T0323716

Difference Equations and Applications

Difference Equations and Applications

Youssef N. Raffoul

ACADEMIC PRESS

An imprint of Elsevier

ISBN: 978-0-443-31492-6

For information on all Academic Press publications
visit our website at https://www.elsevier.com/books-and-journals

Publisher: Peter B. Linsley
Editorial Project Manager: Sara Valentino
Publishing Services Manager: Deepthi Unni
Production Project Manager: Nandhini Thanga Alagu
Cover Designer: Greg Harris

Typeset by VTeX

Working together
to grow libraries in
developing countries

www.elsevier.com • www.bookaid.org

I dedicate this book to my grandson
Everett Russell Wood-Raffoul

Contents

Preface   xi

1. The calculus of difference equations

1.1 Preliminaries   1
   1.1.1 Exercises   7
1.2 Gamma function and the factorial function   8
   1.2.1 Exercises   10
1.3 Shift and difference operators   11
   1.3.1 Exercises   17
1.4 Antidifference or summation   18
   1.4.1 Exercises   24

2. Linear difference equations

2.1 First-order difference equations   27
   2.1.1 Exercises   31
   2.1.2 Applications   32
   2.1.3 Discrete logistic model   35
   2.1.4 Exercises   41
2.2 Variation of parameters   43
   2.2.1 Exercises   47
2.3 Higher-order difference equations   48
   2.3.1 Exercises   53
2.4 Equations with constant coefficients   53
   2.4.1 Distinct roots   54
   2.4.2 Repeated roots   55
   2.4.3 Complex roots   57
   2.4.4 Exercises   59
2.5 The method of undetermined coefficients   61
   2.5.1 Exercises   65
2.6 Variation of parameters   66
   2.6.1 Exercises   71
2.7 Recovery of difference equations   71
   2.7.1 Exercises   73
2.8 From non-linear to solvable   73
   2.8.1 Exercises   79

3. Z-transform

3.1 Introduction   83

3.2	Standard functions	86
	3.2.1 Exercises	89
3.3	Shifting and scalings	90
	3.3.1 Exercises	101
3.4	Convolution	103
	3.4.1 Exercises	110
3.5	Inverse z-transform and applications	110
	3.5.1 Exercises	122
	3.5.2 Frequently used Laplace transforms	125
3.6	Engineering applications	126
3.7	Exercises	135

4. Systems

4.1	Introduction	137
	4.1.1 Exercises	139
4.2	Linear systems	139
4.3	Putzer algorithm	147
	4.3.1 Exercises	155
4.4	Time-varying systems	157
4.5	Non-homogeneous systems	161
	4.5.1 Exercises	164

5. Stability

5.1	Scalar equations	165
	5.1.1 Exercises	171
5.2	Stability of systems	174
	5.2.1 Exercises	179
5.3	$x(n + 1) = A(n)x(n)$	179
	5.3.1 Exercises	184
5.4	$x(n + 1) = Ax(n)$	185
	5.4.1 Exercises	191
5.5	Perturbed linear systems	191
	5.5.1 Exercises	197
5.6	Phase plane analysis	198
	5.6.1 Exercises	205
5.7	Linearization	206
	5.7.1 Exercises	218
5.8	Floquet theory	220
	5.8.1 Non-homogeneous systems	227
	5.8.2 Exercise	230

6. Lyapunov functions

6.1	Introduction to Lyapunov functions	235
	6.1.1 Stability of autonomous systems	237
	6.1.2 Exercises	245
6.2	LaSalle invariance principle	248
	6.2.1 Exercises	252

6.3 Non-autonomous systems   253
    6.3.1   Exercises   260
6.4 Connection between Lyapunov functions and eigenvalues   262
    6.4.1   Exercises   267
6.5 Exponential stability   268
    6.5.1   Exercises   277
6.6 l_p-stability   278
    6.6.1   Exercises   283

7.   **New variation of parameters**

7.1 Introduction   285
7.2 Scalar equations   285
    7.2.1   Contraction versus large contraction   290
    7.2.2   Periodic solutions   295
    7.2.3   Neutral difference equations   299
    7.2.4   Exercises   306

A.   **Banach spaces**

Bibliography   319

Index   321

Preface

Difference equations are widely used by mathematicians, scientists, and engineers. They are used to approximate ordinary and partial differential equations. Difference equations are utilized in the study of electrical, mechanical, thermal, and other systems in which there is a recurrence of identical sections. Even though there are many analogies between differential and integral calculus and the calculus of finite difference, the behaviors of the dynamics between the two subjects widely differ, and hence difference equations earned their place as major fields in mathematics. For many years, the author has been encouraged by the graduate students at the University of Dayton to write a concise and reader-friendly book on the subject of advanced differential equations. So part of this book grew out of lecture notes that the author has been constantly revising and using for a graduate course in difference equations. The book should serve as a two-semester graduate textbook exploring the theory and applications of difference equations. It is intended for students who have basic knowledge of ordinary differential equations and real analysis. While writing this book, the author tried to balance rigor and presenting the most difficult material in an elementary format by adopting easier and friendlier notations that make the book accessible to a wide range of audiences. It was the author's main intention to provide many examples to illustrate the theory conveyed in the theorems. The author made every effort to include contemporary topics such as the use of *Banach spaces and other normed spaces* in several places to prove their existence and uniqueness in the case of non-linear functional equations. What makes the book appealing and distinguished from other books is the addition of Chapter 7 and the addition of advanced topics on stability in Chapter 6. The author devotes a whole chapter instead of a section in order to provide a detailed analysis and study of the Z-transform. The same can be said about stability. Some of the materials in the chapter on stability are new and provide research topics for future study. Chapter 7 is totally new, and the author proposes, at the end of the chapter, a way to generalize the chapter to different systems of difference equations. It is the author's conviction that any student who completes the whole book, especially Chapters 6 and 7, should be ready to carry on with meaningful research in non-linear difference systems.

Much of the pedagogical and mathematical development of this book is influenced by the author's style of presentation. The literature on difference equations is vast and well established and some of the ideas found their way into this book. The author will make every effort to contrast the results on difference equations with their counterparts on differential equations.

Chapter 1 deals with the calculus of finite differences, which is necessary for the development of this book. The chapter includes basic definitions and terminologies. It covers topics on difference operators, shifts, gamma functions, factorial functions, and the anti-difference operator.

Chapter 2 is about preliminary theory and introductory topics on difference equations. The chapter introduces topics on first- and higher-order difference equations, variations of parameters, the method of undetermined coefficients, transforming non-linear equations, and recovering an original difference equation knowing its solution.

Chapter 3 introduces the z-transform and its basic definitions. We begin the chapter by developing the z-transform of basic functions and then move on to stating and proving theorems regarding shifting and scalings. Then we embark on the study of the convolution of two sequences and apply the concept to summation equations. The inverse z-transform is covered in depth, including applications to solving various types of difference equations. The chapter ends with a nice section related to engineering applications.

Chapter 4 introduces systems of difference equations. We briefly discuss how higher-order difference equations are changed to systems. Then we develop the notion of the fundamental matrix as a solution and utilize it to write solutions of non-homogeneous systems so they can be analyzed. In particular, we establish the Putzer algorithm for homogeneous systems with constant coefficients.

Stability theory is the central part of this book after the development of the calculus of finite differences. Chapters 5 and 6 are totally devoted to stability. In Chapter 5, we are mainly concerned with the stability of homogeneous and non-homogeneous systems. We analyze stability via the eigenvalues, when possible, or by linearization. For non-linear, non-homogeneous systems, we employ Gronwall's inequality to obtain qualitative knowledge of the behavior of the solutions. The chapter also includes a nice section on Floquet theory with its application to the stability of periodic systems, and then the concept is generalized to non-homogeneous systems.

Chapter 6 delves deeply into the stability of general systems using Lyapunov functions. We prove general theorems regarding the stability of autonomous and non-autonomous systems by assuming the existence of such a Lyapunov function. We extend the theory of Lyapunov functions to non-autonomous systems. For linear systems, we look into the connection between the eigenvalues of the matrix and the construction of the Lyapunov functions. The chapter ends with nice sections on exponential stability and l_p-stability and most of the materials are new. The literature on Lyapunov functions and functionals in differential equations is vast, and this is not the case for discrete dynamical systems. This chapter contains advanced material that we do not expect to be taught in classroom format. Sections 6.2, 6.3, 6.5, and 6.6 along with Chapter 7 can be used as part of a reading course on special topics in discrete dynamical systems.

Chapter 7 deals with current research concerning the use of a *new variation of parameter formula*. The objective is to introduce a new method for inverting first-order difference equations to obtain a new variation of parameter formula that we use to study stability, boundedness, and periodicity of general equations. The content of the chapter is new and will serve as a starting point for meaningful research. At the end of the chapter, we propose three research projects as extensions of different sections in the chapter.

The book ends with an appendix on Banach spaces that is vital to Chapter 7 and for future research.

Chapters 1–4 can be used to deliver an introductory course on difference equations. On the other hand, Chapters 5, 6, Appendix A, and Chapter 7 can be used either for a graduate course or a reading course on topics in difference equations.

The diagram below shows the interconnection of the chapters.

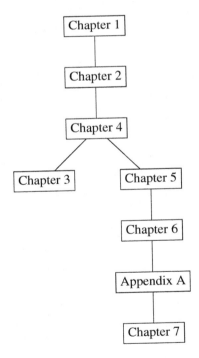

Youssef N. Raffoul
University of Dayton
Dayton, Ohio
October, 2023

Instructor site URL:
https://inspectioncopy.elsevier.com/book/details/9780443314926

Chapter 1

The calculus of difference equations

This chapter is devoted to the calculus of finite differences, which is necessary for the development of important topics in difference equations. The calculus of finite differences is a concept that is parallel to the concept of the calculus on continuous functions.

A difference equation is a mathematical equation that describes the relationship between consecutive values of a sequence or discrete-time function. It is commonly used to model and analyze discrete-time systems in various fields, including mathematics, engineering, physics, economics, and computer science. Difference equations are analogous to differential equations, which describe the relationship between derivatives of a continuous function. However, instead of dealing with derivatives, difference equations involve the differences between successive values of a function. This makes them particularly useful when working with discrete-time systems or processes.

1.1 Preliminaries

Differential equations are extremely useful in modeling scenarios with continuous change. Differential equations do have drawbacks, though, when the change occurs incrementally rather than continuously. Instead, we will make use of difference equations, which are recursively specified sequences that let us represent discontinuous change. Animal populations that reproduce annually, interest that compounds monthly, and models of economic and contagious diseases are a few examples of circumstances where incremental changes occur.

A difference equation, which is a discrete version of a differential equation, symbolizes change in the context of discrete intervals. Difference equations are essential in modeling such time series since the values of these variables can only be recorded at discrete intervals, or it is more convenient to do so. Differential equations are far simpler to study than difference equations in applications. I believe this is the case since differential systems essentially average everything, greatly simplifying the dynamics. Contrarily, discrete systems are more accurate.

However, it's important to note that while the calculus of finite differences shares similarities with calculus, there are also significant differences. A great example of this is the logistic equation. The differential version $x'(t) = rx(1 -$

Difference Equations and Applications. https://doi.org/10.1016/B978-0-44-331492-6.00007-8

1

x) is simple to study and solve, even for students in an introductory course. On the other hand, $x(n + 1) = rx(n)(1 - x(n))$ not only generates interesting dynamics and chaotic behavior but also ties with open theoretical problems. We will touch on the chaos later on in Chapter 2 and Chapter 5.

In this book, the sets $\mathbb{Z}$, $\mathbb{Z}^+$, $\mathbb{N}$, $\mathbb{R}$, and $\mathbb{R}^+$ denote the set of all integers, the set of non-negative integers, the set of natural numbers, the set of real numbers, and the set of nonnegative real numbers, respectively. Throughout this book, the set $\mathbb{N}_{n_0}$ is defined as

$$\mathbb{N}_{n_0} = \{n_0, n_0 + 1, n_0 + 2, n_0 + 3, \cdots\}, \quad n_0 \in \mathbb{Z}.$$

A difference equation, then, is a mathematical equivalence that takes into account the variations between successive values of a function of a discrete variable. A *discrete variable* is one whose definition only applies to values that differ by a certain length of time, typically a constant and frequently 1. In this book our objective is to study sequences of the form

$$x(0), \ x(1), \ x(2), \ x(3), \ldots,$$

that goes on forever. Such a sequence can be abbreviated as $x(n)$, $n \in \mathbb{Z}^+$. On the other hand, the following differences between any two consecutive values in the sequence can be found to be:

$$\triangle x(0) = x(1) - x(0)$$
$$\triangle x(1) = x(2) - x(1)$$
$$\vdots$$
$$\triangle x(n) = x(n + 1) - x(n).$$

This brings us to the notion of delta, or forward difference operator. The forward difference operator, denoted by the symbol $\triangle$, is analogous to the derivative operator in calculus. It measures the difference between consecutive terms of a sequence or the change in a function's value between adjacent points. The forward difference of a sequence $x(n)$ is defined as:

$$\triangle x(n) = x(n + h) - x(n),$$

where h is a step size, which we take to be 1. Thus, we make the following definition.

Definition 1.1.1. Let $x(n)$ be a sequence such that $x : \mathbb{Z} \to \mathbb{R}$. We define the *delta difference operator* of $x(n)$, $\triangle x(n)$ by

$$\triangle x(n) = x(n + 1) - x(n).$$

Also, the *shift operator* E where $E : \mathbb{Z} \to \mathbb{R}$ is defined by

$$Ex(n) = x(n + 1).$$

For example, if $x(n) = 2^n$, $n \in \mathbb{Z}$, then

$$Ex(n) = 2^{n+1},$$

and

$$\Delta x(n) = 2^{n+1} - 2^n = 2^n(2-1) = 2^n.$$

The second order difference of a sequence $x(n)$ is given by

$$\begin{aligned}
\Delta^2 x(n) &= \Delta\big(\Delta x(n)\big) \\
&= \Delta\big(x(n+1) - x(n)\big) \\
&= x(n+2) - x(n+1) - \big(x(n+1) - x(n)\big) \\
&= x(n+2) - 2x(n+1) + x(n).
\end{aligned}$$

As is the case in differential equations, classifications of difference equations with respect to order, linearity, and homogeneity play an important role in guiding us as to what methods and procedures are best suited to deploy in order to solve the difference equation in question. We have the following definition regarding the order of a difference equation.

Definition 1.1.2. The *order* of a difference equation is the difference between the greatest and smallest arguments of the dependent variable that appear in the equation.

Next, we formally define a difference equation.

Definition 1.1.3. A *first order difference equation* is a mathematical expression that relates $\Delta x(n)$ to $x(n)$ and can be expressed in the form

$$x(n+1) = f(n, x(n)), \quad n \in \mathbb{Z} \tag{1.1.1}$$

where n is the independent variable and $x(n)$ is the dependent variable, and the function f is continuous in the second argument. The set of points $\{(n, x(n))\}$ is now a discrete, ordered set of points, and x is a discrete function.

Loosely speaking, by a solution of (1.1.1) on $\mathbb{Z}$, we mean a sequence $x(n) = \varphi(n)$ such that

$$\varphi(n+1) = f(n, \varphi(n)), \quad n \in \mathbb{Z}.$$

Example 1.1. The sequence $x(n+1) = ax(n)$, where a is a nonzero constant, has the solution $x(n) = a^n$, since

$$x(n+1) = a^{n+1} = a^n a = a a^n = ax(n). \quad \square$$

Remark 1.1. Just in the case of differential equations, there are some difference equations that do not have a solution. For example, the difference equation

$$\left(y(n+2) - y(n+1)\right)^2 + y^2(n) = -1$$

has no solution since there is no real-valued function $y(n)$ for which the difference equation is satisfied.

We provide the following example.

Example 1.2. Let n, $k \in \mathbb{Z}^+$ and consider the difference equation

$$a_k x(n+k) + a_{k-1} x(n+k-1) + a_{k-2} x(n+k-2) + \cdots + a_1 x(n+1)$$
$$+ a_0 x(n) = g(n), \tag{1.1.2}$$

where $a_i \neq 0$, $i = 0, 1, \ldots k$ are given coefficients that are independent of the dependent variable $x(n)$, and g is a given function. Then the order of the difference equation given by (1.1.2) is $n + k - (n) = k$. Thus (1.1.2) is a difference equation of order k, or it is a kth order difference equation. Note that by dividing both sides of (1.1.2) by a_k with $a_k \neq 0$, then we may write it in the form

$$x(n+k) = f\left(n, x(n), \ldots, x(n+k-1)\right), \tag{1.1.3}$$

where

$$f\left(n, x(n), \ldots, x(n+k-1)\right)$$
$$= \frac{1}{a_k} \Big[g(n) - a_{k-1} x(n+k-1)$$
$$\quad - a_{k-2} x(n+k-2) - \cdots - a_1 x(n+1) - a_0 x(n) \Big]. \quad \Box$$

Example 1.3. Let $k \in \mathbb{Z}^+$ then the difference equation

$$x(k+2) + 3x(k) + x(k-2) = k+5, \tag{1.1.4}$$

has order $(k+2) - (k-2) = 4$. If we let $n = k - 2$, then we have $k = n + 2$, and under this transformation (1.1.4) can be written as

$$x(n+4) = f\left(n, x(n), x(n+2)\right),$$

where

$$f\left(n, x(n), x(n+2)\right) = n + 7 - x(n) - 3x(n+2). \quad \Box$$

Definition 1.1.4. The kth order difference equation given by (1.1.3) is said to be *linear* in the dependent variable x, if for nonzero constant C, we have that

$$x(n) = Cy(n),$$

then
$$y(n + k) = f(n, y(n), \ldots, y(n + k - 1)).$$

Example 1.4. The difference equation given by (1.1.4) is not linear since

$$Cy(n + 4) = n + 7 + C\left(-y(n) - 3y(n + 2)\right) \neq Cf(n, y(n), y(n + 2)).$$

However, the difference equation

$$x(k + 2) + 3x(k) + x(k - 2) = 0$$

is linear. To see this, we may write is as

$$x(n + 4) = f(n, x(n), x(n + 2)), \quad \text{for } k = n + 2,$$

where

$$f(n, x(n), x(n + 2)) = -x(n) - 3x(n + 2).$$

Thus, for $x(n) = Cy(n)$, we have that

$$Cy(n + 4) = C\left(-y(n) - 3y(n + 2)\right) = Cf(n, y(n), y(n + 2)). \quad \square$$

Example 1.5. The difference equation

$$x(k + 2) + \cos(x(k)) + 3x(k - 2) = 0$$

is not linear. To see this, we may write is as

$$x(n + 4) = f(x(n), x(n + 2)), \quad \text{for } k = n + 2,$$

where

$$f(x(n), x(n + 2)) = -\cos(x(n + 2)) - 3x(n).$$

Thus, for $x(n) = Cy(n)$, we have that

$$Cy(n + 4) = -\cos(Cy(n + 2)) - 3Cy(n) \neq Cf(y(n), y(n + 2)). \quad \square$$

Definition 1.1.5. The difference equation given by (1.1.2) is said to be *homogeneous* if $g(n) = 0$ for all $n \in \mathbb{Z}$. In this case the homogeneous equation associated with (1.1.2) is

$$a_k x(n + k) + a_{k-1} x(n + k - 1) + a_{k-2} x(n + k - 2) + \cdots$$
$$+ a_1 x(n + 1) + a_0 x(n) = 0. \tag{1.1.5}$$

Another way of defining linearity of a difference equation is by the method of operator. But first, we make the following statement. For $k \in \mathbb{Z}^+$ and a sequence $x(n)$, $n \in \mathbb{Z}$ the shift operator E of order k is expressed by

$$
E^k x(n) = \begin{cases} x(n+k), & \text{for } k \geq 1, \\ x(n), & \text{for } k = 0. \end{cases} \tag{1.1.6}
$$

Basically, $E^0 = I$, where I is identity, in the sense that $E^0 x(n) = (Ix)n = x(n)$. With this in mind, the homogeneous Eq. (1.1.5) can be put in the form

$$
a_k E^k x(n) + a_{k-1} E^{k-1} x(n) + a_{k-2} E^{k-2} x(n) + \cdots + a_1 E x(n) + a_0 I x(n) = 0. \tag{1.1.7}
$$

An alternative way of writing (1.1.7) is

$$
\left(a_k E^k + a_{k-1} E^{k-1} + a_{k-2} E^{k-2} + \cdots + a_1 E + a_0 I \right) x(n) = 0. \tag{1.1.8}
$$

Now to better understand linearity we utilize the operator concept $\mathcal{L}$ on an appropriate space, where $\mathcal{L}$ is a *difference operator*. An operator is really just a function that takes a function as an argument instead of numbers as we are used to dealing with in functions. For example, $\mathcal{L}u$ assigns u a new function $\mathcal{L}u$. Another example is if we take

$$
\mathcal{L} = a_k E^k + a_{k-1} E^{k-1} + a_{k-2} E^{k-2} + \cdots + a_1 E + a_0 I
$$

then

$$
\mathcal{L}(x)n = \left(a_k E^k + a_{k-1} E^{k-1} + a_{k-2} E^{k-2} + \cdots + a_1 E + a_0 I \right) x(n).
$$

If we set $\mathcal{L}x(n) = 0$, then we obtain (1.1.5). The next definition gives a precise and convenient way to test for linearity.

Definition 1.1.6. An operator $\mathcal{L}$ is said to be linear if is satisfies

(a)

$$
\mathcal{L}(u_1 + u_2) = \mathcal{L}u_1 + \mathcal{L}u_2,
$$

(b)

$$
\mathcal{L}(cu_1) = c\mathcal{L}u_1,
$$

for any sequences u_1, u_2, and constant c. Moreover, the equation $\mathcal{L}u = 0$ is said to be linear if the operator $\mathcal{L}$ is linear.

By Definition 1.1.6, the difference equation given by (1.1.5) is linear.

Example 1.6. The difference equation

$$x(n+2) + x(n+1) + x^2(n) = 0$$

is not linear. To see this, we define the operator

$$\mathcal{L}(x)n = x(n+2) + x(n+1) + x^2(n).$$

Then it is clear for constant c we have

$$\mathcal{L}(cx)n = cx(n+2) + cx(n+1) + c^2x^2(n) \neq c\mathcal{L}(x)n. \quad \square$$

1.1.1   Exercises

Exercise 1.1. Show that

$$x(n) = c_1(-1)^n + c_2 4^n - \frac{5}{6},$$

is a solution to the second order difference equation

$$x(n+2) - 3x(n+1) - 4x(n) = 5,$$

where c_1, c_2 are constants.

Exercise 1.2. Show that

$$x(n) = c_1 + 2^{n+1},$$

is a solution to the second order difference equation

$$x(n) - x(n-1) = 2^n,$$

where c_1 is constant.

Exercise 1.3. Show that for constants c_1, c_2, and c_3,

$$x(n) = c_1(2)^n + c_2 n 2^n + c_3 3^n,$$

is a solution to the third order difference equation

$$x(n+3) - 7x(n+2) + 16x(n+1) - 12x(n) = 0.$$

Exercise 1.4. Put the given difference equation in the form (1.1.3), where $k \in \mathbb{Z}^+$ is the order. Then classify the order, linearity, and homogeneity of each of the equations.

(a) $x(k+2) + 3x(k) + 2x(k-2) - x(k-3) = 5.$
(b) $(k+1)x(k+2) + 3\sin(x(k)) + 2x(k-1) - x(k-3) = 0.$
(c) $x(k+1) + 3e^{x(k)} + 2x(k-1) - x(k-3) = 0.$
(d) $x(k) + 3x(k-1) = 0.$
(e) $x(k+1) = \dfrac{x(k)}{2 + x(k-1)}.$

1.2 Gamma function and the factorial function

The *Gamma function* plays an important role in the calculus of difference equations.

Definition 1.2.1. We denote the Gamma function by Γ and it is defined by

$$\Gamma(z) = \int_0^\infty e^{-t}t^{z-1}dt, \quad \mathfrak{Re}(z) > 0. \tag{1.2.1}$$

By letting $u = t^{z-1}dt$ and $dv = e^{-t}dt$, then a straight forward integration by parts leads to

$$\begin{aligned}
\Gamma(z) &= \int_0^\infty e^{-t}t^{z-1}dt \\
&= -e^{-t}t^{z-1}\Big|_0^\infty + \int_0^\infty e^{-t}(z-1)t^{z-2}dt \\
&= (z-1)\int_0^\infty e^{-t}t^{z-2}dt \\
&= (z-1)\Gamma(z-1).
\end{aligned}$$

From the above calculations, we arrive at the expression

$$\Gamma(z+1) = z\Gamma(z),$$

which can be used to show that, for every positive integer n

$$\Gamma(n) = (n-1)!, \quad \text{or} \quad \Gamma(n+1) = n!.$$

We will also need the following formula. For positive integer n, we have

$$\Gamma\left(\frac{1}{2}n\right) = \frac{(n-2)!!\sqrt{\pi}}{2^{(n-1)/2}}, \tag{1.2.2}$$

where $n!!$ is a *double factorial*. For example,

$$\Gamma\left(\frac{1}{2}\right) = \sqrt{\pi}, \quad \Gamma\left(\frac{3}{2}\right) = \sqrt{\pi}/2, \quad \Gamma\left(\frac{5}{2}\right) = (3\sqrt{\pi})/4, \text{ etc.}$$

Three other important properties of the Gamma function are

$$\Gamma\left(\frac{1}{2}+n\right) = \frac{(2n-1)!!}{2^n}\sqrt{\pi},$$

$$\Gamma\left(\frac{1}{2}-n\right) = \frac{(-1)^n 2^n}{(2n-1)!!}\sqrt{\pi},$$

and the Euler's reflection formula

$$\Gamma(z)\,\Gamma(1-z) = \frac{\pi}{\sin(\pi z)}, \qquad z \notin \mathbb{Z}.$$

Next we define the *factorial function* $[t]^r$.

Definition 1.2.2. The *factorial function* $[t]^r$ is defined in the following manner:

$$[t]^r = \begin{cases} t(t-1)(t-2)\cdots(t-(r-1)), & \text{for } r \in \mathbb{N} \\ 0, & \text{for } r = 0 \\ \frac{1}{(t+1)(t+2)\cdots(t-r)}, & \text{for } r \in -\mathbb{N} \\ \frac{\Gamma(t+1)}{\Gamma(t-r+1)}, & r \text{ is not an integer.} \end{cases} \tag{1.2.3}$$

For example,

$$[t] = t$$
$$[t]^2 = t(t-1)$$
$$[t]^3 = t(t-1)(t-2)$$
$$[-4]^5 = -4(-4-1)(-4-2)(-4-3)(-4-4).$$

Note that $[1]^r = 0, r \in \mathbb{Z}^+$.

Example 1.7. Compute

(a) $\left[\dfrac{5}{2}\right]^{-\frac{3}{2}}$,

(b) $\left[\dfrac{1}{2}\right]^{-\frac{3}{2}}$.

For (a), we have from (1.2.3) that

$$\begin{aligned} \left[\frac{5}{2}\right]^{-\frac{3}{2}} &= \frac{\Gamma(\frac{5}{2}+1)}{\Gamma(\frac{5}{2}+\frac{3}{2}+1)} \\ &= \frac{\Gamma(\frac{1}{2}7)}{\Gamma(5)} \\ &= \frac{(15\sqrt{\pi})/8}{4!} \\ &= \frac{5\sqrt{\pi}}{60} \end{aligned}$$

Similarly, for (b) we have

$$\left[\frac{1}{2}\right]^{-\frac{3}{2}} = \frac{\Gamma(\frac{1}{2}+1)}{\Gamma(\frac{1}{2}+\frac{3}{2}+1)}$$

$$= \frac{\frac{1}{2}\Gamma(\frac{1}{2})}{\Gamma(3)}$$

$$= \frac{\frac{1}{2}\sqrt{\pi}}{2!}. \quad \square$$

It can be shown that every factorial function to power r can be written as a polynomial of degree r. To explain the idea we provide the following example.

Example 1.8. Express t^3 as a linear combination of $[t]^3$, $[t]^2$ and $[t]$. Since the factorial functions $[t]^3$, $[t]^2$, and $[t]$ produce only one term of t^3 with coefficient one and multiple terms of t^2 and t, we will need to find two constants α and β such that

$$t^3 = [t]^3 + \alpha[t]^2 + \beta[t].$$

Thus, by expanding the factorial functions we arrive at

$$t^3 = t(t-1)(t-2) + \alpha t(t-1) + \beta t$$
$$= t^3 + (\alpha - 3)t^2 + (2 - \alpha + \beta)t.$$

By comparing both sides we see that

$$\alpha - 3 = 0, \quad 2 - \alpha + \beta = 0.$$

This system has the solution $\alpha = 3$, $\beta = 1$. Thus,

$$t^3 = [t]^3 + 3[t]^2 + [t]. \quad \square$$

1.2.1   Exercises

Exercise 1.5. Show that

(a)

$$\int_0^\infty e^{-\lambda t} t^{z-1} dt = \frac{\Gamma(z)}{\lambda^z}, \quad \lambda > 0.$$

(b) Compute

$$\int_0^\infty e^{-5t} t^6 dt$$

Exercise 1.6. Show that

$$\int_0^\infty \frac{\lambda^z e^{-\lambda t} t^{z-1}}{\Gamma(z)} dt = 1.$$

Exercise 1.7. Show that for $a, b > 0$,

$$\Gamma(a)\Gamma(b) = \Gamma(a+b) \int_0^1 \tau^{a-1}(1-\tau)^{b-1} d\tau.$$

Exercise 1.8. Use Exercise 1.7 to show that

$$\Gamma\left(\frac{1}{2}\right) = \sqrt{\pi}.$$

Hint: Use Exercise 1.7 to show that

$$\Gamma\left(\frac{1}{2}\right)\Gamma\left(\frac{1}{2}\right) = \Gamma(1)\int_0^1 \tau^{-1/2}(1-\tau)^{-1/2}d\tau.$$

Then make the substitution $\tau = u^2$. Then final substitution $u = \sin(\theta)$ should do the job.

Exercise 1.9. Verify that

(a) $\Gamma\left(\dfrac{5}{2}\right) = \dfrac{3\sqrt{\pi}}{4}.$

(b) $\Gamma\left(\dfrac{3}{4}\right) = \dfrac{\pi\sqrt{2}}{\Gamma\left(\frac{1}{4}\right)}.$

Exercise 1.10. Compute

(a) $[5]^3.$

(b) $[-7]^{-2}.$

(c) $\left[\dfrac{7}{2}\right]^{-\frac{9}{2}}.$

Exercise 1.11. Express t^2 as a linear combination of $[t]^2$ and $[t]$.

Exercise 1.12. Express t^4 as a linear combination of $[t]^4$, $[t]^3$, $[t]^2$ and $[t]$.

Exercise 1.13. Express $3t^3 - 3t + 1$ as a linear combination of factorial functions.

1.3 Shift and difference operators

In Definition 1.1.1, we defined the *difference operator* and the *shift operator*, acting on a sequence $x(n)$ as

$$\Delta x(n) = x(n+1) - x(n), \tag{1.3.1}$$
$$Ex(n) = x(n+1), \tag{1.3.2}$$

respectively. In this section we are interested in exploring their properties. We begin by noticing that (1.3.1) and (1.3.2) imply that

$$\Delta x(n) = x(n+1) - x(n) = Ex(n) - x(n) = (E - I)x(n).$$

Consequently, we have that

$$\Delta = E - I, \tag{1.3.3}$$

where I is the identity operator. The shift operator is linear in the sense that for any two sequences $x(n)$, $y(n)$, and constants c_1 and c_2 we have that

$$E\left(c_1 x(n) \pm c_2 y(n)\right) = c_1 E\left(x(n)\right) \pm c_2 E\left(y(n)\right).$$

The kth order shifts is defined by

$$E^k x(n) = E\left(E^{k-1} x(n)\right) = x(n+k).$$

Similarly, the operator Δ is linear. That is for any two sequences $x(n)$, $y(n)$ and constants c_1 and c_2 we have that

$$\Delta\left(c_1 x(n) \pm c_2 y(n)\right) = c_1 \Delta\left(x(n)\right) \pm c_2 \Delta\left(y(n)\right).$$

Before, we continue with our development, we pause and give a brief review of the *binomial coefficient function*. The binomial coefficient is a symbol used in combinatorics to represent the variety of ways an object of a particular numerosity can be selected from a bigger collection. Its ability to be used to write the coefficients of a power of a binomial is how it got its name.

Definition 1.3.1. The binomial coefficient is denoted by

$$\binom{k}{i} \tag{1.3.4}$$

and it is read as k choose i, $0 \leq i \leq k$. Moreover, it is defined as follows

$$\binom{k}{i} = \frac{k!}{(k-i)!i!}. \tag{1.3.5}$$

Here is another important property of the binomial coefficient.

$$\binom{k}{i} = \frac{k(k-1)\cdots(k-i+1)}{i!} = \frac{[k]^i}{i!}. \tag{1.3.6}$$

It follows from (1.3.6) that

$$\binom{k}{i} = \frac{[k]^i}{\Gamma(i+1)}. \tag{1.3.7}$$

The following three properties can be proved using (1.3.6).

$$\binom{k}{i} = \binom{k}{k-i}.$$

$$\binom{k}{i} = \frac{k}{i}\binom{k-1}{k-i}.$$

$$\binom{k}{i} = \binom{k-1}{i} + \binom{k-1}{i-1}. \tag{1.3.8}$$

In (1.3.8), if we replace k by $k+1$ and i by $i+1$, then we have

$$\binom{k+1}{i+1} = \binom{k}{i+1} + \binom{k}{i},$$

from which we arrive at

$$\binom{k+1}{i+1} - \binom{k}{i+1} = \binom{k}{i}.$$

Hence,

$$\Delta_k\binom{k}{i+1} = \binom{k+1}{i+1} - \binom{k}{i+1} = \binom{k}{i}. \tag{1.3.9}$$

In algebra, the binomial coefficient is used to expand powers of binomials. According to the binomial theorem,

$$(a+b)^n = \sum_{k=0}^{n} \binom{n}{k} a^k b^{n-k}. \tag{1.3.10}$$

In Section 1.1, we determined the second order difference operator acting on a sequence $x(n)$ to be

$$\Delta^2 x(n) = \Delta\big(\Delta x(n)\big) = x(n+2) - 2x(n+1) + x(n).$$

Similarly,

$$\Delta^3 x(n) = \Delta\big(\Delta^2 x(n)\big) = x(n+3) - 3x(n+2) + 3x(n+1) - x(n).$$

Continuing in this fashion, we see that

$$\Delta^k x(n) = \Delta\big(\Delta^{k-1} x(n)\big)$$
$$= x(n+k) - kx(n+k-1) + \frac{k(k-1)}{2!} x(n+k-2)$$
$$+ \cdots + (-1)^{k-1} kx(k+1) + (-1)^k x(n)$$
$$= \sum_{i=0}^{k} (-1)^i \binom{k}{i} x(n+k-i). \tag{1.3.11}$$

Making use of (1.3.3), we obtain from (1.3.11) that

$$(E - I)^k x(n) = \sum_{i=0}^{k} (-I)^i \binom{k}{i} E^{k-i} x(n)$$

$$= \sum_{i=0}^{k} (-1)^i \binom{k}{i} x(n + k - i). \qquad (1.3.12)$$

One of the major distinctions between calculus on the real line and the calculus of finite differences is the product rule for differentiations and the product rule for differences, as we shall note in the next theorem.

Theorem 1.3.1 (Product rule). *Let $x(n)$, $y(n)$ be given sequences. Then*

$$\triangle (x(n)y(n)) = \triangle x(n) E y(n) + x(n) \triangle y(n).$$

Proof. By the definition of the forward difference operator we have that

$$\triangle (x(n)y(n)) = x(n + 1)y(n + 1) - x(n)y(n).$$

By adding and subtracting the term $x(n)y(n + 1)$, the above equation gives

$$\triangle (x(n)y(n)) = x(n + 1)y(n + 1) - x(n)y(n + 1) + x(n)y(n + 1) - x(n)y(n)$$
$$= [x(n + 1) - x(n)]y(n + 1) + x(n)[y(n + 1) - y(n)]$$
$$= \triangle x(n) E y(n) + x(n) \triangle y(n).$$

This completes the proof. □

Another distinction between calculus on the real line and the calculus of finite differences is the quotient rule.

Theorem 1.3.2 (Quotient rule). *Let $x(n)$, $y(n)$ be given sequences. Then*

$$\triangle \left(\frac{x(n)}{y(n)} \right) = \frac{y(n) \triangle x(n) - x(n) \triangle y(n)}{y(n) E y(n)}$$

provided that $y(n) E y(n) \neq 0$, for all $n \in \mathbb{Z}$.

Proof. The proof is similar to the proof of the product rule. By the definition of the forward difference operator we have that

$$\triangle \left(\frac{x(n)}{y(n)} \right) = \frac{x(n + 1)}{y(n + 1)} - \frac{x(n)}{y(n)}.$$

By taking a common denominator and then adding and subtracting to the numerator the term $x(n)y(n)$, the above equation gives

$$\triangle \left(\frac{x(n)}{y(n)} \right) = \frac{[x(n + 1) - x(n)]y(n) - x(n)[y(n + 1) - y(n)]}{y(n) E y(n)}$$

$$= \frac{y(n)\Delta x(n) - x(n)\Delta y(n)}{y(n)Ey(n)}.$$

This completes the proof. ☐

Recall that the formula of product rule for differentiation is symmetric. That is

$$\frac{d}{dt}(f(t)g(t)) = f'(t)g(t) + f(t)g'(t).$$

Another product of differences is given by the next formula.

$$\Delta(x(n)y(n)) = x(n)\Delta y(n) + y(n)\Delta x(n) + \Delta x(n)\Delta y(n). \qquad (1.3.13)$$

The verification of (1.3.13) is left as an exercise.

Example 1.9. Evaluate the following:

(a) Δe^n, **(b)** $\Delta^2 e^n$, **(c)** $\Delta \tan^{-1}(n)$, **(d)** $\Delta\left(\dfrac{n+1}{n^2 - 3n + 2}\right)$.

We begin with (a).

$$\Delta e^n = e^{n+1} - e^n = (e - 1)e^n.$$

For (b), we have from (a) that

$$\Delta^2 e^n = \Delta(e - 1)e^n = (e - 1)(e - 1)e^n = (e - 1)^2 e^n.$$

Now we turn our attention to (c).

$$\Delta \tan^{-1}(n) = \tan^{-1}(n + 1) - \tan^{-1}(n)$$
$$= \tan^{-1}\left(\frac{n + 1 - n}{1 + (n + 1)n}\right)$$
$$= \tan^{-1}\left(\frac{1}{1 + (n + 1)n}\right).$$

As for (d), we notice that the expression can be written as

$$\frac{n + 1}{n^2 - 3n + 2} = \frac{-2}{n - 1} + \frac{3}{n - 2}.$$

Thus

$$\Delta\left(\frac{n+1}{n^2 - 3n + 2}\right) = \Delta\left(\frac{-2}{n - 1}\right) + \Delta\left(\frac{3}{n - 2}\right)$$
$$= -2\left(\frac{1}{n + 1 - 1} - \frac{1}{n - 1}\right) + 3\left(\frac{1}{n + 1 - 2} - \frac{1}{n - 2}\right)$$
$$= -\frac{n + 4}{n(n - 1)(n - 2)}. \quad ☐$$

Back to the factorial function $[n]^r$ that is defined by (1.2.3). For any variable $x \in \mathbb{R}$, we have that $\frac{d}{dx}x^r = rx^{r-1}$, which is not the case when $x \in \mathbb{Z}$. However, the factorial function gives a similar result.

Theorem 1.3.3 (Factorial and power rule). *Assume $[t]^r$ is defined. Then*

$$\Delta\left([t]^r\right) = r[t]^{r-1}. \tag{1.3.14}$$

Proof. We only do the proof if r is a positive integer. That is $r \in \mathbb{N}$.

$$\begin{aligned}
\Delta([t]^r) &= [t+1]^r - [t]^r \\
&= (t+1)t(t-1)(t-2)\cdots(t-r+2) \\
&\quad - \left[t(t-1)(t-2)\cdots(t-r-2)(t-r+1)\right] \\
&= (t+1-t+r-1)\left[t(t-1)(t-2)\cdots(t-r+2)\right] \\
&= r\left[t(t-1)(t-2)\cdots(t-r+2)\right] \\
&= r[t]^{r-1}.
\end{aligned}$$

This completes the proof. □

In a similar fashion it can be shown that

$$\Delta^2\left([t]^r\right) = r(r-1)[t]^{r-2}$$

$$\vdots$$

$$\Delta^r\left([t]^r\right) = r(r-1)(r-2)\cdots 3.2.1 = r! \tag{1.3.15}$$

$$\Delta^{r+1}\left([t]^r\right) = 0.$$

Now, in some cases, we may be able to use (1.3.15) to solve some difference equations.

Example 1.10. Find a solution to the difference equation

$$3x(n+2) - 6x(n+1) + 3x(n) = n(n-1).$$

By dividing both sides with 3 and noticing that

$$\Delta^2 x(n) = x(n+2) - 2x(n+1) + x(n),$$

the above difference equation can be written in the form

$$\Delta^2 x(n) = \frac{1}{3}[n]^2 = \Delta^2\left(\frac{1}{36}[n]^4\right).$$

Thus, $x(n) = \frac{1}{36}[n]^4$ is a solution of the difference equation. □

As in the case of differential calculus, in the table below, where a is a constant, we list the forward differences of some basic sequences. In the table, $\log(n)$ represents any logarithm of the positive number n.

For convenience, the notation Δ_t means the forward difference is taken with respect to the variable t. For example, $\Delta_t(te^n) = (t+1)e^n - te^n = e^n$.

Theorem 1.3.4. *For $r \neq 0$ the following hold.*

(a) $\Delta_n \binom{n}{r} = \binom{n}{r-1}$.

(b) $\Delta_n \binom{r+n}{n} = \binom{r+n}{n+1}$.

Proof. We make use of (1.3.7). Thus

$$\Delta_n \binom{n}{r} = \Delta_n \frac{[n]^r}{\Gamma(r+1)} = \frac{r[n]^{r-1}}{\Gamma(r+1)}$$

$$= \frac{[n]^{r-1}}{\Gamma(r)} = \binom{n}{r-1}.$$

The proof of (b) follows from (1.3.8). □

TABLE 1.1 Forward differences of some basic sequences.

$f(n)$	$\Delta f(n)$
Constant C	0
a^n	$(a-1)a^n$
$\sin(an)$	$2\sin(\frac{a}{2})\cos\left(a(n+\frac{1}{2})\right)$
$\cos(an)$	$-2\sin(\frac{a}{2})\sin\left(a(n+\frac{1}{2})\right)$
$\log(an)$	$\log(1+\frac{1}{n})$
$\log(\Gamma(n))$	$\log(n)$

1.3.1   Exercises

Exercise 1.14. Verify (1.3.13).

Exercise 1.15. Prove every item of (1.3.15).

Exercise 1.16. Evaluate the following:

(a) $\Delta n e^n$,   **(b)** $\Delta^2 n e^n$,   **(c)** $\Delta \tan^{-1}(n+1)$,   **(d)** $\Delta\left(\dfrac{n+1}{n^2 - 5n + 6}\right)$.

Exercise 1.17. Find a solution to the difference equation

$$x(n+2) - 2x(n+1) + x(n) = 4n(n-1).$$

Exercise 1.18. Find a solution to the difference equation

$$x(n+1) - x(n) = 3^n.$$

Exercise 1.19. Prove each entry of Table 1.1.

Exercise 1.20. Use (1.3.10) and expand $(a+b)^3$.

Exercise 1.21. Show that

$$\sum_{k=0}^{n} \binom{m+k}{k} = \binom{n+m+1}{n}.$$

Exercise 1.22. Prove Theorem 1.3.3 for arbitrary r.

Exercise 1.23. Show that $\displaystyle\sum_{k=0}^{n} \binom{n}{k} = 2^n.$

Exercise 1.24. Show that $E\left[\displaystyle\sum_{i=1}^{k} c_i y_i(n)\right] = \displaystyle\sum_{i=1}^{k} c_i E y_i(n)$, for constants c_i, $i = 1, 2, \cdots k$.

Exercise 1.25. Find a solution to the difference equation

$$x(n+2) - 2x(n+1) + x(n) = n(n-1)(n-2).$$

1.4   Antidifference or summation

We have already seen with the Difference Operator that if f is a real-valued function then the difference operator applied to f is $\triangle f = f(x+1) - f(x)$ which is an analogue to the derivative $\frac{df}{dx}$ of f. We will now look at another calculus analogue. This time with respect to the antiderivative of a real-valued function. Recall that if f is a real-valued function and that F is said to be an antiderivative of f if $\frac{dF}{dx} = f$. We now define the *antidifference* of a sequence.

Definition 1.4.1. Let $x(n)$ be a real-valued sequence. Then if $\triangle F(n) = x(n)$ then F is said to be an *antidifference*, or *indefinite sum* of x using the notation $F = \sum x(n)$ where $\sum$ is the antidifference operator.

A quick example. We have already seen that $\triangle a^n = (a-1)a^n$, $a \neq 1$. Thus, if $x(n) = a^n$, then $F(n) = \frac{a^n}{a-1}$. In other words,

$$\sum a^n = \frac{a^n}{a-1}. \tag{1.4.1}$$

The antidifference operator is linear in the sense that for any two sequences $x(n)$, $y(n)$ and constants c_1 and c_2 we have that

$$\sum (c_1 x(n) \pm c_2 y(n)) = c_1 \sum x(n) \pm c_2 \sum y(n).$$

Example 1.11. Solve

$$x(n+2) - 2x(n+1) + x(n) = a^n, \ a \neq 1.$$

The equation is equivalent to

$$\Delta^2 x(n) = a^n.$$

By applying the antidifference to both sides we get

$$\Delta x(n) = \sum a^n = \frac{a^n}{a-1} + C_1.$$

Applying the antidifference one more time to both sides gives,

$$x(n) = \sum \left(\frac{a^n}{a-1} + C_1 \right) = \frac{a^n}{(a-1)^2} + C_1 n + C_2,$$

for some constants C_1 and C_2. □

We state out first theorem of this section.

Theorem 1.4.1. *If $F(n)$ is the antidifference or indefinite sum of $x(n)$, then so does $F(n) + C(n)$ provided that $C(n)$ has the same domain as x and $\Delta C(n) = 0$. In other words,*

$$\sum x(n) = F(n) + C(n).$$

Proof. Since $F(n)$ is the antidifference of $x(n)$, we have that $\Delta F(n) = x(n)$. Thus,

$$\Delta (F(n) + C(n)) = \Delta F(n) + \Delta C(n) = x(n).$$

This completes the proof. □

In differential calculus, the constant C is found to be a constant irrelevant of the domain. However, this is not the case here. Let's take a closer look on $C(n)$. We begin by assuming $n \in \mathbb{Z}$. Then, $\Delta C(n) = C(n+1) - C(n) = 0$, implies that

$$\cdots C(-2) = C(-1) = C(0) = C(1) = C(2) \cdots$$

We conclude $C(n) = 0$, for all $n \in \mathbb{Z}$. Hence, if the domain is the set of all integers, we write

$$\sum x(n) = F(n) + C,$$

where C is any constant.

On the other hand if the domain of $x(n)$ is set of real numbers, that is $n \in \mathbb{R}$, then $\Delta C(n) = C(n + 1) - C(n) = 0$, is satisfied for any function $C(n)$ with period 1. An example of such a function would be $C(n) = \sin(2\pi n)$.

As is the case for differential calculus, we find the antidifference of some basic sequences. In Theorem 1.3.3 we established that $\Delta_t ([t]^r) = r[t]^{r-1}$. Replacing r with $r + 1$ we have that $\Delta_t ([t]^{r+1}) = (r + 1)[t]^r$. Or $\Delta_t \left(\frac{[t]^{r+1}}{r+1}\right) = [t]^r$. Hence, we have shown, for a sequence $C(t)$ with $\Delta C(t) = 0$, that

$$\sum [t]^r = \frac{[t]^{r+1}}{r + 1} + C(t), \ r \neq -1. \tag{1.4.2}$$

This should remind you of the formula

$$\int x^p dx = \frac{x^{p+1}}{p + 1} + C, \ p \neq -1.$$

Similarly, from Table 1.1, we have that

$$\Delta \sin(at) = 2 \sin(\frac{a}{2}) \cos \left(a(t + \frac{1}{2})\right).$$

Let $u = t + \frac{1}{2}$, then $t = u - \frac{1}{2}$. Thus,

$$\Delta \sin \left(a(u - \frac{1}{2})\right) = 2 \sin(\frac{a}{2}) \cos(au),$$

which implies that

$$\Delta \left(\frac{\sin \left(a(u - \frac{1}{2})\right)}{2 \sin(\frac{a}{2})}\right) = \cos(au).$$

We have shown that for $\Delta C(t) = 0$,

$$\sum \cos(at) = \frac{\sin \left(a(t - \frac{1}{2})\right)}{2 \sin(\frac{a}{2})} + C(t). \tag{1.4.3}$$

As for the binomial coefficient, it follows from (1.3.9) that

$$\sum_k \binom{k}{i} = \binom{k}{i + 1}, \tag{1.4.4}$$

where $\sum_k$ indicates the antidifference with respect to k.

We form a table listing the antidifference of some basic sequences as we did for the forward difference. We assume the domain of all sequences to be the set $\mathbb{Z}$ and take C to be the constant of summation. (See Table 1.2.)

TABLE 1.2 Antidifference of some basic sequences.

$f(n)$	$\sum f(n)$
$[n]^r$	$\frac{[n]^{r+1}}{r+1} + C$
a^n	$\frac{a^n}{a-1} + C, \, a \neq 1$
$\sin(an)$	$-\frac{\cos\left(a(n+\frac{1}{2})\right)}{\sin(\frac{a}{2})} + C, \, a \neq 2k\pi$
$\cos(an)$	$\frac{\sin\left(a(n+\frac{1}{2})\right)}{\sin(\frac{a}{2})} + C, \, a \neq 2k\pi$
$\log(n)$	$\log(\Gamma(n)) + C, \, n \in \mathbb{N}$
$\binom{n}{r}$	$\binom{n}{r+1} + C$

Next, we talk about the antidifference with lower and upper indices of summation. However, to be consistent, we adopt the definition that for $a > b$,

$$\sum_{n=a}^{b} x(n) = 0.$$

The Fundamental Theorem of Difference Calculus, in short FTDC also known as the Summation Theorem or the Discrete Analog of the Fundamental Theorem of Calculus, is a fundamental result in the field of discrete mathematics. It provides a connection between summation and differencing operations.

Theorem 1.4.2 (Fundamental theorem of difference calculus). *Let* $: \mathbb{Z} \to \mathbb{R}$ *and let F be an antidifference of f. That is $\Delta F = f$. Then for $a, b \in \mathbb{Z}$ with $a < b$ we have*

$$\sum_{k=a}^{b} f(k) = F(k)\Big|_{k=a}^{b+1} = F(b+1) - F(a).$$

In particular,

$$\sum_{k=1}^{n} f(k) = F(k)\Big|_{k=1}^{n+1} = F(n+1) - F(1).$$

Proof. We have

$$\sum_{k=a}^{b} f(k) = \sum_{k=a}^{b} \Delta F(k)$$

$$= \sum_{k=a}^{b} \Big(F(k+1) - F(k)\Big) \; (\text{let } u = k + 1)$$

$$= \sum_{u=a+1}^{b+1} F(u) - \sum_{k=a}^{b} F(k)$$

$$= F(b+1) - F(a) + \sum_{u=a+1}^{b} F(u) - \sum_{k=a+1}^{b} F(k)$$

$$= F(b+1) - F(a). \qquad \square$$

We work out a couple of examples.

Example 1.12. Compute

(a) $\displaystyle\sum_{k=1}^{n} a^k, a \neq 1.$

(b) $\displaystyle\sum_{k=1}^{n} [k]^r, r \in \mathbb{Z}^+.$

As for (a), the antidifference is $F(k) = \frac{a^k}{a-1}$. Thus by FTDC, we have

$$\sum_{k=1}^{n} a^k = \frac{a^k}{a-1} \Big|_{k=1}^{n+1}$$

$$= \frac{a^{n+1}}{a-1} - \frac{a}{a-1}.$$

For (b), we have $F(k) = \frac{[k]^{r+1}}{r+1}$, and so

$$\sum_{k=1}^{n} [k]^r = \frac{[k]^{r+1}}{r+1} \Big|_{k=1}^{n+1}$$

$$= \frac{[n+1]^{r+1}}{r+1} - \frac{[1]^{r+1}}{r+1}$$

$$= \frac{[n+1]^{r+1}}{r+1},$$

since $[1]^{r+1} = 0$. $\qquad \square$

Example 1.13. Find a formula for

$$\sum_{k=1}^{n} k^2.$$

By Exercise 1.11, k^2 can be expressed as a combination of $[k]^2$ and $[k]^1$. As a matter of fact, we have that $k^2 = [k]^2 + [k]^1$. Thus, by (b) of Example 1.12 we arrive at

$$\sum_{k=1}^{n} k^2 = \sum_{k=1}^{n} \left([k]^2 + [k]^1 \right)$$

$$= \sum_{k=1}^{n} [k]^2 + \sum_{k=1}^{n} [k]^1$$

$$= \frac{[n+1]^3}{3} + \frac{[n+1]^2}{2}$$

$$= \frac{(n+1)n(n-1)}{3} + \frac{(n+1)n}{2}$$

$$= \frac{2n^3 + 3n^2 + n}{6}$$

$$= \frac{n(n+1)(2n+1)}{6}. \quad \square$$

Remark 1.2. The FTDC implies that

(a) $\displaystyle\sum_{k=n_0}^{n-1} \Delta f(k) = f(n) - f(n_0)$.

However,

(b) $\displaystyle\Delta_n \sum_{k=n_0}^{n-1} f(k) = f(n)$.

To see this,

$$\Delta_n \sum_{k=n_0}^{n-1} f(k) = \sum_{k=n_0}^{n} f(k) - \sum_{k=n_0}^{n-1} f(k)$$

$$= f(n) + \sum_{k=n_0}^{n-1} f(k) - \sum_{k=n_0}^{n-1} f(k)$$

$$= f(n).$$

Note that (b) of Remark 1.2 is the discrete analogue of the second fundamental theorem of calculus, which says that

$$\frac{d}{dx} \left(\int_a^x f(t)dt \right) = f(x).$$

Finally, we present the summation by parts formula, which is the discrete analogue of the integration by parts formula.

Theorem 1.4.3 (Summation by parts). *Let $x(n)$, $y(n)$ be given sequences. Then*

$$\sum_{k=n_0}^{n-1} x(k)\triangle y(k) = x(k)y(k)\Big|_{k=n_0}^{n} - \sum_{k=n_0}^{n-1} \triangle x(k)Ey(k). \tag{1.4.5}$$

Proof. The proof follows by solving for $\triangle x(n)y(n)$ of the product rule given in Theorem 1.3.1 and then summing both sides from n_0 to n. This completes the proof. □

We provide another formula for the summation by parts. It is simply obtained from (1.4.5) by interchanging $x(n)$ and $y(n)$. We make it formal in the next theorem.

Theorem 1.4.4 (Summation by parts). *Let $x(n)$, $y(n)$ be given sequences. Then*

$$\sum_{k=n_0}^{n-1} y(k)\triangle x(k) = x(k)y(k)\Big|_{k=n_0}^{n} - \sum_{k=n_0}^{n-1} \triangle y(k)Ex(k). \tag{1.4.6}$$

Example 1.14. For $a \neq 1$, evaluate

$$\sum_{k=1}^{n} ka^k.$$

We will use Theorem 1.4.3. Let $x(k) = k$ and $\triangle y(k) = a^k$. Then $\triangle x(k) = 1$ and $y(k) = \frac{a^k}{a-1}$. Thus,

$$\sum_{k=1}^{n} ka^k = k\frac{a^k}{a-1}\Big|_{k=1}^{n+1} - \frac{1}{a-1}\sum_{k=1}^{n} 1 \cdot a^{k+1}$$

$$= k\frac{a^k}{a-1}\Big|_{k=1}^{n+1} - \frac{a}{(a-1)^2}a^k\Big|_{k=1}^{n+1}$$

$$= \frac{1}{a-1}\Big[ka^k - \frac{a}{a-1}a^k\Big]\Big|_{k=1}^{n+1}$$

$$= \frac{1}{a-1}\Big[\Big((n+1) - \frac{a}{a-1}\Big)a^{n+1} - \Big(a - \frac{a^2}{a-1}\Big)\Big]. \quad □$$

1.4.1 Exercises

Exercise 1.26. Compute

$$\sum_{k=5}^{64} a^k, \quad a \neq 1.$$

Exercise 1.27. Find a formula for

$$\sum_{k=1}^{n} [k]^3.$$

Exercise 1.28. Prove the Abel's summation formula,

$$\sum_{k=1}^{n} x(k)y(k) = x(n+1)\sum_{k=1}^{n} y(k) - \sum_{k=1}^{n}\left(\Delta x(k)\sum_{i=1}^{k} y(i)\right).$$

Exercise 1.29. Compute

$$\sum_{k=0}^{n-1} (2k+1)a^k, \ a \neq 1.$$

Exercise 1.30. Compute

$$\sum [k]^4 5^k.$$

Exercise 1.31. Compute

$$\sum_{k=1}^{6} \frac{1}{(k+1)(k+2)(k+3)(k+4)(k+5)}.$$

Exercise 1.32. Compute

$$\sum \binom{k}{6}\binom{k}{2}.$$

Hint: Use summation by parts, by taking $x(k) = \binom{k}{2}$, $\Delta y(k) = \binom{k}{6}$.

Exercise 1.33. Compute

$$\sum \binom{k}{5}\binom{k}{3}.$$

Hint: Use summation by parts, by taking $x(k) = \binom{k}{3}$, $\Delta y(k) = \binom{k}{5}$.

Exercise 1.34. Solve

$$\Delta^2 x(n) = [n]^3.$$

Exercise 1.35. Solve

$$\Delta^2 x(n) = n2^{-n}.$$

Chapter 2

Linear difference equations

Linear difference equations are a type of recurrence relation that describe the relationship between successive terms in a sequence. They are analogous to linear differential equations in continuous domain. We will start with homogeneous equations and use their solutions to obtain solutions of non-homogeneous equations. The closed form solution will be called variation of parameters.

2.1 First-order difference equations

Consider the first-order differential equation

$$x'(t) = ax(t), \quad x(t_0) = x_0.$$

Then its solution is known to be

$$x(t) = x_0 e^{a(t-t_0)}.$$

The discrete analogue is $\triangle x(n) = bx(n)$, or $x(n+1) = (b+1)x(n)$. For simplicity and generality, we will always consider, unless it is noted, the first-order difference equation of the form

$$x(n+1) = ax(n), \quad x(0) = x_0, \quad n = 0, 1, 2 \dots \tag{2.1.1}$$

where a is constant. Solving the homogeneous equation (2.1.1) is simple. By varying n over its domain, we arrive at

$$x(1) = ax_0$$
$$x(2) = ax(1) = a^2 x_0$$
$$x(3) = ax(2) = a^3 x_0$$
$$x(4) = ax(3) = a^4 x_0$$
$$\vdots$$
$$x(n) = ax(n-1) = a^n x_0.$$

Hence, it can be easily verified that

$$x(n) = a^n x_0, \quad n = 0, 1, 2 \dots \tag{2.1.2}$$

Difference Equations and Applications. https://doi.org/10.1016/B978-0-44-331492-6.00008-X

is the solution of (2.1.1). To see this,

$$x(n+1) = a^{n+1}x_0 = aa^n x_0 = ax(n).$$

Moreover, $x(0) = a^0 x_0 = x_0$.

It should be clear from (2.1.2) that the solution is $x(n) = x_0$, for $a = 1$. Let us pause for a second and look at the behavior of the sequence solution given by (2.1.2) in terms of the size a for large n. In later chapters, such behavior will be looked at under the topic *stability*. The behavior of $x(n) = a^n x_0$ will be discussed in several cases.

Case 1: $a > 1$. Then, $x(n) \to \pm\infty$ as $n \to \infty$, depending on the sign of x_0.

Case 2: $a \in (0, 1)$. Then, $x(n) \to 0$ as $n \to \infty$. However, if $x_0 > 0$, then the sequence $x(n)$ monotonically and strictly decreases to zero. On the other hand, if $x_0 < 0$, then the sequence $x(n)$ monotonically and strictly increases to zero.

Case 3: $a \in (-1, 0)$. Then, $x(n) \to 0$ as $n \to \infty$. However, $x(n)$ will *oscillate* between negative and positive with the magnitude of oscillate shrinking to zero over time.

Case 4: $a < -1$. Then, as n increases $x(n)$ will oscillate between positive and negative values with the magnitude of oscillate growing to infinity over time.

Case 5: $a = 0$. Then, the system immediately takes the first value $x(1) = 0$, regardless of the location of x_0 and then $x(n) = 0$, $n \geq 1$.

Case 6: $a = -1$. Then, $x(n)$ will oscillate between x_0 and $-x_0$.

Case 7: $a = 1$. Then, $x(n) = x_0$ for all $n \geq 0$.

Fig. 2.1 depicts Case 1 and Case 2.

Notice that $x(t) = x_0 e^{at}$, $t \geq 0$, which is the solution of $x'(t) = ax(t)$, $x(t_0) = x_0$ behaves differently, and its behavior depends on the sign of a and not on its magnitude. This can be seen from the fact that if $a < 0$, then $x(t) \to 0$ as $t \to \infty$, and $|x(t)| \to \infty$ as $t \to \infty$.

We offer the following example.

Example 2.1. Suppose a certain population of owls is growing at the rate of 2 percent per year. If we let x_0 represent the size of the initial population of owls and $x(n)$ the number of owls n years later, then

$$x(n+1) = x(n) + 0.02x(n) = 1.02x(n), \quad n \geq 0.$$

That is, the number of owls in any given year is equal to the number of owls in the previous year plus 2 percent of the number of owls in the previous year. Hence, the number of owls at any time n is the solution of the above difference equation and is given by

$$x(n) = (1.02)^n x_0. \quad \square$$

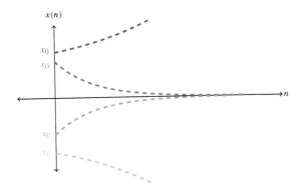

FIGURE 2.1   Diagram for Case 1 and Case 2.

Next we attempt to solve the non-homogeneous linear difference equation

$$x(n+1) = ax(n) + b, \quad x(0) = x_0, \quad n = 0, 1, 2 \cdots \qquad (2.1.3)$$

where a and b are constants. As in the case of the homogeneous equation, we vary n over its domain. Consequently,

$$x(1) = ax_0 + b$$
$$x(2) = ax(1) + b = a^2 x_0 + ab + b$$
$$x(3) = ax(2) + b = a^3 x_0 + a^2 b + ab + b$$
$$x(4) = ax(3) + b = a^4 x_0 + a^3 b + a^2 b + ab + b$$
$$\vdots$$
$$x(n) = a^n x_0 + a^{n-1} b + \cdots + a^2 b + ab + b.$$

Factoring b we arrive at

$$x(n) = a^n x_0 + b \left(1 + a + a^2 + a^3 + \cdots + a^{n-1} \right). \qquad (2.1.4)$$

(i) $a = 1$.

In this case we have that

$$1 + a + a^2 + a^3 + \cdots + a^{n-1} = n,$$

and hence from (2.1.4) we obtain the solution

$$x(n) = x_0 + nb, \quad n = 0, 1, 2, \cdots \qquad (2.1.5)$$

(ii) $a \neq 1$.

This is a more complicated case that requires some calculations as we shall see. The right side of (2.1.4) can be written as

$$x(n) = a^n x_0 + b \sum_{k=0}^{n-1} a^k$$

$$= a^n x_0 + b \frac{a^k}{a-1} \Big|_{k=0}^{n}$$

$$= a^n x_0 + b \left(\frac{a^n}{a-1} - \frac{1}{a-1} \right)$$

$$= a^n x_0 + b \frac{a^n - 1}{a-1}.$$

We conclude that in the case $a \neq 1$, the solution of (2.1.3) is

$$x(n) = a^n x_0 + b \frac{a^n - 1}{a-1}, \quad n = 0, 1, 2, \ldots. \tag{2.1.6}$$

It should have caused no difficulty to start the solution at any initial time n_0, such that $n \geq n_0$, where $n \in \mathbb{Z}$.

Example 2.2. **(a)** The solution of $\Delta x(n) = 2$, $x(0) = 3$ is $x(n) = 3 + 2n$.

(b) The solution of $x(n+1) = \frac{1}{2}x(n) + 4$, $x(0) = 3$ is

$$x(n) = 3.2^{-n} + 4 \frac{2^{-n} - 1}{\frac{1}{2} - 1}$$

$$= 3.2^{-n} + 4 \frac{2^{-n} - 1}{\frac{1}{2} - 1}$$

$$= 3.2^{-n} - 8(2^{-n} - 1).$$

We ask the reader to verify that the obtained solution in part (b) does indeed satisfy the difference equation. □

We end this section by taking a closer look at the relations between differential equations and difference equations. For simplicity we consider the first-order difference equation

$$x(n+1) = ax(n), \quad x(0) = x_0, \quad n \geq 0, \tag{2.1.7}$$

which has solution

$$x(n) = a^n x_0. \tag{2.1.8}$$

For step size h, consider the first-order difference equation

$$t_{n+1} = t_n + h, \quad t_0 = 0.$$

Then its solution is $t_n = nh$. Let $x(n) = x(t_n)$ and rewrite (2.1.7) as

$$x(n+1) - x(n) = (a - 1)x(n).$$

Since $x(n+1) = x(t_{n+1}) = x(t_n + h)$ we have that

$$\frac{x(t_n + h) - x(t_n)}{h} = \frac{a - 1}{h} x(t_n). \tag{2.1.9}$$

For small h

$$\frac{x(t_n + h) - x(t_n)}{h} \approx x'(t)$$

and thus (2.1.9) can be approximated by the differential equation

$$x'(t) = \frac{a - 1}{h} x(t)$$

and has the solution

$$x(t) = x_0 e^{\frac{a-1}{h} t}.$$

Notice that $e^{\frac{a-1}{h} t} = a^n + O(n(a-1)^2)$, and so for $n << \frac{1}{(a-1)^2}$ we obtain good agreement with (2.1.8).

Going from differential equations to difference equations, one would have to use what is called Euler's method.

2.1.1 Exercises

Exercise 2.1. Obtain a formula for the solution similar to (2.1.6) of the difference equation

$$x(n+1) = ax(n) + b, \quad x(n_0) = x_0, \quad n \geq n_0.$$

Make sure to consider $a = 1$, and $a \neq 1$.

Exercise 2.2. Solve.

(a) $x(n+1) = 2x(n) + 5$, $x(0) = 6$, $n \geq 0$.
(b) $x(n+1) = x(n) + 5$, $x(-2) = 6$, $n \geq -2$.
(c) $x(n+1) = 4x(n) + 7$, $x(2) = 9$, $n \geq 2$.

Exercise 2.3. Let $x_1(n)$ be a solution of $x(n+1) = a(n)x(n) + b_1(n)$ and $x_2(n)$ be a solution of $x(n+1) = a(n)x(n) + b_2(n)$. Show that $z(n) = x_1(n) + x_2(n)$ is a solution of

$$z(n+1) = a(n)z(n) + b_1(n) + b_2(n).$$

2.1.2 Applications

In this section we look at some basic applications to linear difference equations that are particular to Section 2.1. We will display a few examples.

Example 2.3 (Bank account). Suppose the balance of a bank account after the nth year is denoted by $x(n)$, $n \geq 0$. Let $x(0) = x_0$ be the initial balance and r be the interest rate earned on the account. Then the annual change in the balance is $\triangle x(n)$, where the interest rate r can be expressed as $r = \dfrac{\triangle x(n)}{x(n)}$. Thus, we have the first-order difference equation $\triangle x(n) = rx(n)$, or

$$x(n+1) = (1+r)x(n).$$

Hence, by the results of Section 2.1 the solution is

$$x(n) = (1+r)^n x_0, \ n \geq 0,$$

which represents the amount in the bank at any time (in years) n. It is clear from the solution that the account is increasing in time since $1 + r > 1$. □

The next application is regarding *cell division*.

Example 2.4 (Cell division). Assume that each member of a population of cells produces a daughter cell during synchronous cell division. Let's use

$$x(1), x(2), \ldots, x(n),$$

to define the number of cells in each generation. The relationship between succeeding generations is then

$$x(n+1) = ax(n),$$

for some positive constant a. Consequently, the nth generation is given by the solution

$$x(n) = a^n x_0, \ n \geq 0,$$

where we assumed x_0 to be the initial number of cells. We remark that if $a > 1$, then $x(n)$ increases over successive generation. On the other hand, if $0 < a < 1$, then $x(n)$ decreases over successive generation. Finally, if $a = 1$, then $x(n)$ remains unchanged. That is $x(n) = x_0$, for all $n \geq 0$. □

Example 2.5 (Fibonacci-Rabbit breeding). The problem of how many pairs of rabbits you will have after 1 year if you start with 1 pair and they each take 1 month to mature and produce 1 other pair each month afterwards is illustrated through a calendar. This is a representation of the classic Fibonacci problem of reproducing rabbits. Take a pair of baby rabbits and put them into a field, how many pairs will there be: a) at the end of each month, and b) at the end of one year? Criteria: Rabbits are fully-grown at 1 month and have another pair of bunnies at 2 months. Each pair is comprised of 1 male and 1 female and no rabbits die or leave the field. This is the classic rabbit problem Fibonacci used to generate the sequence:

$$1, 1, 2, 3, 5, 8, 13, 21, 34, 55, 89, 144, 233, \ldots . \qquad (2.1.10)$$

Let $x(n)$ be the number of pairs at the end of n months, then based on the sequence, this is modeled using the difference equation

$$x(n+2) = x(n) + x(n+1), \quad x(1) = 1, \ x(2) = 1. \qquad (2.1.11)$$

The sequence given by (2.1.10) is called the Fibonacci sequence. We will solve (2.1.11) once we learn how to solve higher-order difference equations.   □

Example 2.6 (Amortization of a Loan). Compound interest computation is one of the simplest economic situations where a difference equation readily appears. Consider the progression of debt. The outstanding debt at the start of period t is denoted by $D(t)$, and the debt in period $t+1$, $D(t+1)$ is computed using the straightforward accounting rule:

$$D(t+1) = D(t) + rD(t) - g(t) = (1+r)D(t) - g(t), \qquad (2.1.12)$$

where $rD(t)$, is the interest that is accruing as of period t's end. At the conclusion of time t, the debt contract is serviced by paying the sum $g(t)$. Normally, this payment covers both the principal payments and the interest. Eq. (2.1.12) is a linear non-homogeneous first-order difference equation. We will revisit its solution once the topic is covered.   □

The next example is regarding *Newton's cooling law*.

Example 2.7 (Newton law of cooling). Newton's law of cooling states that the rate at which a body cools is proportional to the difference between the body's temperature and the medium's temperature, or what is known as the ambient temperature. In discrete time, if $T(n)$ represents the temperature of an object after n units of time, then the change of temperature over one unit of time $\Delta T(n)$ is given by

$$T(n+1) - T(n) = K(T(n) - A), \quad n \geq 0$$

where A is the ambient temperature.

For an application, assume that when a cake is removed from oven, its temperature is measured at 300 °F. Three minutes later, its temperature is 200 °F. How long will it take to cool off to a room temperature of 70 °F? We are given $A = 70$ °F, $T(0) = 300$ °F, and $T(3) = 200$ °F. Hence we must solve the initial value problem

$$T(n+1) - T(n) = K(T(n) - 70), \quad T(0) = 300, \quad n \geq 0 \qquad (2.1.13)$$

and determine the value K using $T(3) = 200$. By rewriting (2.1.13) we arrive at

$$T(n + 1) = (K + 1)T(n) - 70K.$$

Using (2.1.6) we arrive at the solution

$$T(n) = (K + 1)^n T(0) - 70K \frac{(K + 1)^n - 1}{K}$$
$$= 300(K + 1)^n - 70(K + 1)^n + 70$$
$$= 230(K + 1)^n + 70.$$

By making use of $T(3) = 200\,°\text{F}$, we arrive at

$$200 = 230(K + 1)^3 + 70.$$

This is reduced to

$$\frac{13}{23} = (K + 1)^3.$$

From this we obtain

$$K = (\frac{13}{23})^{\frac{1}{3}} - 1 \approx -0.1732.$$

Thus the solution is

$$T(n) = 230(0.827)^n + 70. \tag{2.1.14}$$

Notice that the solution displayed in (2.1.14) provides no finite solution to $T(n) = 70$, since

$$\lim_{n \to \infty} T(n) = \lim_{n \to \infty} [230(0.827)^n + 70] = 70.$$

Yet, intuitively, we expect the cake to reach room temperature after a reasonable amount of time. □

Example 2.8 (The Tower of Hanoi). Legend says that Edouard Lucas, a French mathematician, created the Tower of Hanoi in 1883. It is also known as the Tower of Brahma or the End of the World Puzzle. He was motivated by a myth about a Hindu temple where young priests would have been trained mentally using the pyramid puzzle. According to legend, 64 gold disks, each somewhat smaller than the one below it, were presented to the temple's priests at the beginning of time. Their task was to move the 64 disks between the three poles, but they were given a crucial restriction:

1. Only one disk can be moved at a time.
2. A disk can only be placed on top of a larger disk or an empty rod.
3. No disk can be placed on top of a smaller disk.

Day and night, the priests were working efficiently. According to the myth, the world and the temple would vanish after they had accomplished their work.

What is the smallest number of moves needed, using the guidelines stated, to move n disks from the first pole to the last pole? Let $x(n)$ be this number. The problem is solved if we could find a recurrence relation for this sequence. The bottom disk cannot be moved off the first pole if there are $n+1$ disks until all the others have been moved off of it and onto the other poles. We begin by moving the top n disks to the second pole. That takes $x(n)$ moves. Then we can move the bottom disk to the third pole, requiring one move. We then move the n disks from the second pole onto the third pole, requiring another $x(n)$ moves. Consequently, we arrive at the relation

$$x(n+1) = x(n) + 1 + x(n),$$

or

$$x(n+1) - 2x(n) = 1, \quad x(1) = 1.$$

This has the solution

$$x(n) = 2^n - 1.$$

Keep in mind that according to the narrative, the priests in the temple had 64 disks to move. Thus, it will take

$$x(64) = 18,446,744,073,709,551,615$$

number of moves in order to move these disks from the first pole to the third pole. No wonder the earth has not vanished yet! □

2.1.3   Discrete logistic model

We begin by discussing a general model that can be applied to various types of discrete population models. Let $N(t+1)$ be the population at time $t+1$. Then, a reliable model would relate $N(t+1)$ to $N(t)$ in the sense that

$$N(t+1) = f(N(t)), \quad t > 0, \ t \in \mathbb{N} \qquad (2.1.15)$$

where f is continuous and it is in general non-linear. Equations of the form (2.1.15) are rich in information regarding the growth or decay of the population. However, in most cases, the equation does not possess a closed form solution and hence other means must be employed to extract such important information. One of the most important tools is to look at the *steady state solution*, or *equilibrium solution*, which we will discuss in later chapters. We shall see later that the decay or growth of the population heavily depends on the size of the initial population, besides other factors. If the population one step later is proportional to the current population, then we take $f(N(t)) = rN(t), r > 0$. Then, Eq. (2.1.15) becomes $N(t+1) = rN(t)$, that has the closed solution $N(t) = r^t N_0$, where N_0 is the initial population. It is evident that the population will grow or decay depending on whether $r > 1$ or $0 < r < 1$, respectively.

A more realistic model is discrete Verhulst process given by

$$N(t+1) = rN(t)\left(1 - \frac{N(t)}{K}\right), \quad K > 0, \qquad (2.1.16)$$

which is kind the discrete analogue of the continuous logistic growth model. The constant K is called the *carrying capacity* of the environment due to limited resources. The Verhulst process has its draw back since, for $N(t) > K$, implies that $N(t+1) < 0$.

A more realistic model that $N(t+1)$ stays positive while $N(t) > K$ is given by

$$N(t+1) = N(t)e^{r\left(1 - \frac{N(t)}{K}\right)}, \quad K > 0. \qquad (2.1.17)$$

The mortality factor $e^{-r\frac{N(t)}{K}}$ becomes negligent for large $N(t)$.

In the above discussion, all of the three models can be put in the form

$$N(t+1) = N(t)F(N(t)), \quad t > 0, \quad t \in \mathbb{N} \qquad (2.1.18)$$

for continuous function F.

The function F is non-linear and hence in most cases the solution can not be explicitly found. However, we may search for the steady state solution or equilibrium points. The steady state solutions of (2.1.18) are N^*, where

$$N^* = N^* F(N^*).$$

In other words we set, $N(t+1) = N(t)$. This relation has the two steady state solutions $N^* = 0$, $F(N^*) = 1$. Indeed, the equilibrium points or steady state are the points of intersection of the curves $y = N$ and $y = F(N)$.

From now on we use the independent variable n instead of t. Let r be the growth rate of a population $N(n)$. A reasonable assumption to make is to consider an environmental limitation on the growth on the population. This can be reflected by assuming r depends on the population size. Consequently, by letting $f(N(n)) = r(N(n))N(n)$ the difference equation would be

$$N(n+1) = r(N)N(n), \qquad (2.1.19)$$

where the growth function $r(N)$ decreases as the population increases. We begin our study by considering r to be a linear equation of N. From (2.1.19) the population is at an equilibrium when $r = 1$. To better illustrate the idea we consider different population sizes.

(a) Assume that a population of $N = 100$ has a growth rate of $r(100) = 1.1$. In addition, the population is at equilibrium when $N = 500$. Then an equation

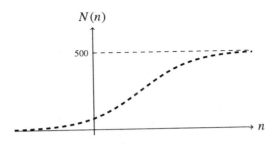

FIGURE 2.2 Growth logistic curve.

of the linear equation that combine the points $(100, 1.1)$ and $(500, 1))$ in the Nr-plane is

$$r(N) = 1.125 - 0.0025N.$$

Thus, the discrete logistic equation given by (2.1.19) is

$$N(n+1) = (1.125 - 0.0025N(n))\,N(n) = 0.0025N(n)\,(450 - N(n)).$$

The graph is represented by the dots on the logistic curve given by Fig. 2.2.
(b) We assume a population $N = 1000$ has a growth rate of $r(1000) = 3.1$. In addition the population is at equilibrium when $N = 6000$. As before, in the Nr-plane we have

$$r(N) = 3.52 - 0.00048N.$$

Thus, the discrete logistic equation given by (2.1.19) yields

$$N(n+1) = (3.52 - 0.00042N(n))N(n).$$

Fig. 2.3(a) below shows the population enter into a periodic cycle of period $T = 4$. One more case.
(c) We assume a population $N = 1000$ has a growth rate of $r(1000) = 3.4$. In addition the population is at equilibrium when $N = 6000$. As before, in the Nr-plane we have

$$r(N) = 3.88 - 0.00048N.$$

Thus, the discrete logistic equation given by (2.1.19) yields

$$N(n+1) = (3.88 - 0.00048N(n))N(n).$$

In this case the graph has an unrecognizable pattern and chaos takes over.
So far we have been considering models in the forms of (2.1.16). By letting $x = \frac{N}{K}$, then using n for t, model (2.1.16) takes the form

$$x(n+1) = rx(n)(1 - x(n)), \tag{2.1.20}$$

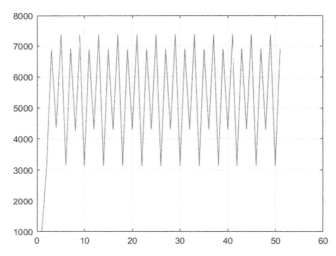

FIGURE 2.3 Periodic cycle of period $T = 4$, with $N_0 = 1000$.

where $r > 0$ is the growth rate.

Notice that comparing $N(n + 1) = 0.0025N(n)(450 - N(n))$ from part (a) with (2.1.20), we see that $r = 1.125$. This is obvious from the fact that

$$N(n + 1) = 0.0025N(n)(450 - N(n)) = 0.0025(450)N(n)(1 - \frac{N(n)}{450}).$$

Now by letting $x = \dfrac{N}{450}$, we obtain

$$x(n + 1) = 1.125x(n)(1 - x(n)).$$

Similarly, for (b) we have

$$x(n + 1) = 3.52x(n)(1 - x(n)),$$

and for (c),

$$x(n + 1) = 3.88x(n)(1 - x(n)).$$

We create the bifurcation diagram as shown below to better understand the complexity. We iterate the appropriate logistic function $x(n + 1) = rx(n)(1 - x(n))$ starting from the point $x = \frac{1}{2}$ by varying the growth rate $2 < r < 4$. We perform over a thousand iterations. Since transient behavior is unimportant to us, we delete the first hundred or so iterations. The remaining points are plotted after that. We refer to the bifurcation diagram in Fig. 2.4(a). We conclude that the discrete logistic equation (2.1.20) event though it looks simple, however, it displays a chaotic behavior as the parameter r changes. In Fig. 5.2 of Chapter 5, we

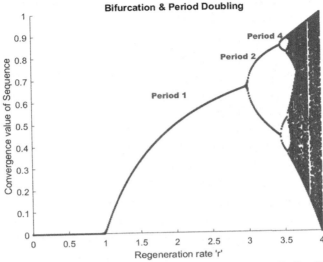

FIGURE 2.4 Bifurcation diagram for $x(n+1) = rx(n)(1 - x(n))$, $x \in [0, 1]$, $x(1) = 0.5$, and $r \in [0, 4]$. We can observe that the two cycle splits into a four cycle at about $r \approx 3.45$ as r increases. As we go through a process known as a period doubling cascade, that keeps happening repeatedly, the pandemonium that results from that infinite cascade ends about $r \approx 3.57$. There are, however, brief periods of calm intermingled, with the most notable one occurring about $r \approx 3.832$.

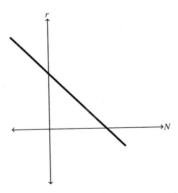

FIGURE 2.5 The growth rate r can be become negative when it is taken to be a linear function in N.

display a graphical method for different values of r, and study the nature of the two fixed points. More on this in Chapter 5. In the previous discussion that lead to chaos in the bifurcation diagram we assumed the growth rate as a linear function of N, as depicted in Fig. 2.5. One of the rules in logistic population models is the fact that the growth rate r decreases, when the population increases, which is not reflected in Fig. 2.5. The logistic model given by (2.1.16) incorporates a

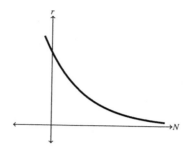

FIGURE 2.6 Non-linear relation between r and the population N.

linear relation between r and the population N. Moreover, model (2.1.16) predicts a negative population once the population grows over a certain threshold. For example, if the population in (a) can grow to a value, say $N(n) = 600$, then the model predicts the next term to be

$$N(n+1) = 0.0025(600)(450 - 600) < 0.$$

However, in this example, the population can never reach 600 with an initial population, $0 < N_0 < 450$. Another alternative to assuming r as a linear function of N is by taking

$$r(N) = \frac{1}{cN + d} \tag{2.1.21}$$

for positive constant c and d. See Fig. 2.6.

Contrasting Fig. 2.5 and Fig. 2.6 we see that in Fig. 2.5 the growth rate r decreases with constant slope until crossing the N-axis and taking on negative values. However, in Fig. 2.6 the graph flattens down as it gets closer to the N-axis, avoiding negative r values.

Substituting (2.1.21) into (2.1.18) and by noting that we are using n for t we arrive at the non-linear difference equation, which we call the *reformed discrete logistic model*

$$N(n+1) = \frac{N(n)}{cN(n) + d}, \quad n \geq 0, \ n \in \mathbb{N}. \tag{2.1.22}$$

It is worth noting that the model given by (2.1.22) will only predict positive population for positive initial population. System (2.1.22) has two equilibrium solutions; namely, $N^* = 0$ and $N^* = \frac{1-d}{c} > 0$, for $0 < d < 1$. If we substitute $N^* = \frac{1-d}{c}$ for N in $r(N) = \frac{1}{cN + d}$ we obtain $r(N) = 1$. We leave it as an exercise to solve the reformed discrete logistic model in (2.1.22).

Concluding remarks. The discrete logistic equation, or model, given by (2.1.20), in most cases cannot be solved in closed form. However, Wolfram

(2002, p. 1098) [41] conjectured that any closed form of the solution of (2.1.20) will most likely be of the form

$$x(n) = \frac{1}{2}\Big[1 - f\left(r^n f^{-1}(1 - 2x_0)\right)\Big], \qquad (2.1.23)$$

for some function f with inverse function f^{-1}. In the same line of thought, M. Trott, did show that smooth solutions can only exist for nonzero and even r. Finally, Wolfram (2002, p. 1098) [41] and R. Germundsson, displayed the type of functions f with a particular value of r for which (2.1.23) is satisfied. Those types of functions are

(a) $f(x) = 2\cos\left(\frac{1}{3}(\pi - \sqrt{3}x)\right)$; $r = -2$,
(b) $f(x) = e^x$; $r = 2$, and
(c) $f(x) = \cos(x)$; $r = 4$.

As we have said before, model (2.1.16) is the discretized model of the logistic equation

$$x'(t) = rN(t)\left(1 - \frac{N(t)}{K}\right), \qquad K \neq N_0 > 0,$$

where N_0 is the initial population at time $t = 0$. Using separation of variables, we obtain the solution

$$x(t) = \frac{ce^{rt}}{1 + \frac{c}{K}e^{rt}}, \quad t \geq 0,$$

where $c = \frac{KN_0}{K - N_0}$. The graph of the solution resembles the graph in Fig. 2.2, where no chaotic behavior occurs, regardless of the value of the growth rate r. This was not the case in the discrete model.

2.1.4   Exercises

Exercise 2.4. A bank account has a yearly interest rate of 3 percent compounded monthly. If you invest $1000, dollars how much money do you have after 5 years?

Exercise 2.5. Your great uncle Jacob invested $100 into a bank account 30 years ago and forgot about it. The account made 5% interest, compounded annually. Your uncle just died, and you inherited the account. How much did you get? Now suppose the bank charges $10 a month to maintain the account. How much will you have in 30 years?

Exercise 2.6. A bank account has a yearly interest rate of 3 percent compounded monthly. If you invest $1000 dollars and add $10 dollars each month, how much money do you have after 5 years?

Exercise 2.7. A thermometer is brought outside, where the air temperature is $5\,^\circ$F, from an interior room. The thermometer shows a reading of $55\,^\circ$F, after one minute and $30\,^\circ$F after five minutes. What was the room's starting temperature?

Exercise 2.8. Suppose a cup of tea, initially at a temperature of 180 °F, is placed in a room which is held at a constant temperature of 80 °F. Moreover, suppose that after one minute the tea has cooled to 175 °F. What will the temperature be after 20 minutes?

Exercise 2.9. Find the equilibrium solution N^* of

$$N(t+1) = rN(t)(1 - N(t))$$

and decide if N^* is a repeller or an attractor.

Exercise 2.10. This exercise generalizes the Tower of Hanoi, in Example 2.8 to four towers. Suppose you have four towers (A, B, C, and D). The goal is to write a difference equation that will give the minimum required moves to move all the disks from one tower, say A to another, say D, using the other towers B and C as intermediate steps, while adhering to the same rules as in Example 2.8.

Exercise 2.11. Assumption: Today's demand for any commodity is a function of present price $p(n)$ while today's supply depends upon yesterday's decisions about the output or supply. Hence, output or supply is naturally influenced by yesterday's price $p(n-1)$. We say the market is at equilibrium when supply is equal to demand. Let S and D be the supply and demand functions of apple, respectively. Assume $S(n) = 4p(n-1) - 10$ and $D(n) = -5p(n) + 35$. Find the price $p(n)$ knowing that $p(0) = 6$ when the market of apple is at equilibrium.

Exercise 2.12. Find $r(N)$ that is given by (2.1.21) in each of the cases (a), (b), and (c) and then find the corresponding $N(n+1)$ given by (2.1.22).

Exercise 2.13. Use the substitution $y(n) = \dfrac{1}{N(n)}$ with $y_0 = \dfrac{1}{N_0}$ to solve the reformed discrete logistic model in (2.1.22).

Answer: $N(n) = \dfrac{1-d}{(y_0(1-c) - c)d^n + c}$.

Exercise 2.14. This exercise is a validation of the bifurcation diagram given in Fig. 2.4(a).

For what values of the growth rate r a 2-cycle begins for the discrete logistic equation $x(n+1) = rx(n)(1 - x(n))$? What happens when $r = 3$?
Hint: To find the branches of the 2-cycle, set $x(n+2) = x(n)$ and solve for $x(n)$.

Exercise 2.15. Let $g(x) = rx(1-x)$. Verify that

$$\frac{g^2(x) - x}{g(x) - x} = \frac{(r+1)(3-r)}{r^2},$$

where $g^2(x) = g(g(x))$. Validate the result of Exercise 2.14 for obtaining the minimum value of r at which 2-cycle begins for $x(n+1) = g(x(n))$.

Exercise 2.16. Verify that for each of the functions f given in (a), (b), and (c), when substituted into (2.1.23), the resulting equation satisfies the discrete logistic equation given by (2.1.20).

2.2 Variation of parameters

In this section we look at the non-homogeneous difference equation

$$x(n+1) = a(n)x(n) + g(n, x(n)), \quad x(n_0) = x_0, \quad n \geq n_0, \quad n \in \mathbb{Z}^+. \quad (2.2.1)$$

The linearity of (2.2.1) depends on the given function $g(n, x(n))$. Throughout this book, we adopt the following two conventions:

1. $\displaystyle\sum_{s=a}^{b} y(s) = 0$ for all $a > b$ and

2. the product of $y(t)$ from $t = a$ to b is denoted by $\displaystyle\prod_{t=a}^{b} y(t)$ such that $\displaystyle\prod_{t=a}^{b} y(t) = 1$ for all $a > b$.

In differential calculus, the first-order differential equation

$$x'(t) = a(t)x(t), \quad x(t_0) = x_0, \quad t \geq t_0$$

has the solution

$$x(t) = x_0 e^{\int_0^t a(s)ds}. \quad (2.2.2)$$

The discrete analogue is evident from the solution of the first-order difference equation

$$x(n+1) = a(n)x(n), \quad x(n_0) = x_0, \quad n \geq n_0. \quad (2.2.3)$$

Note that the domain of (2.2.3) is the set

$$\mathbb{N}_{n_0} = \{n_0, n_0 + 1, n_0 + 2, n_0 + 3, \cdots\}.$$

We assume $a(n) \neq 0$, for all $n \in \mathbb{N}_{n_0}$. To obtain the discrete analogue of (2.2.2) we vary $n \in \mathbb{N}_{n_0}$, in a consecutive way beginning with n_0. Thus, we iterate starting with $n = n_0$ and obtain the following:

$$x(n_0 + 1) = a(n_0)x_0$$
$$x(n_0 + 2) = a(n_0 + 1)x(n_0 + 1) = a(n_0 + 1)a(n_0)x_0$$
$$x(n_0 + 3) = a(n_0 + 2)x(n_0 + 2) = a(n_0 + 2)a(n_0 + 1)a(n_0)x_0$$

$$\vdots$$

$$x(n_0 + k) = a(n_0)a(n_0 + 1) \cdots a(n_0 + k - 1)x_0 = x_0 \prod_{i=n_0}^{n_0+k-1} a(i).$$

Thus, by replacing $n_0 + k$ with n, we arrive at the closed form solution of (2.2.3)

$$x(n) = x_0 \prod_{i=n_0}^{n-1} a(i), \quad n \geq n_0. \quad (2.2.4)$$

Next we verify that (2.2.4) satisfies (2.2.3). First we note that $x(n_0) = x_0 \prod_{i=n_0}^{n_0-1} a(i) = x_0(1) = x_0$.

Now,

$$x(n+1) = x_0 \prod_{i=n_0}^{n} a(i) = a(n) \left(x_0 \prod_{i=n_0}^{n-1} a(i) \right) = a(n)x(n).$$

We have the following example.

Example 2.9. Consider the difference equation

$$x(n+1) = nx(n), \quad n(1) = 2.$$

Then using (2.2.4), its solution is found to be

$$x(n) = 2 \prod_{s=1}^{n-1} s = 2[1.2.3\cdots(n-2)(n-1)] = 2(n-1)! \quad \square$$

Next we give an expression that defines the solution of (2.2.1). First we make the following definition regarding the product factor.

Definition 2.2.1. Suppose $a(n) \neq 0$ for all $n \in \mathbb{N}$. The *product factor* $\mu(n)$ of (2.2.1) is

$$\mu(n) = \prod_{s=n_0}^{n} a^{-1}(s). \tag{2.2.5}$$

The product factor given by (2.2.5) is the discrete analogue of the integrating factor in differential equations.

Theorem 2.2.1. *Suppose $a(n) \neq 0$ for all $n \in \mathbb{N}_{n_0}$. Then $x(n)$ is a solution of* (2.2.1) *if and only if*

$$x(n) = x_0 \prod_{s=n_0}^{n-1} a(s) + \sum_{r=n_0}^{n-1} \prod_{s=r+1}^{n-1} a(s)g(r, x(r)). \tag{2.2.6}$$

Proof. Let $x(n)$ be a solution of (2.2.1). Multiply both sides of (2.2.1) by the product factor and get

$$\prod_{s=n_0}^{n} a^{-1}(s)x(n+1) - \prod_{s=n_0}^{n} a^{-1}(s)a(n)x(n) = g(n, x(n)) \prod_{s=n_0}^{n} a^{-1}(s).$$

Combining the middle term we may rewrite the above expression in the form

$$\prod_{s=n_0}^{n} a^{-1}(s)x(n+1) - \prod_{s=n_0}^{n-1} a^{-1}(s)x(n) = g(n, x(n)) \prod_{s=n_0}^{n} a^{-1}(s). \tag{2.2.7}$$

Thus, Eq. (2.2.7) is equivalent to

$$\Delta\left(x(n)\prod_{s=n_0}^{n-1}a^{-1}(s)\right)=g(n,x(n))\prod_{s=n_0}^{n}a^{-1}(s). \qquad (2.2.8)$$

By summing (2.2.8) from $r=n_0$ to $r=n-1$, we have that

$$\sum_{r=n_0}^{n-1}\Delta\left(x(r)\prod_{s=n_0}^{r-1}a^{-1}(s)\right)=\sum_{r=n_0}^{n-1}g(r,x(r))\prod_{s=n_0}^{r}a^{-1}(s).$$

Applying Theorem 1.4.2 FTDC, we arrive at the expression

$$x(n)\prod_{s=n_0}^{n-1}a^{-1}(s)=x_0+\sum_{r=n_0}^{n-1}g(r,x(r))\prod_{s=n_0}^{r}a^{-1}(s).$$

Multiplying the above expression with $\prod_{s=n_0}^{n-1}a(s)$ yields

$$x(n)=x_0\prod_{s=n_0}^{n-1}a(s)+\sum_{r=n_0}^{n-1}g(r,x(r))\frac{\prod_{s=n_0}^{n-1}a(s)}{\prod_{s=n_0}^{r}a(s)}$$

$$=x_0\prod_{s=n_0}^{n-1}a(s)+\sum_{r=n_0}^{n-1}\prod_{s=r+1}^{n-1}a(s)g(r,x(r)).$$

Now suppose $x(n)$ satisfies (2.2.6). Then using (2.2.6) we arrive at

$$x(n+1)=x_0\prod_{s=n_0}^{n}a(s)+\sum_{r=n_0}^{n}\prod_{s=r+1}^{n}a(s)g(r,x(r))$$

$$=a(n)x_0\prod_{s=n_0}^{n-1}a(s)+\sum_{r=n_0}^{n}\prod_{s=r+1}^{n}a(s)g(r,x(r))$$

$$=a(n)x_0\prod_{s=n_0}^{n-1}a(s)+\sum_{r=n_0}^{n-1}\prod_{s=r+1}^{n}a(s)g(r,x(r))+\prod_{s=n+1}^{n}a(s)g(n,x(n))$$

$$=a(n)\left[x_0\prod_{s=n_0}^{n-1}a(s)+\sum_{r=n_0}^{n-1}\prod_{s=r+1}^{n-1}a(s)g(r,x(r))\right]+g(n,x(n))$$

$$=a(n)x(n)+g(n,x(n)).$$

This completes the proof. $\qquad\qquad\qquad\qquad\qquad\qquad\qquad\Box$

Note that if $a(n)=0$, then (2.2.6) is not a solution of (2.2.1). Special forms of (2.2.6) can be easily obtained depending on the functions a and g. For example:

if $a(n) = a = $ **constant**, then (2.2.6) reduces to

$$x(n) = a^{n-n_0}x_0 + \sum_{r=n_0}^{n-1} a^{n-r-1}g(r, x(r)). \tag{2.2.9}$$

We have the following example.

Example 2.10. Consider the difference equation

$$x(n+1) = 3x(n) + 5^n, \quad x(1) = 2.$$

Then by (2.2.9) we have

$$x(n) = 3^{n-1}(2) + \sum_{r=1}^{n-1} 3^{n-r-1}5^r$$

$$= 3^{n-1}(2) + 3^{n-1}\sum_{r=1}^{n-1} 3^{-r}5^r$$

$$= 3^{n-1}(2) + 3^{n-1}\sum_{r=1}^{n-1}\left(\frac{5}{3}\right)^r = 3^{n-1}(2) + 3^{n-1}\frac{\left(\frac{5}{3}\right)^r}{\frac{5}{3}-1}\Big|_{r=1}^{n}$$

$$= 3^{n-1}(2) + \frac{3^n}{2}\left[\left(\frac{5}{3}\right)^n - \frac{5}{3}\right]. \quad \square$$

Comparing the variation of parameters obtained for (2.2.1) with the one for the non-linear differential equation

$$x'(t) = a(t)x(t) + g(t, x(t)), \quad x(t_0) = x_0, \ t \geq t_0$$

where $g \in C(\mathbb{R} \times \mathbb{R}, \mathbb{R})$ and $a \in C(\mathbb{R}, \mathbb{R})$, we obtain the formula

$$x(t) = x_0 e^{\int_{t_0}^{t} a(u)du} + \int_{t_0}^{t} g(s, y(s))e^{\int_{s}^{t} a(u)du}ds, \quad t \geq t_0. \tag{2.2.10}$$

We end this section with existence and uniqueness results.

Corollary 2.1. *The solution of the linear difference equation*

$$x(n+1) = a(n)x(n) + f(n), \quad x(n_0) = x_0, \ n \geq n_0, \ n \in \mathbb{N}_{n_0}, \tag{2.2.11}$$

is unique.

Proof. The existence of the solution is implied from (2.2.6) by replacing g with $f(n)$. Left to show uniqueness. Assume there are two solutions $x_1(n)$ and $x_2(n)$ that satisfy (2.2.11) along with the given initial condition. Let $n^* \in \mathbb{N}_{n_0}$ be the smallest in the set. Then we must have $n^* \geq n_0$ since $x_1(n_0) = x_2(n_0) = x_0$. On the other hand, by the definition of n^* we have $x_1(n^* - 1) = x_2(n^* - 1)$. Consequently,

$$
\begin{aligned}
x_1(n^*) &= a(n^* - 1)x_1(n^* - 1) + f(n^* - 1) \\
&= a(n^* - 1)x_2(n^* - 1) + f(n^* - 1) \\
&= x_2(n^*),
\end{aligned}
$$

which is a contradiction. Thus, $n^* = n_0$. Since $x_1(n_0) = x_2(n_0)$, we must have $x_1(n) = x_2(n)$ for all $n \in \mathbb{N}_{n_0}$ and consequently, the solution is unique. This completes the proof. $\square$

2.2.1 Exercises

Exercise 2.17. Show that if $a(n) = a$ and $g(n) = b$, where a and b are constants, then (2.2.9) reduces to

$$
x(n) = a^{n-n_0} x_0 + \frac{ba^n}{1-a} \left(a^{-n} - a^{-n_0} \right), \quad n = n_0, n_0 + 1, n_0 + 2 \dots.
$$

Compare this answer with (2.1.6) when $n_0 = 0$.

Exercise 2.18. Let

$$
x(n + 1) = a(n)x(n) + g(n, x(n)), \quad n \in \mathbb{Z}. \tag{2.2.12}
$$

Suppose

$$
a(n + T) = a(n), \quad g(n + T, x(\cdot)) = g(n, x(\cdot)).
$$

Also assume that $a(n) \neq 0$ for $n \in \mathbb{Z}$ and $1 - \prod_{s=n-T}^{n-1} a(s) \neq 0$, $n \in \mathbb{Z}$. Let P_T be the space of all periodic sequences of period T. That is

$$
P_T = \{\phi : \mathbb{Z} \to \mathbb{R} : \phi(n + T) = \phi(n)\}.
$$

Show that if $x \in P_T$ is a solution of (2.2.12), then it satisfies

$$
x(n) = \left(\prod_{s=n-T}^{n-1} a(s) \right) x(n - T) + \sum_{u=n-T}^{n-1} \left(g(u, x(u)) \prod_{s=u+1}^{n-1} a(s) \right). \tag{2.2.13}
$$

In addition, show that (2.2.13) is equivalent to

$$
x(n) = \frac{1}{1 - \prod_{s=n-T}^{n-1} a(s)} \sum_{u=n-T}^{n-1} \left(g(u, x(u)) \prod_{s=u+1}^{n-1} a(s) \right).
$$

Exercise 2.19. Solve.

(a) $x(n+1) = 2x(n) + n$, $x(0) = 5$, $n \geq 0$. Ans: $5.2^n + 2^n - n - 1$.
(b) $x(n+1) = (n-7)x(n)$, $x(-2) = 6$, $n \geq -2$.
(c) $x(n+1) = 4x(n) + 7$, $x(2) = 9$, $n \geq 2$.
(d) $x(n+1) = \dfrac{1}{n+1}x(n)$, $x(0) = 1$, $n \geq 0$.
(e) $x(n+1) = \dfrac{n-4}{n+1}x(n)$, $x(0) = 1$, $n \geq 0$.
(f) $x(n+1) = (n+1)x(n) + 2^n(n+1)!$, $x(0) = 1$, $n \geq 0$. (Ans: $n!2^n$)
(g) $x(n+1) = 2^n x(n) + [n]^3$, $x(1) = 1$, $n \geq 1$.
(h) $x(n+1) = nx(n) + [n]^3$, $x(1) = 1$, $n \geq 1$.
(i) $x(n+1) = x^2(n)$, $x(0) = x_0$, $n \geq 0$.

Exercise 2.20. Consider the non-linear difference equation

$$y(n+1) = \frac{y(n)}{1 + y(n)}, \quad y(0) = y_0 \neq 0, \quad n \in \mathbb{N}_0. \tag{2.2.14}$$

(a) Show that $y(n) \neq 0$, for all $n \in \mathbb{N}_0$.
(b) Use the transformation $x(n) = \frac{1}{y(n)}$ and transform (2.2.14) to the linear difference equation

$$x(n+1) - x(n) = 1.$$

(c) Find the solution of (2.2.14).

2.3   Higher-order difference equations

In Section 1.1 we introduced higher-order linear difference equations, by considering, for $n \in \mathbb{N}_{n_0}$, the kth-order linear difference equation

$$a_k x(n+k) + a_{k-1}x(n+k-1) + a_{k-2}x(n+k-2) + \cdots$$
$$+ a_1 x(n+1) + a_0 x(n) = F(n), \tag{2.3.1}$$

where the coefficients a_i, $i = 1, 2, \ldots, k$ are given along with $F(n)$ and $a_k \neq 0$. The coefficients a_i, $i = 1, 2, \ldots, k$ may depend on n. If the function $F(n)$ vanishes for all $n \in \mathbb{N}_{n_0}$, then we call (2.3.1) a *homogeneous* linear equation; otherwise, it is *non-homogeneous*. Thus the *homogeneous* linear equation associated with (2.3.1) is

$$a_k x(n+k) + a_{k-1}x(n+k-1) + a_{k-2}x(n+k-2) + \cdots$$
$$+ a_1 x(n+1) + a_0 x(n) = 0. \tag{2.3.2}$$

Definition 2.3.1. Let $f_1, f_2, \ldots, f_n$ be a set of functions defined on the set $\mathbb{N}_{n_0}$. We say the set $\{f_1, f_2, \ldots, f_j\}$ is *linearly dependent* on $\mathbb{N}_{n_0}$ if there exists constants $c_1, c_2, \ldots, c_j$ not all zero, such that

$$c_1 f_1 + c_2 f_2 + \ldots + c_j f_j = 0$$

for every $n \in \mathbb{N}_{n_0}$. If the set of functions is not linearly dependent on the interval $\mathbb{N}_{n_0}$, it is said to be *linearly independent*.

Definition 2.3.2. (*Fundamental set of solutions*). A set of k solutions of the linear difference equation (2.3.2) all defined on the same set $\mathbb{N}_{n_0}$, is called a *fundamental set of solutions* on $\mathbb{N}_{n_0}$ if the solutions are linearly independent functions on $\mathbb{N}_{n_0}$.

We have the following corollary.

Corollary 1. *If* $\{\phi_1(n), \phi_2(n), \ldots, \phi_k(n)\}$ *form a fundamental set of solutions on* $\mathbb{N}_{n_0}$, *then the general solution of* (2.3.2) *is given by*

$$x(n) = c_1\phi_1(n) + c_2\phi_2(n) + \ldots + c_k\phi_k(n),$$

for constants c_i, $i = 1, 2, \ldots, k$.

Example 2.11. The second order difference equation

$$x(n+2) + x(n+1) - 2x(n) = 0, \tag{2.3.3}$$

has the two solutions $\varphi_1(n) = 1$ and $\varphi_2(n) = (-2)^n$ and they are linearly independent on $\mathbb{Z}$ and hence the fundamental solution is

$$x(n) = c_1\varphi_1(n) + c_2\varphi_2(n) = c_1 + c_2(-2)^n,$$

for constants c_1 and c_2. $\qquad\qquad\square$

To completely describe the solution of either (2.3.1) or (2.3.2), we impose the initial conditions

$$x(n_0) = d_0, \quad x(n_0 + 1) = d_1, \quad \ldots, \quad x(n_0 + k - 1) = d_{k-1}, \tag{2.3.4}$$

for an initial point n_0, and constants d_i, $i = 0, 1, 2, \ldots, k - 1$.

Theorem 2.3.1 (Existence and uniqueness). *Consider the (IVP) defined by* (2.3.1) *and* (2.3.4), *where F and $a_i(n)$, $i = 1, 2, \ldots, k$ are defined on $\mathbb{N}_{n_0}$. Assume $a_i(n)$, $\neq 0$ for $i = 0, k$ and for all $n \in \mathbb{N}_{n_0}$. For $n_0 \in \mathbb{N}_{n_0}$ and given the constants d_i, $i = 0, 1, 2, \ldots, k - 1$, the difference equation (2.3.1) has a unique solution on the entire set $\mathbb{N}_{n_0}$ satisfying the initial condition (2.3.4).*

Proof. Since $a_k(n) \neq 0, a_0(n) \neq 0$ for all $n \in \mathbb{N}_{n_0}$, we may solve for $x(n+k)$ in terms of all other terms and then it follows from iteration on n that

$$x(n_0 + k) = \frac{F(n_0) - a_{k-1}(n_0)d_{k-1} - \cdots - a_0(n_0)d_0}{a_k(n_0)}.$$

In a similar way we are able to solve for $x(n)$ in (2.3.1) for $n > n_0 + k$. The same process can be performed for $n \leq n_0 - 1$. $\qquad\qquad\square$

Another way of determining whether a set of functions is linearly independent or not is to look at the *Casoratian*.

Definition 2.3.3 (Casoratian). Given two functions f and g, the Casoratian of f and g is the determinant

$$W = \begin{vmatrix} f(n) & g(n) \\ f(n+1) & g(n+1) \end{vmatrix} = f(n)g(n+1) - f(n+1)g(n).$$

We write $W(f, g)$ to emphasize the dependence on the functions. Consider the two functions given in Example 2.11. Then

$$W(\varphi_1, \varphi_2) = \begin{vmatrix} 1 & (-2)^n \\ 1 & (-2)^{n+1} \end{vmatrix} = (-2)^{n+1} - (-2)^n = -3(-2)^n \neq 0,$$

for all $n \in \mathbb{Z}$.

This is an example of a linearly independent pair of functions. Note that the *Casoratian* is nonzero everywhere. On the other hand, if the functions f and g are linearly dependent, with $g = cf$ for a nonzero constant c, then

$$W(f, g) = \begin{vmatrix} f(n) & cf(n) \\ f(n+1) & cf(n+1) \end{vmatrix} = cf(n)f(n+1) - cf(n)f(n+1) = 0.$$

Thus the Casoratian of two linearly dependent functions is zero. This will be made formal in Theorem 2.3.2. The above Casoratian discussion can be easily extended to the set of functions $f_1, f_2, \ldots, f_k$, where

$$W(f_1, f_2, \ldots, f_k) = \begin{vmatrix} f_1(n) & f_2(n) & \cdots & f_k(n) \\ f_1(n+1) & f_2(n+1) & \cdots & f_k(n+1) \\ \vdots & \vdots & \ddots & \vdots \\ f_1(n+k-1) & f_2(n+k-1) & \cdots & f_k(n+k-1) \end{vmatrix}.$$

For better illustration, we consider the second-order difference equation

$$x(n+2) + b_1(n)x(n+1) + b_0(n)x(n) = 0, \quad n \in \mathbb{N}_{n_0}. \tag{2.3.5}$$

Theorem 2.3.2. *Suppose x_1 and x_2 are solutions of (2.3.5) on $\mathbb{N}_{n_0}$. Then x_1 and x_2 are linearly independent if and only if*

$$W(x_1(n), x_2(n)) \neq 0, \quad \text{for all } n \in \mathbb{N}_{n_0}.$$

Proof. Let x_1 and x_2 be linearly independent. Then,

$$c_1 x_1(n) + c_2 x_2(n) = 0, \quad \text{for all } n \in \mathbb{N}_{n_0},$$

is true only when $c_1 = c_2 = 0$. Since this is true for all $n \in \mathbb{N}_{n_0}$, it implies that

$$c_1 x_1(n) + c_2 x_2(n) = 0 \text{ and } c_1 x_1(n+1) + c_2 x_2(n+1) = 0 \qquad (2.3.6)$$

for all $x \in I$. Now (2.3.6) can only be true for some c_1, and c_2 not both zero if and only if $W(x_1(n), x_2(n)) = 0$, for all $x \in \mathbb{N}_{n_0}$. If (2.3.6) holds for some point $n^* \in \mathbb{N}_{n_0}$, then the function $x = c_1 x_1 + c_2 x_2$ is a solution (2.3.5) and satisfies the initial conditions, $x(n^*) = x(n^* + 1) = 0$. On the other hand the zero function $x = 0$ is also a solution and satisfies the initial conditions. This violates the uniqueness of the solution unless x and the zero solution, $x = 0$ are the same. Now $x = 0$ implies (2.3.6) is true for all $n \in \mathbb{N}_{n_0}$. This shows that $W(x_1(n), x_2(n)) = 0$, for all $n \in \mathbb{N}_{n_0}$ if and only if $W(x_1(n), x_2(n)) = 0$, for at least one $n^* \in \mathbb{N}_{n_0}$. This completes the proof. $\qquad \square$

We have the following lemma regarding $W(x_1(n), x_2(n))$. Sometimes we adopt the following notation:

$$W(x_1(n+1), x_2(n+1)) := W(n+1).$$

Lemma 2.1. *Suppose* x_1 *and* x_2 *are solutions of* (2.3.5) *on* $\mathbb{N}_{n_0}$. *Then for* $n_0 \in \mathbb{N}_{n_0}$, *we have that for all* $n \geq n_0$

$$W(n) = W(n_0) \prod_{k=n_0}^{n-1} b_0(k). \qquad (2.3.7)$$

Proof. Let x_1 and x_2 be solutions of (2.3.5) on $\mathbb{N}_{n_0}$. Then we have

$$x_1(n+2) = -b_1(n) x_1(n+1) - b_0(n) x_1(n)$$

and

$$x_2(n+2) = -b_1(n) x_2(n+1) - b_0(n) x_2(n).$$

By the definition of the Casoratian we obtain

$$
\begin{aligned}
W(n+1) &= \begin{vmatrix} x_1(n+1) & x_2(n+1) \\ x_1(n+2) & x_2(n+2) \end{vmatrix} \\[2mm]
&= \begin{vmatrix} x_1(n+1) & x_2(n+1) \\ -b_1(n)x_1(n+1) - b_0(n)x_1(n) & -b_1(n)x_2(n+1) - b_0(n)x_2(n) \end{vmatrix} \\[2mm]
&= \begin{vmatrix} x_1(n+1) & x_2(n+1) \\ -b_0(n)x_1(n) & -b_0(n)x_2(n) \end{vmatrix} \\[2mm]
&= b_0(n) \begin{vmatrix} x_1(n+1) & x_2(n+1) \\ -x_1(n) & -x_2(n) \end{vmatrix}
\end{aligned}
$$

$$= b_0(n)\big(x_1(n)x_2(n+1) - x_2(n)x_1(n+1)\big)$$
$$= b_0(n)W(n).$$

Thus, solving the first-order difference equation

$$W(n+1) = b_0(n)W(n),$$

with initial condition $W(n_0)$ we arrive at the result. This completes the proof.
□

In the proof of the above lemma, to get from the second step to the third step in the string of equalities, we multiplied the first row by b_1 and added the resulting row to the second row. This does not change the value of the determinant.

Next, we define the general solution of the non-homogeneous difference equation (2.3.1). The proof of the next theorem will be left as an exercise.

Theorem 2.3.3. *Let $F(n)$ and $a_i(n)$, $i = 1, 2, \ldots, k$ be defined on $\mathbb{N}_{n_0}$. Suppose*

$$\{\phi_1(n), \ \phi_2(n), \ \ldots, \ \phi_k(n)\}$$

form a fundamental set of solutions on $\mathbb{N}_{n_0}$ of the homogeneous difference equation (2.3.2). Denote such solution with

$$y_h(c) = c_1\phi_1(n) + c_2\phi_2(n) + \ldots + c_k\phi_k(n),$$

for constants c_i, $i = 1, 2, \ldots, k$. Let $x_p(n)$ be a particular solution of the non-homogeneous difference equation (2.3.1). Then the general solution of (2.3.1) on $\mathbb{N}_{n_0}$ is given by

$$x(n) = x_h(n) + x_p(n).$$

Example 2.12. Consider the second order difference equation

$$y(n+2) + 3y(n+1) + 2y(n) = 6, \quad y(0) = 1, \quad y(1) = -3. \tag{2.3.8}$$

Clearly, each of the functions $\varphi_1(n) = (-1)^n$, and $\varphi_2(n) = (-2)^n$ is a solution of the homogeneous equation $y(n+2) + 3y(n+1) + 2y(n) = 0$. Also, they are linearly independent since

$$W(\varphi_1, \varphi_2) = \begin{vmatrix} (-1)^n & (-2)^n \\ (-1)^{n+1} & (-2)^{n+1} \end{vmatrix} = -3(2)^n \neq 0, \text{ for all } x \in \mathbb{N}_0$$

Thus, the homogeneous solution of $y(n+2) + 3y(n+1) + 2y(n) = 0$ is given by

$$y_h(x) = c_1(-1)^n + c_2(-2)^n$$

for constants c_1 and c_2. Moreover, $y_p(n) = 1$ is a particular solution of (2.3.8), since

$$y_p(n+2) + 3y_p(n+1) + 2y_p(n) = 6.$$

Thus, the general solution of (2.3.8) is

$$y(n) = y_h(n) + y_p(n) = c_1(-1)^n + c_2(-2)^n + 1.$$

Applying the initial conditions we arrive at $c_1 = -4$, and $c_2 = 4$.

Later on in the chapter we will look at different techniques for finding the particular solution $y_p(n)$. $\square$

2.3.1 Exercises

Exercise 2.21. Use Definition 2.3.1 to show that for any nonzero constant λ and $k \in \mathbb{N}_1$ the set

$$\{\lambda^n,\ n\lambda^n,\ n^2\lambda^n,\ \ldots,\ n^k\lambda^n\}$$

is linearly independent.

Exercise 2.22. Decide if the sequences listed below are independent or not.
(a) $\{n, 1+n\}$, (b) $\{(n+2)^2 - n, 1+n^2\}$, $n \in \mathbb{N}$.

Exercise 2.23. Decide whether or not the solutions given determine a fundamental set of solutions for the equation.

(a) $y(n+3) - 3y(n+2) - y(n+1) + 3y(n) = 0$, $y_1 = 3^n + 1$, $y_2 = 1 - (-1)^n$, $y_3 = 3^n + (-1)^n$.
(b) $y(n+3) - 2y(n+2) - y(n+1) + 2y(n) = 0$, $y_1 = 1$, $y_2 = (-1)^n$, $y_3 = 2^n$.
(c) $y(n+2) - 6y(n+1) + 9y(n) = 0$, $y_1 = 3^n$, $y_2 = n3^n$.
(d) $y(n+2) - 25y(n) = 0$, $y_1 = 5^n$, $y_2 = (-5)^n$.

2.4 Equations with constant coefficients

Now we examine the solutions of the homogeneous nth-order difference equation with constant coefficients

$$a_k x(n+k) + a_{k-1} x(n+k-1) + a_{k-2} x(n+k-2) + \cdots$$
$$+ a_1 x(n+1) + a_0 x(n) = 0, \tag{2.4.1}$$

where the coefficients a_i, $i = 0, 1, 2 \cdots k$ are constants. In the previous sections we noticed that solutions to some of the difference equations that were considered, were some constants raised to power n. Thus, we search for solutions of (2.4.1) of the form

$$x(n) = \lambda^n, \quad n \in \mathbb{N}_0$$

where λ is a parameter to be determined. We begin with the observation that

$$E^k x(n) = E^k \lambda^n = \lambda^{n+k} \quad \text{for } k = 0, 1, 2, \ldots \tag{2.4.2}$$

A substitution of (2.4.2) into (2.4.1) leads to

$$\lambda^n \left(a_k \lambda^k + a_{k-1} \lambda^{k-1} + \ldots + a_2 \lambda^2 + a_1 \lambda + a_0 \right) = 0.$$

Since, $\lambda^n \neq 0$ for any finite λ or n, we must have that

$$a_k \lambda^k + a_{k-1} \lambda^{k-1} + \ldots + a_2 \lambda^2 + a_1 \lambda + a_0 = 0. \tag{2.4.3}$$

Eq. (2.4.3) is referred to as the *characteristic equation* or the *auxiliary equation*.

2.4.1 Distinct roots

The fundamental solution or general solution of (2.4.1) can be written down provided that the roots of (2.4.3) can be found. The simplest case is when all the roots λ_i, $i = 1, 2, \ldots, k$ are real and distinct. That is, no two roots are the same, or

$$\lambda_i \neq \lambda_j, \quad i, j = 1, 2, \ldots, k.$$

We have the following theorem.

Theorem 2.4.1 (Distinct real roots). *Suppose all the roots of (2.4.3) λ_i, $i = 1, 2, \ldots, n$ are real and distinct. Then the general solution of (2.4.1) is given by*

$$x(n) = \sum_{i=1}^{k} c_i \lambda_i^n, \tag{2.4.4}$$

for constants c_i, $i = 1, 2, \ldots, k$.

Proof. Since the roots are real and distinct, the set $\{\lambda_i^n, \ i = 1, 2, \ldots, k\}$ is linearly independent. Moreover, each function in the set is a solution of (2.4.1) and hence they form a fundamental set of solutions. Then, by Theorem 2.3.3, the solution is given by (2.4.4). $\qquad \square$

Example 2.13. Consider the third-order difference equation

$$y(n + 3) + 2y(n + 2) - y(n + 1) - 2y(n) = 0.$$

Its characteristic equation is found to be $\lambda^3 + 2\lambda^2 - \lambda - 2 = 0$, which factors into $(\lambda^2 - 1)(\lambda + 2) = 0$. Thus the three roots are -2, -1, and 1 and they are real and distinct and hence by Theorem 2.4.1 the general solution is

$$y = c_1(-2)^n + c_2(-1)^n + c_3 1^n$$
$$= c_1(-2)^n + c_2(-1)^n + c_3. \quad \square$$

Example 2.14 (Fibonacci-Rabbit breeding). Consider the Fibonacci sequence that we formed its corresponding difference equation in Example 2.5

$$x(n+2) = x(n) + x(n+1), \quad x(1) = 1, \; x(2) = 1. \tag{2.4.5}$$

Its auxiliary equation is

$$\lambda^2 - \lambda - 1 = 0,$$

which has the solution $\lambda = \frac{1 \pm \sqrt{5}}{2}$. Consequently, the general solution of (2.4.5) is given by

$$x(n) = c_1 \left(\frac{1 + \sqrt{5}}{2} \right)^n + c_2 \left(\frac{1 - \sqrt{5}}{2} \right)^n.$$

Applying the initial conditions we see that $c_1 = \frac{1}{\sqrt{5}}$, and $c_2 = -\frac{1}{\sqrt{5}}$. Thus the solution is

$$x(n) = \frac{1}{\sqrt{5}} \left(\frac{1 + \sqrt{5}}{2} \right)^n - \frac{1}{\sqrt{5}} \left(\frac{1 - \sqrt{5}}{2} \right)^n, \quad n \ge 0.$$

We make the following two observations:

1. Sine $\lim\limits_{n \to \infty} \left(\frac{1 - \sqrt{5}}{2} \right)^n = 0$, the rabbit population grows as the term $\frac{1}{\sqrt{5}} \left(\frac{1 + \sqrt{5}}{2} \right)^n$ increases with n.

2. The ratio of two consecutive rabbit generations, for large n is given by

$$\lim_{n \to \infty} \frac{x(n+1)}{x(n)} = \frac{1 + \sqrt{5}}{2} \approx 1.618,$$

which is the famous *golden ratio*. $\qquad\qquad\square$

2.4.2 Repeated roots

Now we turn our attention to the case when the characteristic Eq. (2.4.3) has some of its roots repeated. In such cases, we are not able to produce k linearly independent solutions using Theorem 2.4.1. For example, if the characteristic equation of a given difference equation has the roots 3, 5, 2, and 2, then we can only produce the three linearly independent functions 3^n, 5^n, and 2^n. The problem is then to find a way to obtain the linearly independent solutions. As we have noted before, Eq. (2.4.1) can take the form

$$a_k E^k + a_{k-1} E^{k-1} + a_{k-2} E^{k-2} + \cdots + a_1 E + a_0 I. \tag{2.4.6}$$

We remind ourselves that if we have $(E - r_1)x(n) = 0$ then $x(n) = r_1^n$ is its solution. With this in mind, suppose $\lambda_1, \lambda_2, \cdots \lambda_j$ are roots of the auxiliary

Eq. (2.4.3). Then, Eq. (2.4.6) can be written as

$$(E - \lambda_1)^{\beta_1}(E - \lambda_2)^{\beta_2} \cdots (E - \lambda_j)^{\beta_j} x(n) = 0, \qquad (2.4.7)$$

where $\sum_{i=1}^{j} \beta_i = k$. Suppose that $\beta_1 > 1$. In other word, the root λ_1 has multiplicity at least 2. We already know one solution of $(E - \lambda_1)x(n) = 0$, which is $x(n) = \lambda_1^n$. Thus, we search for other solutions of the form $x(n) = y(n)\lambda_1^n$. Then using (1.3.12) we obtain

$$
\begin{aligned}
(E - \lambda_1)^{\beta_1} \lambda_1^n y(n) &= \sum_{i=0}^{\beta_1} \binom{\beta_1}{i}(-\lambda_1)^{\beta_1 - i} E^i \lambda_1^n y(n) \\
&= \sum_{i=0}^{\beta_1} \binom{\beta_1}{i}(-\lambda_1)^{\beta_1 - i} \lambda_1^{n+i} E^i y(n) \\
&= \lambda_1^{\beta_1 + n} \sum_{i=0}^{\beta_1} \binom{\beta_1}{i}(-1)^{\beta_1 - i} E^i y(n) \\
&= \lambda_1^{\beta_1 + n}(E - 1)^{\beta_1} y(n) \\
&= \lambda_1^{\beta_1 + n} \Delta^{\beta_1} y(n) \\
&= 0
\end{aligned}
$$

only when $y(n)$ is of the form $y(n) = 1, n, n^2, \cdots n^{\beta_1 - 1}$. Thus, $(E - \lambda_1)^{\beta_1}$ has the solutions $1.\lambda_1, n\lambda_1, n^2\lambda_1, \cdots n^{\beta_1 - 1}\lambda_1$. We arrived at the following theorem.

Theorem 2.4.2 (Repeated roots). *Suppose a root λ of the characteristic equation (2.4.3) has multiplicity $j > 1$. Then the root λ contributes to the general solution of (2.4.1) the term*

$$\left(c_0 + c_1 n + c_2 n^2 + \ldots + c_j n^{j-1}\right)\lambda^n,$$

for constants c_i, $i = 0, 1, 2, \ldots, j$.

To see the set

$$\{\lambda^n, n\lambda^n, n^2\lambda^n, \ldots, n^j\lambda^n\}$$

is linearly independent, we refer to Exercise 2.21.

Example 2.15. Consider the fourth order difference equation

$$y(n + 4) - 7y(n + 3) + 18y(n + 2) - 20y(n + 1) + 8y(n) = 0.$$

Its characteristic equation is found to be $\lambda^4 - 7\lambda^3 + 18\lambda^2 - 20\lambda + 8 = 0$, which factors into $(\lambda - 2)^3(\lambda - 1) = 0$. Thus, we have a simple root 1 and another root 2 of multiplicity 3. Now by Theorem 2.4.2 the root 2 contributes the term $\left(c_0 + c_1 n + c_2 n^2\right) 2^n$. Consequently, the general solution is

$$y = \left(c_0 + c_1 n + c_2 n^2\right) 2^n + c_3. \quad \square$$

2.4.3 Complex roots

We discuss the situation when one of the roots is complex. That is, if (2.4.3) has a simple complex root then, it appears in complex conjugate pairs $\lambda = \alpha \pm i\beta$, where α, and β are real and $i = \sqrt{-1}$. Recall

$$e^t = \sum_{n=0}^{\infty} \frac{t^n}{n!} = 1 + t + \frac{t^2}{2!} + \frac{t^3}{3!} + \frac{t^n}{n!} + \cdots.$$

If we let $t = i\theta$, then the above series becomes

$$e^{i\theta} = \sum_{n=0}^{\infty} \frac{(i\theta)^n}{n!} = 1 + i\theta - \frac{\theta^2}{2!} - \frac{i\theta^3}{3!} + \frac{\theta^4}{4!} + \frac{i\theta^5}{5!} - \cdots$$

$$= \left(1 - \frac{\theta^2}{2!} + \frac{\theta^4}{4!} - \cdots\right) + i\left(\theta - \frac{\theta^3}{3!} + \frac{\theta^5}{5!} + \cdots\right)$$

$$= \cos(\theta) + i \sin(\theta).$$

Let

$$\rho = \sqrt{\alpha^2 + \beta^2}, \quad \text{and} \quad \theta = \tan^{-1} \frac{\beta}{\alpha}.$$

Then by Euler's formula we have that

$$\alpha \pm i\beta = \rho e^{\pm i\theta} = \rho\left(\cos(\theta) \pm i \sin(i\theta)\right).$$

Therefore, for arbitrary constants c_1 and c_2 that particular part of the solution can be written as

$$x(n) = c_1 (\alpha + i\beta)^n + c_2 (\alpha - i\beta)^n$$

$$= c_1\left(\rho e^{i\theta}\right)^n + c_2\left(\rho e^{-i\theta}\right)^n$$

$$= \rho^n \left(c_1 e^{in\theta} + c_2 e^{-in\theta}\right)$$

$$= \rho^n \left((c_1 + c_2) \cos(n\theta) + i(c_1 - c_2) \sin(n\theta)\right)$$

$$= \rho^n \left(c_3 \cos(n\theta) + c_4 \sin(n\theta)\right),$$

where $c_3 = c_1 + c_2$ and $c_4 = i(c_1 - c_2)$.

We briefly discuss the angle θ. Suppose we have a complex number denoted by $a + ib$. Then the real part a lies along the x-axis, and the imaginary part b lies along the y-axis. Then the point $a + ib$ is associated with the ordered pair (a, b), and from this we have a triangle with side lengths a, b and hypotenuse ρ. The value of θ is always measured from the positive real axis. To remove any ambiguities introduced by the inverse tangent, we could take this to be the four quadrant inverse tangent. For example, if $(a, b) = (1, 1)$, then $\theta = \frac{\pi}{4}$. On the other hand, if $(a, b) = (-1, 1)$ then $\theta = \frac{3\pi}{4}$. In addition, if $(a, b) = (-1, -1)$, then $\theta = \frac{5\pi}{4}$. Lastly, if $(a, b) = (1, -1)$, then $\theta = \frac{7\pi}{4}$. Note that, it should not matter if you use $a - ib$, since the values of the arbitrary constants will vary and consequently the solutions will match. We arrived at the following theorem.

Theorem 2.4.3 (Complex simple roots). *Suppose the characteristic equation* (2.4.3) *has a nonrepeated pair of complex conjugate roots* $\alpha \pm i\beta$. *Then the corresponding part of the general solution of* (2.4.1) *is*

$$\rho^n \left(c_1 \cos(n\theta) + c_2 \sin(n\theta) \right)$$

for constants c_1, *and* c_2, *where* $\rho = \sqrt{\alpha^2 + \beta^2}$ *and* $\theta = \tan^{-1} \frac{\beta}{\alpha}$.

Example 2.16. Consider the second order difference equation

$$x(n + 2) - 2x(n + 1) + 2x(n) = 0, \quad x(0) = 1, \ x(1) = -2.$$

The characteristic equation is $\lambda^2 - 2\lambda + 2 = 0$, which has the roots $\lambda = 1 \pm i$. Hence, if we set $\lambda = 1 + i$, then we have $\rho = \sqrt{2}$ and $\theta = \frac{\pi}{4}$. Thus, the solution is

$$x(n) = (\sqrt{2})^n \left(c_1 \cos(n\frac{\pi}{4}) + c_2 \sin(n\frac{\pi}{4}) \right).$$

Applying the given initial conditions we obtain

$$c_1 = 1, \ c_2 = -3.$$

Now if we make use of $\lambda = 1 - i$, then we have $\rho = \sqrt{2}$ and $\theta = -\frac{\pi}{4}$, and the solution

$$x(n) = (\sqrt{2})^n \left(c_1 \cos(n\frac{\pi}{4}) - c_2 \sin(n\frac{\pi}{4}) \right).$$

Applying the given initial conditions we obtain

$$c_1 = 1, \ c_2 = 3.$$

Plugging c_1 and c_2 back into their corresponding solutions, we see the solution in either case is

$$x(n) = (\sqrt{2})^n \left(\cos(n\frac{\pi}{4}) - 3 \sin(n\frac{\pi}{4}) \right). \quad \square$$

The next example summarizes all three cases.

Example 2.17. Suppose the characteristic equation of a given difference equation is found to be

$$(r^2 - 2r + 5)(r^2 - 9)(r + 2)^2 = 0.$$

We are interested in finding its general solution. The term $r^2 - 2r + 5 = 0$, has the pairs of complex conjugate roots $1 - 2i$, and $1 + 2i$, and therefore its contribution to the general solution is given by $(\sqrt{5})^n(c_1 \cos(n\theta) + c_2 \sin(n\theta))$, where $\theta = \tan^{-1}(2)$. Similarly, $r^2 - 9 = 0$ makes the contribution $c_3(3)^n + c_4(-3)^n$. Finally, $c_5(-2)^n + c_6 n(-2)^n$ is the contribution corresponding to $(r + 2)^2 = 0$. Hence, the general solution is

$$y = (\sqrt{5})^n \Big(c_1 \cos(n\theta) + c_2 \sin(n\theta) \Big) + c_3(3)^n + c_4(-3)^n$$
$$+ c_5(-2)^n + c_6 n(-2)^n. \quad \square$$

2.4.4 Exercises

Exercise 2.24. Find the solution of

$$x(n + 2) - 4x(n + 1) + 4x(n), \quad x(0) = 1, \; x(1) = -3.$$

Exercise 2.25. Consider the difference equation

$$x(n + 2) + \alpha x(n + 1) + \beta x(n) = \gamma. \tag{2.4.8}$$

A fixed point of (2.4.8) is a real-valued number d such that

$$d + \alpha d + \beta d = \gamma.$$

Now use the transformation $y(n) = x(n) - d$ to transform the non-homogeneous equation (2.4.8) to the homogeneous equation

$$y(n + 2) + \alpha y(n + 1) + \beta y(n) = 0.$$

In Exercises 2.26–2.27 the characteristic equation of a certain difference equation is given. Find the corresponding general solution.

Exercise 2.26.

$$(r^2 + 4)(r^2 - 9)^2(r + 2)^2 = 0.$$

Exercise 2.27.

$$(r^2 + 4)^2(r^2 + 3r + 2)^3(r + 5)^2 = 0.$$

In Exercises 2.28–2.32 solve the given difference equation.

Exercise 2.28.

$$y(n+2) + y(n+1) - 2y(n) = 0, \quad y(0) = 1, \quad y(1) = 4.$$

Exercise 2.29.

$$y(n+3) - y(n+2) - 4y(n+1) + 4y(n) = 0.$$

Exercise 2.30.

$$y(n+4) - 2y(n+3) + 3y(n+2) + 2y(n+1) - 4y(n) = 0.$$

Exercise 2.31.

$$y(n+4) + 2y(n+3) + 3y(n+2) + 2y(n+1) + y(n) = 0.$$

Exercise 2.32.

$$9y(n+2) - 6y(n+1) + y(n) = 0, \quad y(0) = 0, \quad y(1) = 1.$$

Exercise 2.33. Let λ_1 and λ_2 be the roots of the characteristic equation of the second order-difference equation

$$x(n+2) + \alpha x(n+1) + \beta x(n) = 0. \tag{2.4.9}$$

(a) Write (2.4.9) using the operator E and get

$$\left(E^2 + \alpha E + \beta\right)x(n) = 0.$$

(b) Show that

$$E^2 + \alpha E + \beta = (E - \lambda_1)(E - \lambda_2).$$

(c) In this notation, show that the difference equation (2.4.9) takes the form

$$(E - \lambda_1)(E - \lambda_2)x(n). \tag{2.4.10}$$

(d) Let $y(n)$ be the solution of $(E - \lambda_2)x(n) = 0$. In other words,

$$(E - \lambda_2)x(n) = y(n).$$

Then (2.4.10) becomes

$$(E - \lambda_1)y(n). \tag{2.4.11}$$

(e) Show that (2.4.11) has the solution $y(n) = \lambda_1^n$.
(f) Show that with this $y(n)$, the difference equation in part (d) reduces to

$$x(n+1) - \lambda_2 x(n) = \lambda_1^n. \tag{2.4.12}$$

(g) Refer to Section 2.1 to show that solution of (2.4.12) is

$$x(n) = \begin{cases} c_1\lambda_1^n + c_2\lambda_2^n, & \text{for } \lambda_1 \neq \lambda_2 \\ c_1\lambda_1^n + c_2 n\lambda_2^n, & \text{for } \lambda_1 = \lambda_2. \end{cases}$$

Exercise 2.34.

$$1, 2, 3, 5, 8, 13, 21, 34, 55, 89, 144, 233, \ldots. \tag{2.4.13}$$

Write down a second order difference equation in $x(n)$ that represents (2.4.13) with the proper initial conditions and find its solution $x(n)$ for $n \geq 0$. Compute

$$\lim_{n \to \infty} \frac{x(n+1)}{x(n)}$$

and compare your answer with Example 2.14.

2.5 The method of undetermined coefficients

In the previous section, our work mainly was about solving homogeneous difference equations. In this section we learn how to solve non-homogeneous kth-order difference equations with constant coefficients, of the form

$$\begin{aligned} a_k x(n+k) + a_{k-1}x(n+k-1) + a_{k-2}x(n+k-2) + \cdots \\ + a_1 x(n+1) + a_0 x(n) = f(n), \end{aligned} \tag{2.5.1}$$

and the associated homogeneous equation

$$\begin{aligned} a_k x(n+k) + a_{k-1}x(n+k-1) + a_{k-2}x(n+k-2) + \cdots \\ + a_1 x(n+1) + a_0 x(n) = 0, \end{aligned} \tag{2.5.2}$$

where the function $f(n)$ is referred to as the *forcing function*. We express Eqs. (2.5.1) and (2.5.2) in terms of the operator $\mathcal{L}$ by defining

$$\mathcal{L} = a_k E^k + a_{k-1}E^{k-1} + a_{k-2}E^{k-2} + \cdots + a_1 E + a_0 I$$

then

$$\mathcal{L}(x)n = \left(a_k E^k + a_{k-1}E^{k-1} + a_{k-2}E^{k-2} + \cdots + a_1 E + a_0 I\right)x(n).$$

Consequently, if we set $\mathcal{L}x(n) = 0$, then we obtain (2.5.2).

In terms of the operator $\mathcal{L}$ Eqs. (2.5.1) and (2.5.2) take the form

$$\mathcal{L}x = f, \quad \mathcal{L}x = 0,$$

respectively. If x_p is a given particular solution of (2.5.1), then $\mathcal{L}x_p = f$. In addition, if z is any other solution of (2.5.1), then we have $\mathcal{L}z = f$. Consequently, due to the linearity of $\mathcal{L}$ we see that

$$\mathcal{L}(x_p - z) = \mathcal{L}x_p - \mathcal{L}z = f - f = 0.$$

Thus, $x_h = z - x_p$ is a solution of the associated homogeneous equation given by (2.5.2). It follows from Theorem 2.3.3 that

$$x_h(x) = c_1\phi_1(n) + c_2\phi_2(n) + \ldots + c_k\phi_k(n)$$

for constants c_i, $i = 1, 2, \ldots, k$, where the functions $\varphi_i(n)$, $i = 1, 2, \ldots, k$ are linearly independent solutions of (2.5.2). Finding the particular solution x_p depends on two things:

(a) The type of function $f(n)$ in (2.5.1), and
(b) the nature of the homogeneous solution x_h of (2.5.2).

The method of Undetermined Coefficients only applies to functions $f(n)$ that are polynomial in n, combinations of *sine* or *cosine*, exponentials in n or combinations of the after-mentioned forms of $f(n)$. We illustrate the idea with several examples.

Example 2.18. Consider the difference equation

$$x(n + 2) + x(n + 1) - 6x(n) = 8.$$

Then the characteristic equation of the homogeneous part of the equation is $\lambda^2 + \lambda - 6 = 0$, which has the two roots $\lambda_1 = 2$ and $\lambda_2 = -3$. Consequently, the homogeneous solution is $x_h(n) = c_1(2)^n + c_2(-3)^n$. Since $f(n) = 4$ is a constant we consider a particular solution of the form $x_p = A$, where A is to be determined. Substituting x_p into the difference equation results into $A + A - 6A = 8$ from which we have $A = -2$. Thus the general solution is

$$x(n) = x_h(n) + x_p = c_1(2)^n + (-3)^n - 2. \quad \square$$

Example 2.19. For

$$x(n + 2) + x(n + 1) - 6x(n) = 4(3)^n$$

we have $x_h(n) = c_1(2)^n + c_2(-3)^n$. Since $f(x) = 4(3)^n$, we consider a particular solution of the form $x_p = A(3)^n$, where A is to be determined. Note that the term 4 is already included in A. We substitute x_p into the difference equation and obtain

$$A3^{n+2} + A3^{n+1} - 6A3^n = 4(3)^n.$$

Factoring the term 3^n yields,

$$3^n \left[A3^2 + 3A - 6A \right] = 4(3)^n.$$

Equating coefficients on both sides we obtain $9A + 3A - 6A = 4$, which implies $A = \frac{2}{3}$. Hence, the general solution is

$$x(n) = x_h(n) + x_p = c_1(2)^n + c_2(-3)^n + \frac{2}{3}(3)^n. \quad \square$$

Example 2.20. The equation

$$x(n+2) + x(n+1) - 6x(n) = 2^n$$

has $x_h(n) = c_1(2)^n + c_2(-3)^n$. Since $f(n) = 2^n$, let us try as before, the particular solution $x_p = A2^n$, where A is to be determined. Substituting x_p into the difference equation we arrive at the relation $A2^{n+2} + 2^{n+1} - 6A2^n = 2^n$. Factoring out 2^n we obtain

$$(4A + 2A - 6A)2^n = 2^n.$$

Or, $0.2^n = 1.2^n$, which can only imply that

$$0 = 1.$$

So what went wrong? Well, we said in the beginning that x_h depends on the form of $f(n)$ too. Let's start over. Now $f(n) = 2^n$, and so we try a particular solution of the form $x_p = A2^n$. But the term 2^n is already present in x_h and so we try to multiply $A2^n$ by n. Thus, we end up with the particular solution of the form $x_p = An2^n$. Substituting x_p into the difference equation we arrive at the relation $10A(2)^n = 2^n$, from which we obtain $A = \frac{1}{10}$. Thus, $x_p(n) = \frac{1}{10}n(2)^n$, and the general solution is

$$x(n) = x_h(n) + x_p(n) = c_1(2)^n + c_2(-3)^n + \frac{1}{10}n(2)^n. \quad \square$$

Example 2.21. In this example the emphasis is on writing the particular solution. Consider the difference equation

$$x(n+2) + 9x(n) = f(n).$$

Find the particular solution $x_p(n)$ when:

(a) $f(n) = 1 + 5^n + \sin(n\frac{\pi}{3})$,
(b) $f(n) = 1 + 5^n + (3)^n \sin(n\frac{\pi}{2})$.

We begin by finding the homogeneous solution $x_h(n)$. The characteristic equation is $\lambda^2 + 9 = 0$, which has the roots $\lambda = \pm 3i$. Thus, if we $\lambda_1 = 3i$, then we have $\rho = 3$ and $\theta = \frac{\pi}{2}$. Consequently,

$$x_h(n) = (3)^n\left(c_1 \cos(n\frac{\pi}{2}) + c_2 \sin(n\frac{\pi}{2})\right).$$

(a) Since none of the terms in $f(n)$ makes its presence in the particular solution, we construct $x_p(n)$ in the usual way. For the term 1, the particular solution must include the parameter A. As for the term 5^n, the particular solution must include $B(5)^n$. Finally, for the term $\sin(n\frac{\pi}{3})$, the particular solution must include $C\sin(n\frac{\pi}{3}) + D\cos(n\frac{\pi}{3})$. Thus, putting these terms together we obtain the general form of the particular solution to be

$$x_p(n) = A + B(5)^n + C\sin(n\frac{\pi}{3}) + D\cos(n\frac{\pi}{3}).$$

For part (b), the first two terms in $f(n)$ do not appear in the homogeneous solution $x_h(n)$, and hence their contribution to the particular solution is already known from part (a). Now that the third term $(3)^n \sin(n\frac{\pi}{2})$ of $f(n)$ is present in the homogeneous solution $x_h(n)$ its contribution to the particular solution is in the form

$$(3)^n \left[C n \sin(n\frac{\pi}{2}) + D n \cos(n\frac{\pi}{2}) \right].$$

Thus the general form of the particular solution $x_p(n)$ is

$$x_p(n) = A + B(3)^n + (3)^n \left[C n \sin(n\frac{\pi}{2}) + D n \cos(n\frac{\pi}{2}) \right]. \quad \square$$

One more example in which we make use of the trigonometric identities $\sin(u + v) = \sin u \cos v + \cos u \sin v$, $\cos(u + v) = \cos u \cos v - \sin u \sin v$.

Example 2.22. Consider the difference equation $x(n + 2) + 9x(n) = \sin(3n)$, $n \in \mathbb{N}$.

Then,

$$x_h(n) = (3)^n \left(c_1 \cos(n\frac{\pi}{2}) + c_2 \sin(n\frac{\pi}{2}) \right).$$

Let

$$x_p(n) = A \sin(3n) + B \cos(3n).$$

Substituting into the difference equation we obtain the string of equalities

$$x(n + 2) + 9x(n) = A \sin(3n + 6) + B \cos(3n + 6) + 9A \sin(3n) + 9B \cos(3n)$$

$$= A \Big(\sin(3n) \cos(6) + \cos(3n) \sin(6) \Big) + 9A \sin(3n)$$

$$+ B \Big(\cos(3n) \cos(6) - \sin(3n) \sin(6) \Big) + 9B \cos(3n)$$

$$= \Big(A \cos(6) + 9A - B \sin(6) \Big) \sin(3n)$$

$$+ \Big(A \sin(6) + B \cos(6) + 9B) \Big) \cos(3n)$$

$$= \sin(3n).$$

Equating coefficients on both sides we arrive at the system of two unknowns,

$$(\cos(6) + 9)A - B\sin(6) = 1,$$
$$A\sin(6) + (\cos(6) + 9)B = 0.$$

Solving for A in the second equation and then substituting it into the first equation we arrive at

$$B = -\frac{\sin(6)}{(\cos(6) + 9)^2 + \sin^2(6)} = -\frac{\sin(6)}{82 + 18\cos(6)}.$$

As a result, we have

$$A = -\frac{\cos(6) + 9}{82 + 18\cos(6)}.$$

Finally, the general solution is

$$x(n) = x_h(n) + x_p(n)$$
$$= (3)^n\left(c_1\cos(n\frac{\pi}{2}) + c_2\sin(n\frac{\pi}{2})\right)$$
$$-\frac{\cos(6) + 9}{82 + 18\cos(6)}\sin(3n) - \frac{\sin(6)}{82 + 18\cos(6)}\cos(3n). \quad \square$$

From the above examples, we observed that the form of a particular solution may often be inferred from the form of the function $f(n)$. Moreover, the method of undetermined coefficients can be applied successfully when the function $f(n)$ is a linear combination of sums and products of functions of the form a^n, $\sin(bn)$, $\cos(bn)$, and n^k, where a and b are any constants and k is any nonnegative integer.

On the other hand, if the particular solution $x_p(n)$ includes a term which is present in the homogeneous solution $x_h(n)$, then $x_p(n)$ should first be multiplied by n and the resulting $x_p(n)$ used. This process can be repeated until no term in the evolving $x_p(n)$ is present in $x_h(n)$.

Table 2.1 provides guidance on how to construct x_p in the case that none of the terms present in x_p are parts of the forcing function $f(n)$.

2.5.1 Exercises

In Exercises 2.35–2.36 the characteristic equation and the *forcing function* of a certain difference equation are given. Write down the particular solution without solving for the coefficients.

Exercise 2.35.

$$(\lambda^2 + 4)(\lambda^2 - 9)^2(\lambda + 2)^2 = 0; \quad f(n) = \sin(2n) + (-5)^n + n^2.$$

TABLE 2.1 Shows how to find $x_p(n)$.

$f(n)$	$x_p(n)$
Constant C	A
a^n	Aa^n
Cn^k, $k = 0, 1, 2, \ldots$	$A_0 + A_1 n + A_2 n^2 + \ldots + A_k n^k$
$\cos(bn)$ or $\sin(bn)$	$A_1 \cos(bn) + A_2 \sin(bn)$
$n^k \cos(bn)$ or $n^k \sin(bn)$	$(A_0 + A_1 n + A_2 n^2 + \ldots + A_k n^k) \cos(bn) +$ $(B_0 + B_1 n + B_2 n^2 + \ldots + B_k n^k) \sin(bn)$
$n^k a^n$	$(A_0 + A_1 n + A_2 n^2 + \ldots + A_k n^k) a^n$

Exercise 2.36.

$$(\lambda - 1)(\lambda + 4)(\lambda^2 - 2\lambda + 4)(\lambda + 5)^2 = 0,$$

$$f(n) = \cos(n\frac{\pi}{4}) + 2^n \sin(n\frac{\pi}{3}) + n(-5)^n + 1.$$

Exercise 2.37. Solve each of the given difference equations.

(a) $x(n + 2) + 2x(n + 1) - 3x(n) = 4(2)^n$, $x(0) = 0$, $x(1) = 2$.

(b) $x(n + 2) + 2x(n + 1) - 3x(n) = 1 + 4(-3)^n$.

(c) $x(n + 2) + 4x(n) = \sin(n)$, $n \in \mathbb{N}$.

(d) $4x(n + 2) + 4x(n + 1) + x(n) = n^2$, $x(0) = 0$, $x(1) = 0$.

Answer: $x(n) = \dfrac{1}{27}\left[-4(-\dfrac{1}{2})^n + 2n(-\dfrac{1}{2})^n + 3n^2 - 8n + 4\right]$.

(e) $x(n + 2) - 4x(n + 1) + 16x(n) = 4(2)^n + 26$, $x(0) = 1$, $x(1) = 1$.

Answer: $x(n) = 4^n\left[-\dfrac{4}{3}\cos(n\dfrac{\pi}{3}) + \dfrac{1}{2\sqrt{3}}\sin(n\dfrac{\pi}{3})\right] + \dfrac{2^n}{3} + 2$.

(f) $x(n + 2) - 7x(n + 1) + 10x(n) = 12(4)^n$, $x(0) = 0$, $x(1) = 0$.

Answer: $x(n) = 2(2)^n + 4(5)^n - 6(4)^n$.

(g) $x(n + 2) - 7x(n + 1) + 10x(n) = 12(5)^n$, $x(0) = 0$, $x(1) = 0$.

Answer: $x(n) = \dfrac{4}{3}\left[(2)^n - (5)^n\right] + \dfrac{4}{5}(5)^n$.

(h) $x(n + 3) - 5x(n + 2) + 9x(n + 1) - 5x(n) = 4$, $x(0) = x(1) = -2$, $x(2) = 0$.

Answer: $x(n) = (\sqrt{5})^n\left[\sin(n \tan^{-1}(\theta)) - 3\cos(n \tan^{-1}(\theta))\right] + 2n + 1$, $\theta = \frac{1}{2}$.

2.6 Variation of parameters

Our aim is to solve non-homogeneous difference equations with variable coefficients. In particular, we consider the non-homogeneous kth-order difference equation with variable coefficients a_i, $i = 0, 1, \cdots k - 1$

$$x(n + k) + a_{k-1}(n)x(n + k - 1) + a_{k-2}(n)x(n + k - 2) + \cdots$$

$$+ a_1(n)x(n+1) + a_0(n)x(n) = f(n). \tag{2.6.1}$$

We already know from Section 2.5, whenever a fundamental set of solutions

$$\{\phi_1(n), \ \phi_2(n), \ \ldots, \ \phi_k(n)\}$$

of the corresponding homogeneous equation

$$x(n+k) + a_{k-1}(n)x(n+k-1) + a_{k-2}(n)x(n+k-2) + \cdots$$
$$+ a_1(n)x(n+1) + a_0(n)x(n) = 0, \tag{2.6.2}$$

is known, then a general solution of (2.6.2) is given by

$$x_h(n) = c_1\phi_1(n) + c_2\phi_2(n) + \ldots + c_k\phi_k(n).$$

The method of undetermined coefficients that we studied in Section 2.5 will not work for equations when the forcing function $f(n)$ is not one of the functions in Table 2.1. Thus, in this section, we will develop a method called the *variation of parameters* to deal with equations similar to (2.6.1) by using the known solution of (2.6.2). To better illustrate the method of variations of parameters, we carry out the derivations for $k = 2$. Thus, we consider the general linear second-order difference equation

$$x(n+2) + p(n)x(n+1) + q(n)x(n) = f(n), \tag{2.6.3}$$

where the functions $p(n)$, $q(n)$, and $f(n)$ are defined on $\mathbb{N}_{n_0}$. Assume $x_1(n)$ and $x_2(n)$ are two linearly independent known solutions on $\mathbb{N}_{n_0}$ of the corresponding homogeneous equation

$$x(n+2) + p(n)x(n+1) + q(n)x(n) = 0. \tag{2.6.4}$$

Then the homogeneous solution of (2.6.4) is

$$x_h(n) = c_1 x_1(n) + c_2 x_2(n),$$

for constants c_1 and c_2. Assume a particular solution $x_p(n)$ of (2.6.3) of the form

$$x_p(n) = u_1(n)x_1(n) + u_2(n)x_2(n) \tag{2.6.5}$$

where the functions u_1 and u_2 are to be determined. We assume u_1 and u_2 satisfy the natural condition

$$\Delta(u_1(n))x_1(n+1) + \Delta(u_2(n))x_2(n+1) = 0. \tag{2.6.6}$$

We will need to substitute $x_p(n+1)$ and $x_p(n+2)$ into (2.6.3). First, we perform the following computations.

$$x_p(n+1) = u_1(n+1)x_1(n+1) + u_2(n+1)x_2(n+1)$$

$$= u_1(n)x_1(n+1) + u_2(n)x_2(n+1)$$
$$+ \Delta(u_1(n))x_1(n+1) + \Delta(u_2(n))x_2(n+1).$$

Using the imposed condition (2.6.6) we compute $x_p(n+2)$ and obtain

$$x_p(n+2) = u_1(n)x_1(n+2) + u_2(n)x_2(n+2)$$
$$+ \Delta(u_1(n))x_1(n+2) + \Delta(u_2(n))x_2(n+2).$$

Substitute $x_p(n)$, $x_p(n+1)$, $x_p(n+2)$ into (2.6.5) and making use of the fact that $x_i(n+2) + p(n)x_i(n+1) + q(n)x_i(n) = 0$, for $i = 1$ and $i = 2$, we arrive at the simplified expression

$$\Delta(u_1(n))x_1(n+2) + \Delta(u_2(n))x_2(n+2) = f(n).$$

To determine $\Delta(u_1(n))$ and $\Delta(u_2(n))$ we must solve the system

$$\Delta(u_1(n))x_1(n+1) + \Delta(u_2(n))x_2(n+1) = 0,$$
$$\Delta(u_1(n))x_1(n+2) + \Delta(u_2(n))x_2(n+2) = f(n).$$

Multiply the first equation with $-x_2(n+2)$ and the second equation with $x_2(n+1)$ and adding the resulting two equations and obtain

$$\Delta(u_1(n)) = -\frac{x_2(n+1)f(n)}{x_1(n+1)x_2(n+2) - x_2(n+1)x_1(n+2)}.$$

Similarly,

$$\Delta(u_2(n)) = \frac{x_1(n+1)f(n)}{x_1(n+1)x_2(n+2) - x_2(n+1)x_1(n+2)}.$$

The Casoratian of $x_1(n+1)$ and $x_2(n+1)$ denoted with $W(n+1)$ is given by

$$W(n+1) = \begin{vmatrix} x_1(n+1) & x_2(n+1) \\ x_1(n+2) & x_2(n+2) \end{vmatrix}$$
$$\neq 0$$

since $x_1(n)$ and $x_2(n)$ are linearly independent. Hence the denominator in the expressions for $\Delta(u_1(n))$ and $\Delta(u_2(n))$ is not zero, which makes $\Delta(u_1(n))$ and $\Delta(u_2(n))$ well defined on $\mathbb{N}_{n_0}$. Using the Casoratian notation, we may rewrite

$$\Delta(u_1(n)) = -\frac{x_2(n+1)f(n)}{W(n+1)},$$

and

$$\Delta(u_2(n)) = \frac{x_1(n+1)f(n)}{W(n+1)}.$$

Next we claim, using the above expressions for $\Delta(u_1(n))$ and $\Delta(u_2(n))$ in terms of the Casoratian, that the functions $u_1(n)$ and $u_2(n)$ are given for $n \in \mathbb{N}_{n_0}$ by

$$u_1(n) = -\sum_{s=n_0}^{n-1} \frac{x_2(s+1)f(s)}{W(s+1)}, \tag{2.6.7}$$

and

$$u_2(n) = \sum_{s=n_0}^{n-1} \frac{x_1(s+1)f(s)}{W(s+1)}. \tag{2.6.8}$$

Now we make sure $u_1(n)$ and $u_2(n)$ given by (2.6.7) and (2.6.8) satisfy $\Delta u_1(n)$ and $\Delta u_2(n)$, respectively. To see this,

$$\Delta(u_2(n)) = \sum_{s=n_0}^{n} \frac{x_1(s+1)f(s)}{W(s+1)} - \sum_{s=n_0}^{n-1} \frac{x_1(s+1)f(s)}{W(s+1)}$$

$$= \frac{x_1(n+1)f(n)}{W(n+1)} + \sum_{s=n_0}^{n-1} \frac{x_1(s+1)f(s)}{W(s+1)} - \sum_{s=n_0}^{n-1} \frac{x_1(s+1)f(s)}{W(s+1)}$$

$$= \frac{x_1(n+1)f(n)}{W(n+1)}.$$

The proof of $\Delta u_1(n)$ follows along the same lines. We summarize the above discussion in the following theorem.

Theorem 2.6.1. *Suppose $x_1(n)$ and $x_2(n)$ are two linearly independent solutions of (2.6.4) on $\mathbb{N}_{n_0}$. Then the general solution of the non-homogeneous equation (2.6.3) on $\mathbb{N}_{n_0}$ is given by*

$$x(n) = x_h(n) + x_p(n)$$

where

$$x_h(n) = c_1 x_1(n) + c_2 x_2(n),$$

for constants c_1, c_2 and

$$x_p(n) = -x_1(n) \sum_{s=n_0}^{n-1} \frac{x_2(s+1)f(s)}{W(s+1)} + x_2(n) \sum_{s=n_0}^{n-1} \frac{x_1(s+1)f(s)}{W(s+1)}.$$

Example 2.23. Solve by the method of variation of parameters the difference equation

$$x(n+2) - 3x(n+1) + 2x(n) = 5^n, \quad n \in \mathbb{N}_{n_0}.$$

By the method of the previous section we can easily see that $x_1(n) = 1$ and $x_2(n) = 2^n$ are two linearly independent solutions of the corresponding homogeneous equation. Left to find

$$x_p(n) = u_1(n)x_1(n) + u_2(n)x_2(n) = u_1(n) + u_2(n)(2)^n.$$

The Casoratian,

$$W(n+1) = \begin{vmatrix} 1 & 2^{n+1} \\ 1 & 2^{n+2} \end{vmatrix} = 2^{n+1} \neq 0.$$

Left to compute $u_1(n)$ and $u_2(n)$.

$$u_1(n) = -\sum_{s=n_0}^{n-1} \frac{2^{s+1}.5^s}{2^{s+1}} = -\sum_{s=n_0}^{n-1} 5^s$$

$$= \left[\frac{5^s}{5-1}\right]\Bigg|_{s=n_0}^{n} = -\frac{1}{4}\left[5^n - 5^{n_0}\right].$$

Moreover,

$$u_2(n) = \sum_{s=n_0}^{n-1} \frac{5^s}{2^{s+1}} = \frac{1}{2}\sum_{s=n_0}^{n-1}\left(\frac{5}{2}\right)^s$$

$$= \frac{1}{2}\left[\frac{\left(\frac{5}{2}\right)^s}{\frac{5}{2}-1}\right]\Bigg|_{s=n_0}^{n} = \frac{1}{3}\left[\left(\frac{5}{2}\right)^n - \left(\frac{5}{2}\right)^{n_0}\right].$$

Finally, the general solution is

$$x(n) = c_1 + c_2.2^n - \frac{1}{4}\left[5^n - 5^{n_0}\right] + \frac{2^n}{3}\left[\left(\frac{5}{2}\right)^n - \left(\frac{5}{2}\right)^{n_0}\right]. \quad \square$$

Theorem 2.6.1 can be easily generalized to (2.6.1).

Theorem 2.6.2. *Suppose $x_1(n), x_2(n), \cdots x_k(n)$ are linearly independent solutions of (2.6.2). Then the particular solution $x_p(n)$ of (2.6.1) is given by*

$$x_p(n) = u_1(n)x_1(n) + u_2(n)x_2(n) + \cdots u_k(n)x_k(n),$$

where $u_1(n), u_2(n), \cdots u_k(n)$ satisfy the relations

$$\begin{bmatrix} x_1(n+1) & x_2(n+1) & \cdots & x_k(n+1) \\ x_1(n+2) & x_2(n+2) & \cdots & x_k(n+2) \\ \vdots & \vdots & \ddots & \vdots \\ x_1(n+k) & x_2(n+k) & \cdots & x_k(n+k) \end{bmatrix} \begin{bmatrix} \Delta u_1(n) \\ \Delta u_2(n) \\ \vdots \\ \Delta u_k(n) \end{bmatrix} = \begin{bmatrix} 0 \\ 0 \\ \vdots \\ f(n) \end{bmatrix}.$$

Then the general solution of (2.6.1) is

$$x(n) = x_h(n) + x_p(n)$$

where

$$x_h(n) = c_1 x_1(n) + \cdots c_k x_k(n),$$

for constants c_i, $i = 1, 2, \cdots k$.

2.6.1 Exercises

Exercise 2.38. Solve by the method of variation of parameters each of the given difference equations.

(a) $x(n+2) - 4x(n+1) + 3x(n) = 5^n$, $n \in \mathbb{N}_{n_0}$.
(b) $x(n+2) - 2x(n+1) + x(n) = n$, $n \in \mathbb{N}_{n_0}$.
(c) $x(n+2) - 4x(n) = 2^n$, $n \in \mathbb{N}_{n_0}$.
(d) $x(n+3) - 2x(n+2) - x(n+1) + 2x(n) = 3^n$, $n \in \mathbb{N}_{n_0}$.

Exercise 2.39. Solve by the method of variation of parameters.
$x(n+2) - 3x(n+1) + 2x(n) = 5^n$, $x(0) = 1$, $x(1) = 2$.

Exercise 2.40. Solve by the method of variation of parameters.
$x(n+2) - 5x(n+1) + 6x(n) = 2 + 3n + n^2$, $n \geq 0$, $n \in \mathbb{N}$.
Answer: $x(n) = c_1 2^n + c_2 3^n + \frac{1}{2} \left[\frac{23}{2} + 6n + n^2 \right]$.

Exercise 2.41. Solve by the method of variation of parameters.
$x(n+2) - 7x(n+1) + 6x(n) = -1 + 2n$, $n \geq 0$, $n \in \mathbb{N}$.
Answer: $x(n) = c_1 + c_2 6^n + \frac{1}{5} \left[\frac{8}{5} - n^2 \right]$.

2.7 Recovery of difference equations

Suppose we are given a sequence and asked to find the recursive equation that the sequence satisfies. We have seen in previous sections that a solution to a kth-order difference equation contains k constants that are to be determined. In other words, the order of the difference equation is equal to the number of arbitrary constants in its solution. Mathematically speaking, suppose we have the relationship

$$x(n) = f(n, A),$$

where A is some constant. Then

$$x(n+1) = f(n+1, A).$$

Upon the elimination of A between these two equations, we obtain a relation between n, $x(n)$, and $x(n+1)$ that we denote with

$$F(n, x(n), x(n+1)) = 0,$$

which represents the difference equation of interest. This process can be repeated to accommodate solution that contains k constants of the form

$$x(n) = f(n, A_1, A_2, \cdots A_k).$$

We illustrate the idea by several examples.

Example 2.24. Find the difference equation that its solution is given by

$$x(n) = A5^n + n2^{n-1}.$$

Since the solution contains one constant, the corresponding difference equation is of order one. Replacing n with $n + 1$ produces the new relation

$$x(n + 1) = A5^{n+1} + (n + 1)2^n$$
$$= 5A5^n + 2(n + 1)2^{n-1}.$$

Solving for A in the solution and then substituting it into the right-side of $x(n + 1)$ gives

$$A = \frac{x(n) - n2^{n-1}}{5^n}, \quad \text{and}$$
$$x(n + 1) = 5\left[\frac{x(n) - n2^{n-1}}{5^n}\right]5^n + 2(n + 1)2^{n-1}$$
$$= 5x(n) - 5n2^{n-1} + 2(n + 1)2^{n-1} = 5x(n) + (-3n + 2)2^{n-1}.$$

Thus, the difference equation is

$$x(n + 1) - 5x(n) = (2 - 3n)2^{n-1}.$$

We note that this can be worked out by computing $\triangle x(n)$ and then manipulating the two equations. The reason for only taking the first order $\triangle$ is due to the existence of only one constant. $\square$

Example 2.25. Find the difference equation that its solution is given by

$$x(n) = An^2 - Bn.$$

Since the solution contains two constants, the corresponding difference equation is of order two. Thus, we must compute $x(n + 1)$ and $x(n + 2)$. Replacing n with $n + 1$ and then n with $n + 2$ produce the new relations

$$x(n + 1) = A(n + 1)^2 - B(n + 1)$$
$$x(n + 2) = A(n + 2)^2 - B(n + 2).$$

Eliminating A and B from the three equations given by $x(n)$, $x(n+1)$, and $x(n+2)$ is equivalent to

$$\begin{vmatrix} x(n) & n^2 & -n \\ x(n+1) & (n+1)^2 & -(n+1) \\ x(n+2) & (n+2)^2 & -(n+2) \end{vmatrix} = 0.$$

This simplifies to the second-order difference equation

$$n^3(n+1)x(n+2) - 2n^2(n+2)x(n+1) - (n^2+3n+2)(n^2-2)x(n) = 0. \quad \square$$

2.7.1 Exercises

Exercise 2.42. Find the difference equation that its solution is given by

$$x(n) = A2^n + n3^{n-1}.$$

Answer: $x(n+1) - 2x(n) = (n+3)3^{n-1}$.

Exercise 2.43. Redo Example 2.25 using Δ, Δ^2, and Δ^3.

Exercise 2.44. Find the difference equation that its solution is given by

$$x(n) = \frac{A}{n} + B, \ n \in \mathbb{N}.$$

Answer: $(n+2)x(n+2) - 2(n+1)x(n+1) - nx(n) = 0$.

Exercise 2.45. Find the difference equation that its solution is given by

$$x(n) = A\cos(n\theta) + B\sin(n\theta).$$

Answer: $x(n+2) - 2\cos(\theta)x(n+1) + x(n) = 0$.

Exercise 2.46. Find the difference equation that its solution is given by

$$x(n) = A2^n + B3^n + \frac{1}{2}.$$

Answer: $x(n+2) - 5x(n+1) + 6x(n) = 1$.

2.8 From non-linear to solvable

Some non-linear difference equations can be transformed into linear equations or to non-linear equations but solvable, when the right transformation is used. One of the well-known non-linear equations that can be transformed into equations that can be solved is the *Riccati equation*.

Riccati.

We begin by considering the non-homogeneous Riccati equation

$$x(n+1) = \frac{a + bx(n)}{c + dx(n)}, \quad n \in \mathbb{N}_0 \tag{2.8.1}$$

where a, b, c, and d are real constants. In order for (2.8.1) to be well defined, we ask that when the initial point $x(0)$ is assigned that

$$x(0) = x_0 \neq -\frac{c}{d}. \tag{2.8.2}$$

Later on in our discussion it will become clearer why we must ask for

$$b + c \neq 0. \tag{2.8.3}$$

With minor calculations, we may see that (2.8.1) is equivalent to

$$dx(n)x(n+1) + cx(n+1) - bx(n) = a, \quad n \in \mathbb{N}_0. \tag{2.8.4}$$

Eq. (2.8.4) is known as the *non-homogeneous Riccati equation* with constant coefficients. Of course (2.8.4) is non-linear. To transform (2.8.1) to a linear difference equation we consider the transformation

$$x(n) = \frac{b+c}{d} \cdot \frac{y(n+1)}{y(n)} - \frac{c}{d}, \quad n \in \mathbb{N}_0. \tag{2.8.5}$$

Substituting into (2.8.4) we obtain

$$dx(n)x(n+1) + cx(n+1) - bx(n) - a$$
$$= x(n+1)\left[dx(n) + c\right] - bx(n) - a$$
$$= \left[\frac{b+c}{d} \cdot \frac{y(n+2)}{y(n+1)} - \frac{c}{d}\right]\left[(b+c)\frac{y(n+1)}{y(n)}\right] - \frac{b}{d}(b+c)\frac{y(n+1)}{y(n)} - a + \frac{bc}{d}$$
$$= \left[\frac{b+c}{d}\right]\left[(b+c)\cdot\frac{y(n+2)}{y(n)} - (b+c)\cdot\frac{y(n+1)}{y(n)}\right] - a + \frac{bc}{d}$$
$$= \left[\frac{(b+c)^2}{d}\right]\left[\frac{y(n+2)}{y(n)} - \frac{y(n+1)}{y(n)}\right] + \frac{bc - ad}{d} = 0.$$

Thus, we have arrived at the expression

$$\left[\frac{(b+c)^2}{d}\right]\left[\frac{y(n+2)}{y(n)} - \frac{y(n+1)}{y(n)}\right] + \frac{bc - ad}{d} = 0.$$

A multiplication of both sides of the above equation by $\dfrac{d}{(b+c)^2} y(n)$ leads to the second-order linear difference equation in y

$$y(n+2) - y(n+1) + \alpha y(n) = 0, \quad n \in \mathbb{N}_0, \tag{2.8.6}$$

where

$$\alpha = \frac{bc - ad}{(b+c)^2}.$$

In (2.8.5), if we choose $y(0) = 1$, then we have

$$y(1) = \frac{d(0)}{b(0) + c(0)} \left(x(0) + \frac{c(0)}{d(0)} \right).$$

The passage from an equation of first-order to second-order was due to the presence of the term a. We would have used a different transformation and obtained a linear first-order difference equation if we had $a = 0$. In other words, it is the difference between having a homogeneous difference equation and a non-homogeneous difference equation.

The transformation $x(n) = \dfrac{1}{y(n)}$ transforms the homogeneous Riccati equation

$$x(n)x(n+1) + c(n)x(n+1) + b(n)x(n) = 0, \quad n \in \mathbb{N}_0 \tag{2.8.7}$$

to

$$b(n)y(n+1) + c(n)y(n) + 1 = 0, \quad n \in \mathbb{N}_0. \tag{2.8.8}$$

On the other hand, the transformation $x(n) = \dfrac{y(n+1)}{y(n)} - c(n)$, will transform the non-homogeneous Riccati equation

$$x(n)x(n+1) + c(n)x(n+1) + b(n)x(n) = a(n), \quad n \in \mathbb{N}_0 \tag{2.8.9}$$

to

$$y(n+2) + (b(n) - c(n+1)) y(n+1) - (a(n) + c(n)b(n)) y(n) = 0, \quad n \in \mathbb{N}_0. \tag{2.8.10}$$

Rationally homogeneous.

Next, we discuss equations that are called rationally homogeneous in the sense of the definition that follows.

Definition 2.8.1. A kth-order difference equation is said to be *rationally homogeneous,* if it can be put in the form

$$\sum_{i=1}^{j} \alpha_i(n) \left(\frac{x(n+k)}{x(n+l)} \right)^i = f(n), \quad l < k = 1, 2, \cdots \tag{2.8.11}$$

The functions $f(n)$ and $\alpha_i(n)$ are defined on $\mathbb{N}_{n_0}$ and may be constants. To solve equations of the form (2.8.11), we make use of the substitution

$$y(n) = \frac{x(n+k)}{x(n+l)}$$

and obtain a polynomial in y of degree j.

Example 2.26. The difference equation

$$x^2(n+1) - 2x^2(n) = 0, \quad n \in \mathbb{N}_{n_0}$$

is rationally homogeneous, since it can be written as

$$x^2(n+1) - 2x^2(n) = \left[\frac{x(n+1)}{x(n)} \right]^2 = 2,$$

which is of the form of (2.8.11) with $\alpha_1 = 0$, $\alpha_2 = 1$ and $f(n) = 2$. $\qquad\square$

Example 2.27. Consider the difference equation in Example 2.26 with $x(n_0) = x_0 \neq 0$. We let $y(n) = \frac{x(n+1)}{x(n)}$, which transforms the difference equation to the equation

$$y^2(n) - 2 = 0.$$

We have either $y(n) = \sqrt{2}$, or $y(n) = -\sqrt{2}$. In terms of $x(n)$ this implies that

$$x(n+1) = \sqrt{2}x(n), \text{ or } x(n+1) = -\sqrt{2}x(n).$$

This results into the two solutions with the same initial condition

$$x(n) = (\sqrt{2})^{n-n_0} x_0, \text{ or } x(n) = (-\sqrt{2})^{n-n_0} x_0, \quad n \in \mathbb{N}_{n_0}. \quad \square$$

Example 2.28. For $n \in \mathbb{N}_0$, consider the difference equation

$$x^2(n+2) - \frac{5}{2}x(n)x(n+2) + x^2(n) = 0, \quad x(0) = x_0 \neq 0, \ x(1) = x_1 \neq 0.$$

Dividing with $x^2(n)$ the equation takes the form

$$\left[\frac{x(n+2)}{x(n)} \right]^2 - \frac{5}{2}\frac{x(n+2)}{x(n)} + 1 = 0.$$

Thus, it is of the form (2.8.11) with $\alpha_1 = -\frac{5}{2}$, $\alpha_2 = 1$ and $f(n) = -1$.

We let $y(n) = \frac{x(n+2)}{x(n)}$, which transforms the difference equation to

$$y^2(n) - \frac{5}{2}y(n) + 1 = 0.$$

We have either $y(n) = \frac{1}{2}$, or $y(n) = 2$. In terms of $x(n)$ this implies that

$$x(n+2) = \frac{1}{2}x(n), \text{ or } x(n+2) = 2x(n).$$

The equation $x(n+2) = \frac{1}{2}x(n)$, has two linearly independent solutions $\left(\frac{1}{\sqrt{2}}\right)^n$, $\left(-\frac{1}{\sqrt{2}}\right)^n$. Hence its general solution is given by

$$x(n) = c_1\left(\frac{1}{\sqrt{2}}\right)^n + c_2\left(-\frac{1}{\sqrt{2}}\right)^n. \qquad (2.8.12)$$

Similarly, the general solution of $x(n+2) = 2x(n)$ is

$$x(n) = c_3(\sqrt{2})^n + c_4(-\sqrt{2})^n. \qquad (2.8.13)$$

We remark that both solutions (2.8.12) and (2.8.13) have the same initial conditions. This does not violate uniqueness of solutions since the original equation in non-linear. $\qquad \square$

Example 2.29. The difference equation

$$x^2(n+3) - 9x^2(n+2) = 0, \quad n \in \mathbb{N}_{n_0}$$

is rationally homogeneous, since it can be written as

$$x^2(n+3) - 9x^2(n+2) = \left[\frac{x(n+3)}{x(n+2)}\right]^2 - 9 = 0.$$

Thus, it is of the form (2.8.11) with $\alpha_1 = 0$, $\alpha_2 = 1$ and $f(n) = 9$. We let $y(n) = \frac{x(n+3)}{x(n+2)}$, and we obtain the polynomial

$$y^2(n) - 9 = 0,$$

with solutions $y(n) = 3$, or $y(n) = -3$. In terms of $x(n)$, we have the two equations

$$x(n+3) = 3x(n+2) \quad \text{or} \quad x(n+3) = -3x(n+2).$$

The first equation has the characteristic equation $\lambda^3 - 3\lambda^2 = 0$, with roots $\lambda_1 = \lambda_2 = 0$, $\lambda_3 = 3$. Hence its corresponding general solution is

$$x(n) = c_1 3^n.$$

Similarly, the corresponding general solution of the second equation is

$$x(n) = c_2(-3)^n.$$

If we assign initial conditions $x(i) = x_i$, $x_i \neq 1$, $i = 0, 1, 2$, then the constants can be computed. $\qquad \square$

Applying log.
In some cases, we are able to apply log to both sides of the equation and then make a transformation of the form $y(\cdot) = \log(x(\cdot))$. Suppose a given difference equation of order k can be written as

$$\prod_{s=0}^{k} x^{r_s}(n+s) = a(n), \quad n \in \mathbb{N}_{n_0}. \tag{2.8.14}$$

Then applying log to both sides of (2.8.14) we arrive at

$$\sum_{s=0}^{k} r_s \log x(n+s) = \log(a(n)), \quad n \in \mathbb{N}_{n_0}.$$

Finally, we make the substitution

$$y(n) = \log(x(n)).$$

Example 2.30. Consider the difference equation

$$\frac{x(n+1)x^{\frac{1}{n}}(n+1)}{x(n)} = e, \quad x(1) = x_0 \neq 0, \ n \in \mathbb{N}_{n_0}.$$

Taking logarithms of both sides leads to

$$\log x(n+1) + \frac{1}{n}\log x(n+1) - \log x(n) = 1.$$

Let $y(n) = \log(x(n))$. After combining likewise terms, the above difference equation transfers to the linear non-homogeneous difference equation

$$(n+1)y(n+1) - ny(n) = n,$$

or

$$\Delta(ny(n)) = n, \quad n \in \mathbb{N}_1.$$

Summing both sides of the above equation from $s = 1$ to $s = n - 1$ leads to

$$ny(n) = \sum_{s=1}^{n-1} s = \sum_{s=1}^{n-1}[s] = \frac{1}{2}[s]^2\Big|_{s=1}^{n}$$

$$= \frac{1}{2}\left[[n]^2 - [n]^1\right] = \frac{n(n-1)}{2} + C,$$

for a constant C. Divide both sides by n and then take the exponential on both sides to arrive at

$$x(n) = e^{\frac{1}{2}(n-1)+\frac{C}{n}}, \quad n \in \mathbb{N}_1.$$

Applying the initial condition gives $e^C = x_0$. Hence, the solution of the original difference equation is

$$x(n) = x_0^{\frac{1}{n}} e^{\frac{1}{2}(n-1)}, \quad n \in \mathbb{N}_1. \quad \square$$

2.8.1 Exercises

Exercise 2.47. Verify (2.8.8).

Exercise 2.48. Verify (2.8.10).

Exercise 2.49. Solve the difference equation

$$x(n)x(n+1) - x(n+1) + x(n) = 0, \quad n \in \mathbb{N}_0.$$

Exercise 2.50. Solve the difference equation

$$x(n+1) = \frac{1 + 2x(n)}{2 + x(n)}, \quad n \in \mathbb{N}_0.$$

Exercise 2.51. Discuss the nature of solutions of (2.8.6) in terms of α.

Exercise 2.52. Solve the difference equation

$$x^2(n+1) - 5x(n+1)x(n) + 6x^2(n) = 0, \quad x(1) = 4, \ n \in \mathbb{N}_1.$$

Exercise 2.53. Solve the difference equation

$$x^2(n+2) - 5x(n+2)x(n) + 6x^2(n) = 0, \quad x(1), = 4, \ n \in \mathbb{N}_1.$$

Exercise 2.54. Solve the difference equation

$$x^2(n+2) - \frac{5}{2}x(n+1)x(n+2) + x^2(n+1) = 0,$$
$$x(0) = x_0 \neq 0, \ x(1) = x_1 \neq 0, \ x(2) = x_2 \neq 0 \, n \in \mathbb{N}_0.$$

Exercise 2.55. Find the general solution of the difference equation

$$x^3(n+4) - 8x^3(n+2) = 0, \quad n \in \mathbb{N}_{n_0}$$

$x(i) = x_i, \ x_i \neq 0, i = 0, 1, 2, 3$. Do not compute the four constants.

Exercise 2.56. Find the general solution of the difference equation

$$x^3(n+4) - 8x^3(n+1) = 0, \quad n \in \mathbb{N}_{n_0},$$

$x(i) = x_i, \ x_i \neq 0, i = 0, 1, 2$. Do not compute the constants.

Exercise 2.57. Solve

$$\left[x(n+2) - 2^{n+1}x(n+1)\right] - 2^n\left[x(n+1) - 2^n x(n)\right] = 0, \quad n \in \mathbb{N}_{n_0}.$$

Hint: Use the proper transformation. $y(n+1) = x(n+1) - 2^n x(n)$.

Exercise 2.58. Solve

$$\left[\frac{x(n+2)}{x(n+1)} - (n+1)\right] - n\left[\frac{x(n+1)}{x(n)} - n\right] = 0, \quad n \in \mathbb{N}_{n_0}.$$

Answer: $x(n) = \left(A\sum_{s=1}^{n-1}\frac{1}{s} + B\right)(n-1)!$, where $\Delta A = 0$, $\Delta B = 0$.

Exercise 2.59. Consider the general Riccati difference equation

$$x(n+1) = \frac{a(n) + b(n)x(n)}{c(n) + d(n)x(n)}, \quad n \in \mathbb{N}_0 \tag{2.8.15}$$

such that $d(n) \neq 0$, $b(n)c(n) - a(n)d(n) \neq 0$ for all $n \in \mathbb{N}_{n_0}$. Use the transformation

$$x(n) = \frac{1}{d(n)}\left[\frac{y(n+1)}{y(n)} - c(n)\right]$$

and transform (2.8.15) to the

$$y(n+2) + p(n)y(n+1) + q(n)y(n) = 0,$$
$$y(0) = 0, \quad y(1) = d(0)x(0) + c(0),$$

where

$$p(n) = -\frac{d(n)c(n+1) + b(n)d(n+1)}{d(n)},$$

$$q(n) = -\frac{d(n+1)}{d(n)}\left(b(n)c(n) - a(n)d(n)\right).$$

Exercise 2.60. Use substitution to solve

$$x(n+1) - 2\sqrt{1 - x^2(n)} = 0, \quad n \in \mathbb{N}_0.$$

Hint. Let $x(n) = \sin(\theta(n))$ and get $x(n+1) = \sin(\theta(n+1))$. Or $\sin(\theta(n+1)) = \sin(2\theta(n))$. Hence $\theta(n+1) = (-1)^r 2\theta(n) + r\pi$. Now take $r = 2s$ even and $r = (2s+1)$, odd.

Exercise 2.61. Solve

$$\frac{x(n+2)x(n)}{x(n+1)} = 1, \quad n \in \mathbb{N}_{n_0}.$$

Exercise 2.62. Solve

$$\frac{x(n+2)x^2(n)}{x^2(n+1)} = 1, \quad n \in \mathbb{N}_{n_0}.$$

Exercise 2.63. Solve

$$\frac{x(n+2)x^2(n)}{x^2(n+1)} = e^n, \quad n \in \mathbb{N}_{n_0}.$$

Exercise 2.64. For a positive number P, solve

$$x(n+1) = \frac{1}{2}\left(x(n) + \frac{P}{x(n)}\right), \quad n \geq 0.$$

Hint: Use the transformation $x(n) = P^{\frac{1}{2}}\cot(y(n))$.

Exercise 2.65. Solve

$$x(n+1) = 2x(n)\,(1 - x(n)), \quad n \geq 0.$$

Hint: Use the transformation $x(n) = \frac{1-y(n)}{2}$.

Exercise 2.66. Solve

$$x(n+1) - nx(n) = 1, \quad n \geq 1.$$

Hint: Divide both sides by $n!$.

Exercise 2.67. Solve the difference equation

$$x(n+1) = \frac{3 + 2x(n)}{2 + 3x(n)}, \quad n \in \mathbb{N}_0.$$

Chapter 3

Z-transform

This chapter is devoted to the study and the development of the z-transform. The z-transform is a powerful mathematical instrument that makes it possible to analyze, create, and synthesize discrete-time systems. Its applications range from signal processing to control systems analysis, making it a valuable tool in various engineering and scientific fields.

We can represent discrete-time signals and systems in the frequency domain using the z-transform. We may examine signals' frequency characteristics, such as their magnitude response, phase response, and poles and zeros, by translating them from the time domain to the z-domain. Understanding system behavior, stability, and performance is made easier by this analysis. The z-transform is heavily used in the applications of digital signal processing. It offers a framework for analyzing and creating digital filters, which are necessary for operations like noise cancellation, signal amplification, and frequency modification. Analysis of filter characteristics including frequency response, phase response, and group delay is possible thanks to the z-transform.

3.1 Introduction

In this section we state some definitions that are essential for the development of the z-transform.

Definition 3.1.1. A sequence $x(n)$ is said to be of *exponential order* if there exists a number $r > 0$, and an integer $N > 0$ such that

$$|x(n)| \leq rc^n \quad \text{for all } n > N,$$

where $c \geq 0$ is some suitable constant.

Definition 3.1.2. The z-transform $Z[x(n)]$ of a sequence $x(n)$ of exponential order is defined by

$$Z[x(n)] = \sum_{n=0}^{\infty} x(n)z^{-n}, \quad |z| > \rho, \tag{3.1.1}$$

where z is a complex number, and ρ is the radius of convergence of $Z[x(n)]$.

We remark a sequence $u(n)$ in Definition 3.1.2 is defined for $n \in \mathbb{N}_0$ and $u(n) = 0$, for $n \leq 0$. In addition, we note that the z-transform is a special kind of Laurent series; but really it behaves more like a Taylor series in $\frac{1}{z}$.

Difference Equations and Applications. https://doi.org/10.1016/B978-0-44-331492-6.00009-1

Definition 3.1.2 is the discrete analogue of the Laplace transform for functions that are defined on continuous intervals. The definition of the Laplace transform is stated as follows.

Definition 3.1.3. Let f be a function defined for $t \geq 0$. The Laplace transform of f is denoted by $F(s)$ and given by

$$\mathcal{L}[f(t)] = F(s) = \int_0^\infty e^{-st} f(t)\, dt, \quad \text{for } s > 0. \tag{3.1.2}$$

The Laplace transform of f is said to exist if the improper integral (3.1.2) converges.

Note that the z-transform defined by Definition 3.1.2 is considered as one sided transform and for two sided z-transform,

$$Z[x(n)] = \sum_{n=-\infty}^{\infty} x(n) z^{-n}.$$

In this book, we restrict ourselves to Definition 3.1.2, unless it is noted otherwise.

Later on, in the development of the z-transform, it is logical to adopt the notation that the z-transform of a sequence $u(n)$ is $U(z)$. We have our first theorem regrading the existence of the z-transform of a sequence.

Theorem 3.1.1. *Suppose the sequence $x(n)$ is of exponential order. Then its z-transform exists.*

Proof. Since $x(n)$ is of exponential order, then there exists a number $r > 0$, and an integer $N^* > 0$ such that

$$|x(n)| \leq rc^n \quad \text{for all } n > N^*.$$

Then from Definition 3.1.2 we have

$$\sum_{n=0}^{\infty} x(n) z^{-n} = \sum_{n=0}^{N^*-1} x(n) z^{-n} + \sum_{n=N^*}^{\infty} x(n) z^{-n}.$$

This implies that

$$\left| \sum_{n=0}^{N^*-1} x(n) z^{-n} \right| \leq \sum_{n=0}^{N^*-1} \left| \frac{x(n)}{z^n} \right| < \infty,$$

since it is summed over finite points. On the other hand,

$$\left| \sum_{n=N^*}^{\infty} x(n) z^{-n} \right| \leq \sum_{n=N^*}^{\infty} \left| \frac{x(n)}{z^n} \right|$$

$$\leq r \sum_{n=N^*}^{\infty} \left| \frac{c}{z} \right|^n,$$

and the last sum converges since $|z| > c$. This completes the proof. $\square$

For any sequence $\langle a(n) \rangle_{n \geq 0}$ there will be an $R \in [0, \infty]$ such that $Z[a(n)]$ is defined for $|z| > R$ and not for $|z| < R$. ($\frac{1}{R}$ is the radius of convergence of the series $\sum_{n=0}^{\infty} x(n) z^n$.) On the open set $\{z : |z| > R\}$, $Z[a(n)]$ will be an analytic function and its derivative will be

$$Z'[a(n)] = A'(z) = -\sum_{n=0}^{\infty} \frac{na(n)}{z^{n+1}} = -\sum_{n=1}^{\infty} \frac{(n-1)a(n-1)}{z^n}$$

$$= -\sum_{n=2}^{\infty} \frac{(n-1)a(n-1)}{z^n}.$$

Moreover, if R is finite, then $a(0) = \lim_{|z| \to \infty} A(z)$, and $a(1) = \lim_{z \to \infty} z^2 A'(z) = \lim_{|z| \to \infty} z(A(z) - a(0))$. We say more about it later on in the chapter. We address uniqueness of z-transform in the next theorem.

Theorem 3.1.2 (Uniqueness Theorem). *Suppose $a(n)$, $b(n)$ are two sequences with z-transforms $A(z)$ and $B(z)$, and suppose that $A(z) = B(z)$ at least whenever $|z|$ is large enough. Then $A(z) = B(z)$ wherever either is defined, and $a(n) = b(n)$ for every n.*

We end this section by comparing the two-sided z-transform to discrete Fourier transform. Consider the two-sided z-transform

$$X(z) = \sum_{n=-\infty}^{\infty} x(n) z^{-n}.$$

By setting $z = e^{j\omega}$, we obtain

$$X(e^{j\omega}) = \sum_{n=-\infty}^{\infty} x(n) e^{-j\omega n},$$

which is the discrete Fourier transform. Thus, one of the similarities is

$$X(z) = X(e^{j\omega}).$$

It is clear that the concept of the z-transform is a general one since z is any complex valued function, unlike the case for the discrete Fourier transform, where $z = e^{j\omega}$, that is 2π-periodic. Moreover, the z-transform is defined everywhere on $|z| > R$, whereas the discrete Fourier transform is only defined on the unit circle.

3.2 Standard functions

In this section we begin the derivation of the z-transform of some basic functions or sequences. First, we list the linearity property of the z-transform.

Theorem 3.2.1 (Linearity). *For constants α and β and sequences $u(n)$ and $v(n)$ we have*

$$Z[\alpha u(n) \pm \beta v(n)] = \alpha U(z) \pm \beta V(z).$$

Proof. We make a direct use of Definition 3.1.2 and obtain

$$Z[\alpha u(n) \pm \beta v(n)] = \sum_{n=0}^{\infty} [\alpha u(n) \pm \beta v(n)] z^{-n}$$

$$= \alpha \sum_{n=0}^{\infty} u(n) z^{-n} \pm \beta \sum_{n=0}^{\infty} v(n) z^{-n}$$

$$= \alpha U(z) \pm \beta V(z). \qquad \square$$

Remark 3.1. Suppose the z-transform of two sequences $u(n)$ and $v(n)$ is defined on $\{z : |z| > R_1\}$, $\{z : |z| > R_2\}$, respectively, then the z-transform of $\alpha u(n) a_n \pm \beta v(n)$ is defined on $\{z : |z| > \max(R_1, R_2)\}$. This is true since if $u(n)$ has z-transform defined on $\{z : |z| > R\}$, and $c \in \mathbb{C}$, then $cu(n)$ has its z-transform defined on $\{z : |z| > R\}$.

In the next theorem we list and obtain the z-transform of some standard sequences.

Theorem 3.2.2. *Let a and b be constants. Then*

(a) $Z[a^n] = \dfrac{z}{z - a}$, $|z| > |a|$

(b) $Z[b] = \dfrac{bz}{z - 1}$, $|z| > 1$

(c) $Z[\cos(n\theta)] = \dfrac{z(z - \cos(\theta))}{z^2 - 2z\cos(\theta) + 1}$, $|z| > 1$

(d) $Z[\sin(n\theta)] = \dfrac{z \sin(\theta)}{z^2 - 2z\cos(\theta) + 1}$, $|z| > 1$.

Proof. Let a be a constant. Then applying Definition 3.1.2 we have

$$Z[a^n] = \sum_{n=0}^{\infty} \left(\frac{a}{z}\right)^n$$

$$= 1 + \frac{a}{z} + \frac{a^2}{z^2} + \frac{a^3}{z^3} + \cdots \frac{a^n}{z^n} + \cdots$$

$$= \frac{1}{1 - \frac{a}{z}}, \quad \left|\frac{a}{z}\right| < 1$$

$$= \frac{z}{z-a}, \quad |z| > |a|.$$

Now we prove (b). Again, we make use of Definition 3.1.2.

$$Z[b] = \sum_{n=0}^{\infty} \left(\frac{b}{z}\right)^n = b \sum_{n=0}^{\infty} \left(\frac{1}{z}\right)^n$$

$$= b\left[1 + \frac{1}{z} + \frac{1}{z^2} + \frac{1}{z^3} + \cdots \frac{1}{z^n} + \cdots\right]$$

$$= b\frac{1}{1 - \frac{1}{z}}, \quad \left|\frac{1}{z}\right| < 1$$

$$= \frac{bz}{z-1}, \quad |z| > 1.$$

For part (c) and (d), we make use of Euler formula

$$e^{i\zeta} = \cos(\zeta) + i\sin(\zeta),$$

by letting $\zeta = -in\theta$, we have by using part (a) that

$$Z[e^{-in\theta}] = Z[(e^{-i\theta})^n]$$

$$= \frac{z}{z - e^{-i\theta}} = \frac{z(z - e^{i\theta})}{(z - e^{-i\theta})(z - e^{i\theta})}$$

$$= \frac{z(z - \cos(\theta) - i\sin(\theta))}{z^2 - z(e^{i\theta} + e^{-i\theta}) + 1}$$

$$= \frac{z(z - \cos(\theta) - i\sin(\theta))}{z^2 - 2z\cos(\theta) + 1},$$

where we have used

$$\cos(\theta) = \frac{e^{i\theta} + e^{-i\theta}}{2}$$

in the fourth equality to get to the last equality. Hence we have so far that

$$Z[e^{-in\theta}] = \frac{z(z - \cos(\theta))}{z^2 - 2z\cos(\theta) + 1} - i\frac{z\sin(\theta)}{z^2 - 2z\cos(\theta) + 1},$$

or

$$Z[\cos(n\theta) - i\sin(n\theta)] = \frac{z(z - \cos(\theta))}{z^2 - 2z\cos(\theta) + 1} - i\frac{z\sin(\theta)}{z^2 - 2z\cos(\theta) + 1}.$$

Using Theorem 3.2.1, the above equality readily gives

$$Z[\cos(n\theta)] = \frac{z(z - \cos(\theta))}{z^2 - 2z\cos(\theta) + 1},$$

and

$$Z\big[\sin(n\theta)\big] = \frac{z\sin(\theta))}{z^2 - 2z\cos(\theta) + 1}.$$

This completes the proof. □

Example 3.1. Find

(a) $Z[1]$,
(b) $Z[(-1)^n]$,
(c) $Z[\sin(\frac{n\pi}{4}) + 4a^7]$.

For (a) we set $b = 1$ in Theorem 3.2.2 and get $Z[1] = \dfrac{z}{z-1}$. For (b) we set $a = -1$ in Theorem 3.2.2 and get $Z[(-1)^n 1] = \frac{z}{z+1}$. Now we consider (c) by making use of (b) and (d) of Theorem 3.2.2.

$$Z\left[\sin\left(\frac{n\pi}{4}\right) + 4a^7\right] = Z\left[\sin\left(\frac{n\pi}{4}\right)\right] + 4Z[a^7]$$

$$= \frac{\frac{z}{\sqrt{2}}}{z^2 - z\sqrt{2} + 1} + \frac{4a^7 z}{z - 1}. \qquad \square$$

Definition 3.2.1.

(a) The *unit step sequence* $\mathcal{U}(n)$ is defined by

$$\mathcal{U}(n) = \begin{cases} 0, & \text{for } n < 0 \\ 1, & \text{for } n \geq 0 \end{cases} \tag{3.2.1}$$

(b) The *unit impulse sequence* $\delta(n)$ is defined by

$$\delta(n) = \begin{cases} 1, & \text{for } n = 0 \\ 0, & \text{for } n \neq 0. \end{cases} \tag{3.2.2}$$

Thus, as a consequence of the above definition we have that

$$\sum_{k=-\infty}^{\infty} \delta(k) = \delta(0) = 1,$$

and in general

$$\sum_{k=-\infty}^{\infty} a(k)\delta(k) = a(0).$$

Application wise, suppose we have an arbitrary signal $a(n)$. If we multiply $a(n)$ by the time-shifted impulse function $\delta_k(n)$, (see Exercise 3.1) then we get a signal that is zero everywhere except at $n = k$, where it takes the value $a(k)$. This is known as the sampling or sifting property of the impulse function:

$$a(n)\delta_k(n) = a(k)\delta_k(n).$$

In general, for any signal $a(n)$, we have

$$a(n) = \sum_{k=-\infty}^{\infty} a(k)\delta_k(n). \qquad (3.2.3)$$

In other words, any signal $a(n)$ can be written as a sum of scaled and shifted impulse functions. We have the following theorem.

Theorem 3.2.3. *Let a and b be constants. Then*

(a) $Z[\mathcal{U}(n)] = \dfrac{z}{z-1}$,

(b) $Z[\delta(n)] = 1$.

Proof. Proof of (a). Using Definition 3.1.2 we have

$$Z[\mathcal{U}(n)] = \sum_{n=0}^{\infty} \mathcal{U}(n)z^{-n}$$

$$= \sum_{n=0}^{\infty} z^{-n}$$

$$= \left[1 + \frac{1}{z} + \frac{1}{z^2} + \frac{1}{z^3} + \cdots \frac{1}{z^n} + \cdots\right]$$

$$= \frac{z}{z-1}.$$

As for (b),

$$Z[\delta(n)] = \sum_{n=0}^{\infty} \delta(n)z^{-n} = 1 + 0 + \cdots = 1. \qquad \square$$

3.2.1 Exercises

Exercise 3.1. Let

$$\delta_k(n) = \begin{cases} 1, & \text{for } n = k \\ 0, & \text{for } n \neq k. \end{cases}$$

Show that the $Z[\delta_k(n)] = \frac{1}{z^k}$, and find its region of convergence.

Exercise 3.2. [two-sided z-transform] Let

$$x(n) = \begin{cases} \left(\frac{1}{3}\right)^n, & \text{for } n \geq 0 \\ \left(\frac{1}{2}\right)^{-n}, & \text{for } n < 0. \end{cases}$$

Find $Z[x(n)]$ and its region of convergence.

Answer: $\frac{(5/3)z}{(z-1/3)(2-z)}$, $1/3 < |z| < 2$.

Exercise 3.3 (Two-sided z-transform). Explain why the two-sided z-transform of

$$x(n) = \begin{cases} 2^n, & \text{for } n < 0 \\ 3^n, & \text{for } n \geq 0, \end{cases}$$

does not exist.

Answer: $\frac{z}{z^2-5z+6}$, $3 < |z| < 2$, which is not a feasible region of convergence.

Exercise 3.4. Draw $\mathcal{U}(n-1), \mathcal{U}(n-2), \mathcal{U}(n+1)$.

3.3   Shifting and scalings

It is not convenient to use Definition 3.1.2 each time we want to find the z-transform. There are theorems that can be useful in instances where the sequence is in a specific form. In the discussion that follows, we present several labor-saving theorems; these, in turn, enable us to obtain the z-transform of many complicated functions without using Definition 3.1.2.

Theorem 3.3.1 (Scaling). *Let* $Z[u(n)] = U(z)$. *Then*

(a)

$$Z\left[a^{-n}u(n)\right] = U(az),$$

(b)

$$Z\left[a^n u(n)\right] = U\left(\frac{z}{a}\right).$$

Proof.

$$Z\left[a^{-n}u(n)\right] = \sum_{n=0}^{\infty} a^{-n}u(n)z^{-n}$$

$$= \sum_{n=0}^{\infty} u(n)(az)^{-n}$$

$$= U(az).$$

This proves (a). The proof of (b) is similar, but for completeness we will do it.

$$Z[a^n u(n)] = \sum_{n=0}^{\infty} a^n u(n) z^{-n}$$

$$= \sum_{n=0}^{\infty} u(n)(z/a)^{-n}$$

$$= U\left(\frac{z}{a}\right). \qquad \square$$

Example 3.2.

$$Z[2^n \sin(n\theta)] = \frac{\frac{z}{2}\sin(\theta)}{(z/2)^2 - z\cos(\theta) + 1}. \qquad \square$$

Theorem 3.3.2 (Derivative). *For a positive integer p we have*

$$Z[n^p] = -z\frac{d}{dz}Z[n^{p-1}].$$

Proof. By definition we have

$$Z[n^p] = \sum_{n=0}^{\infty} n^p z^{-n}, \qquad (3.3.1)$$

and hence

$$Z[n^{p-1}] = \sum_{n=0}^{\infty} n^{p-1} z^{-n}. \qquad (3.3.2)$$

A differentiation of (3.3.2) with respect to z gives

$$\frac{d}{dz}Z[n^{p-1}] = \sum_{n=0}^{\infty} n^{p-1}(-n)z^{-n-1}$$

$$= -z^{-1}\sum_{n=0}^{\infty} n^p z^{-n}$$

$$= -z^{-1}Z[n^p] \text{ by}(3.3.2).$$

Thus, we have obtained

$$\frac{d}{dz}Z[n^{p-1}] = -z^{-1}Z[n^p].$$

Solving for $Z[n^p]$ implies the result. This completes the proof. $\qquad \square$

We have the following example.

Example 3.3. Find

$$Z[n] \text{ and } Z[n^2].$$

Set $p = 1$ in Theorem 3.3.2. Then

$$Z[n] = -z\frac{d}{dz}Z[1] = -z\frac{d}{dz}\left(\frac{z}{z-1}\right) = \frac{z}{(z-1)^2}.$$

Next, we set $p = 2$ and make use of $Z[n]$.

$$Z[n^2] = -z\frac{d}{dz}Z[n] = -z\frac{d}{dz}\left(\frac{z}{(z-1)^2}\right) = \frac{z^2+z}{(z-1)^3}. \quad \Box$$

Theorem 3.3.3 (Multiplication by n). *Let $u(n)$ be a sequence with $Z[u(n)] = U(z)$. Then*

$$Z[nu(n)] = -z\frac{d}{dz}U(z).$$

Proof. The proof is based on the manipulation of the sum. By definition we have

$$Z[nu(n)] = \sum_{n=0}^{\infty} nu(n)z^{-n}$$

$$= -z\sum_{n=0}^{\infty} u(n)(-n)z^{-n-1}$$

$$= -z\sum_{n=0}^{\infty} \frac{d}{dz}\left(u(n)z^{-n}\right)$$

$$= -z\frac{d}{dz}\sum_{n=0}^{\infty} u(n)z^{-n}$$

$$= -z\frac{d}{dz}U(z). \quad \Box$$

Theorem 3.3.4 (Shifting). *Let $u(n)$ be a sequence with $Z[u(n)] = U(z)$. Then, for a positive integer k, we have*

(a) $Z[u(n-k)] = z^{-k}[U(z) + \sum_{m=-k}^{-1} u(m)z^{-m}], n \geq k$ *(right shifting property)*.

(b) $Z[u(n+k)] = z^{k}[U(z) - u(0) - \dfrac{u(1)}{z} - \dfrac{u(2)}{z^2} - \cdots - \dfrac{u(k-1)}{z^{k-1}}]$ *(left shifting property)*.

Proof. We use Definition 3.1.2. We begin with part (b).

$$Z[u(n+k)] = \sum_{n=0}^{\infty} u(n+k)z^{-n} \quad (j = n + k)$$

$$= \sum_{j=k}^{\infty} u(j)z^{k-j}$$

$$= z^k \left[\sum_{j=0}^{\infty} u(j)z^{-j} - \sum_{m=0}^{k-1} u(m)z^{-m} \right]$$

$$= z^k \left[U(z) - \sum_{m=0}^{k-1} u(m)z^{-m} \right].$$

This completes the proof of (b). The proof of part (a) is left as an exercise. □

We remark that if the sequence $u(n) = 0$, $n < 0$, then part (a) of Theorem 3.3.4 reduces to

$$Z[u(n-k)] = z^{-k}U(z) \quad n \geq k. \tag{3.3.3}$$

We have the immediate corollary.

Corollary 3.1. *Let $x(n)$ be a sequence for $n \geq 0$. Then for $Z[x(n)] = X(z)$ we have*

(a) $Z[x(n+1)] = z[X(z) - x(0)]$

(b) $Z[x(n+2)] = z^2[X(z) - x(0) - \dfrac{x(1)}{z}]$

(c) $Z[x(n+3)] = z^3[X(z) - x(0) - \dfrac{x(1)}{z} - \dfrac{x(2)}{z^2}].$

Proof. The proofs for parts (a), (b), and (c) follow by setting $k = 1, 2, 3$ in Theorem 3.3.4, respectively. □

The graph below shows shifting the sequence one unit to the right.

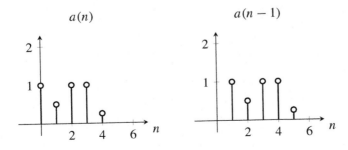

$a(n)$ $a(n-1)$

In Definition 3.2.1 the unit step sequence $\mathcal{U}(n)$ had its jump at zero and stayed there for all $n = 0, 1, 2, \ldots$. However, if the jump does start at a later time, say k, then the jump is being delayed by k units. Such a unit step sequence, we call *delayed unit step sequence* of delay k and denote it by $\mathcal{U}_n(k)$. Thus, we have the following definition.

Definition 3.3.1. For positive integers k, the *unit step sequence* $\mathcal{U}_n(k)$ is defined by

$$\mathcal{U}_n(k) = \begin{cases} 0, & \text{for } 0 \leq n \leq k-1 \\ 1, & \text{for } n \geq k \end{cases} \tag{3.3.4}$$

We have the following theorem.

Theorem 3.3.5. *Let $a(n)$ be a sequence with $Z[a(n)] = A(z)$. Then, for a positive integer k, we have*

$$Z\big[a(n-k)\mathcal{U}_n(k)\big] = z^{-k}A(z).$$

Proof.

$$Z\big[a(n-k)\mathcal{U}_n(k)\big] = \sum_{n=0}^{\infty} a(n-k)\mathcal{U}_n(k)z^{-n}$$

$$= \sum_{n=k}^{\infty} a(n-k)z^{-n} \quad (j=n-k)$$

$$= \sum_{j=0}^{\infty} a(j)z^{-j-k}$$

$$= z^{-k}A(z).$$

This completes the proof. $\square$

Example 3.4. Find

$$Z\big[\mathcal{U}_n(k)\big].$$

Set $a(n) = 1$, in Theorem 3.3.5.

$$Z\big[\mathcal{U}_n(k)\big] = z^{-k}Z[1] = z^{-k}\frac{z}{z-1} = \frac{z^{1-k}}{z-1}. \quad \square$$

Example 3.5. For positive integers k, find $Z[a(n)]$, where

$$a(n) = \begin{cases} 3, & \text{for } 0 \leq n \leq 10 \\ 4, & \text{for } n \geq 11 \end{cases}$$

We may express the sequence $a(n)$ by

$$a(n) = 3 + \mathcal{U}_n(11).$$

Thus,

$$\begin{aligned}
Z[a(n)] &= Z[3 + \mathcal{U}_n(11)] \\
&= Z[3] + Z[\mathcal{U}_n(11)] \\
&= \frac{3z}{z-1} + \frac{z^{-10}}{z-1}. \quad \square
\end{aligned}$$

The next theorem deals with the z-transform of factorial functions.

Theorem 3.3.6 (Factorial). *Let $A(n)$ be a sequence with $Z[a(n)] = A(z)$, on $|z| > \rho$. Then, for a positive integer k, we have*

(a) $Z[[n + k - 1]^k a(n)] = (-1)^k z^k \dfrac{d^k A}{dz^k}(z).$

(b) $Z[[n + k - 1]^k] = \dfrac{z! z^k}{(z-1)^{k+1}}.$

Proof. We know $A(z) = \sum_{n=0}^{\infty} a(n) z^{-n}$. Applying the derivative k times with respect to z gives

$$\begin{aligned}
\frac{d^k A}{dz^k}(z) &= \frac{d^k}{dz^k}\left(\sum_{n=0}^{\infty} n a(n) z^{-n}\right) \\
&= (-1)^k \sum_{n=0}^{\infty} n(n+1)\cdots(n+k-1) a(n) z^{-n-k} \\
&= (-1)^k z^{-k} \sum_{n=0}^{\infty} [n+k-1]^k a(n) z^{-n} \\
&= (-1)^k z^{-k} Z[[n+k-1]^k a(n)].
\end{aligned}$$

Multiplying both sides with $(-1)^k z^k$ gives the desired result of (a). For (b), we set $a(n) = 1$ in (a). Then

$$\begin{aligned}
Z[[n+k-1]^k] &= (-1)^k z^k \frac{d^k}{dz^k}\left(\frac{z}{z-1}\right) \\
&= (-1)^k z^k \frac{(-1)^k z!}{(z-1)^{k+1}} \\
&= \frac{z! z^k}{(z-1)^{k+1}}. \quad \square
\end{aligned}$$

This completes the proof. The next theorem is called *initial value theorem.*

Theorem 3.3.7 (Initial Value Theorem). *Let $A(z) = Z[a(n)]$ for $|z| > \rho$. Then*

(a)

$$a(0) = \lim_{z \to \infty} A(z),$$

(b)

$$a(1) = \lim_{z \to \infty} z[A(z) - a(0)],$$

(c)

$$a(2) = \lim_{z \to \infty} z^2 \left[A(z) - a(0) - \frac{a(1)}{z} \right],$$

$$\vdots$$

Proof. The proof is a direct consequence of the convergence of the infinite series in the definition of the z-transform. That is, using Definition 3.1.2 we have

$$\lim_{z \to \infty} A(z) = \lim_{z \to \infty} \sum_{n=0}^{\infty} a(n)z^{-n}$$

$$= a(0) + \sum_{n=1}^{\infty} \lim_{z \to \infty} \left(a(n)z^{-n} \right)$$

$$= a(0) + 0.$$

For (b), we also have

$$A(z) = \sum_{n=0}^{\infty} a(n)z^{-n}$$

$$= a(0) + a(1)z^{-1} + \sum_{n=2}^{\infty} \left(a(n)z^{-n} \right).$$

Multiplying both sides with z and rearranging the terms we obtain

$$z[A(z) - a(0)] = a(1) + \sum_{n=2}^{\infty} \left(a(n)z^{-n+1} \right).$$

Taking the limit as $z \to \infty$,

$$\lim_{z \to \infty} z[A(z) - a(0)] = a(1) + \sum_{n=2}^{\infty} \lim_{z \to \infty} \left(a(n)z^{-n+1} \right)$$

$$= a(1) + 0.$$

This proves part (b). The proof of part (c) is left as an exercise. $\square$

The next theorem is called the *final value theorem.*

Theorem 3.3.8 (Final value Theorem). *Let $A(z) = Z[a(n)]$ for $|z| > \rho$. Then*

$$\lim_{n \to \infty} a(n) = \lim_{z \to 1} (z - 1)A(z).$$

Proof. Again the proof relies on the convergence of the infinite series. This is crucial to obtaining the result since it implies the tail of the infinite series goes to zero as $n \to \infty$. During the proof, we employ Corollary 3.1. Since z-transform is linear we have

$$Z[a(n+1)] - Z[a(n)] = \sum_{n=0}^{\infty} (a(n+1) - a(n))z^{-n}.$$

By Corollary 3.1 we get

$$z[A(z) - a(0)] - A(z) = \sum_{n=0}^{\infty} (a(n+1) - a(n))z^{-n},$$

or

$$(z - 1)A(z) - za(0) = \sum_{n=0}^{\infty} (a(n+1) - a(n))z^{-n}.$$

Taking the limit as $z \to 1$ we obtain

$$\lim_{z \to 1} (z - 1)A(z) - za(0) = \sum_{n=0}^{\infty} (a(n+1) - a(n)).$$

Consequently,

$$\lim_{z \to 1} (z - 1)A(z) - a(0)$$
$$= \lim_{n \to \infty} \left[(a(1) - a(0)) + (a(2) - a(1)) + \cdots + (a(n+1) - a(n)) \right]$$
$$= \lim_{n \to \infty} \left[a(n+1) - a(0) \right]$$
$$= a(\infty) - a(0).$$

Simplifying with $a(0)$ from both sides gives

$$\lim_{z \to 1} (z - 1)A(z) = \lim_{n \to \infty} a(n) = a(\infty).$$

This completes the proof. □

We remark that Theorems 3.3.7 and 3.3.8 can be used to recover a sequence from its z-transform, as the next example shows.

Example 3.6. Compute $Z[a(n + 1)]$ given that $A(z) = \frac{z^2+z}{(z-1)^3}$. From Corollary 3.1 we have

$$Z[a(n+1)] = z[A(z) - a(0)].$$

By Theorem 3.3.7

$$a(0) = \lim_{z \to \infty} A(z) = \lim_{z \to \infty} \frac{z^2 + z}{(z - 1)^3} = 0.$$

Consequently,

$$Z[a(n + 1)] = zA(z) = \frac{z^3 + z^2}{(z - 1)^3}. \quad \square$$

Example 3.7. Compute $Z[a(n + 2)]$ given that $A(z) = \frac{z}{z-1} + \frac{z}{(z-1)^2}$. From Corollary 3.1 we have

$$Z[a(n+2)] = z^2\left[A(z) - a(0) - \frac{a(1)}{z}\right].$$

By Theorem 3.3.7

$$a(0) = \lim_{z \to \infty} A(z) = \lim_{z \to \infty}\left[\frac{z}{z - 1} + \frac{z}{(z - 1)^2}\right] = 1.$$

On the other hand

$$a(1) = \lim_{z \to \infty} z[A(z) - a(0)] = \lim_{z \to \infty} z\left[\frac{z}{z - 1} + \frac{z}{(z - 1)^2} - 1\right],$$

or

$$a(1) = \lim_{z \to \infty}\left[\frac{2z^2 - z}{(z - 1)^2}\right] = 2.$$

Finally,

$$Z[a(n + 2)] = z^2\left[A(z) - a(0) - \frac{a(1)}{z}\right]$$
$$= \frac{3z^2 - 2z}{(z - 1)^2}. \quad \square$$

The next theorem gives a formula for the z-transform of the binomial coefficient.

Theorem 3.3.9. *For $0 \leq r \leq n$, we have*

$$Z\left[\binom{n}{r}\right] = (1 + z^{-1})^n.$$

Proof. We make use of (1.3.10), which states that

$$(a+b)^n = \sum_{k=0}^{n} \binom{n}{k} a^k b^{n-k}.$$

Thus, applying the definition of z-transform we have

$$Z\left[\binom{n}{r}\right] = \sum_{r=0}^{n} \binom{n}{r} z^{-r}$$

$$= \sum_{r=0}^{n} \binom{n}{r} (z^{-1})^r = (1+z^{-1})^n,$$

where we have taken $b = 1$ and $a = z^{-1}$. This completes the proof. $\square$

Theorem 3.3.10. *We have the following z-transforms:*

(a) $Z[\frac{1}{n!}] = e^{\frac{1}{z}}$

(b) $Z[\frac{1}{(n-r)!}]$.

Proof. We begin with (a).

$$Z\left[\frac{1}{n!}\right] = \sum_{n=0}^{\infty} \frac{1}{n!} z^{-n} = \sum_{n=0}^{\infty} \frac{(z^{-1})^n}{n!} = e^{\frac{1}{z}}.$$

For (b) we make use of the right-shifting property of Theorem 3.3.4. That is

$$Z[a(n-k)] = z^{-k} Z[a(n)].$$

Thus, by letting $a(n) = \frac{1}{n!}$, we get by using part (a) that

$$Z\left[\frac{1}{(n-r)!}\right] = z^{-r} e^{\frac{1}{z}}.$$

This completes the proof. $\square$

We end this section with the following theorem regarding the z-transform of a periodic sequence.

Theorem 3.3.11 (Periodic sequence). *Suppose the sequence $a(n)$ is periodic of period $T \in \mathbb{N}_1$, where T is the smallest number in $\mathbb{N}_1$ such that $a(n+T) = a(n)$. If $Z[a(n)] = A(z)$ for $|z| > \rho$, then*

$$Z[a(n)] = \frac{z^T}{z^T - 1} \sum_{n=0}^{T-1} a(n) z^{-n},$$

for all $|z| > \max\{1, \rho\}$.

Proof. The idea of the proof is to decompose the infinite sum into the addition of sums, each summed over the length of the period T. By definition of the z-transform we have

$$Z[a(n)] = \sum_{n=0}^{\infty} a(n)z^{-n}$$

$$= \sum_{n=0}^{T-1} a(n)z^{-n} + \sum_{n=T}^{2T-1} a(n)z^{-n} + \sum_{n=2T}^{3T-1} a(n)z^{-n} + \cdots$$

$$+ \sum_{n=kT}^{(k+1)T-1} a(n)z^{-n} + \cdots$$

By making the substitutions $k = n - T$ in the second summation and $k = n - 2T$ in the third summation and so on, and then by switching back to n we arrive at

$$Z[a(n)] = \sum_{n=0}^{T-1} a(n)z^{-n} + \sum_{n=0}^{T-1} a(n+T)z^{-n-T} + \sum_{n=0}^{T-1} a(n+2T)z^{-n-2T} \cdots$$

$$= \sum_{n=0}^{T-1} a(n)z^{-n} + \sum_{n=0}^{T-1} a(n)z^{-n-T} + \sum_{n=0}^{T-1} a(n)z^{-n-2T} + \cdots$$

$$= \sum_{n=0}^{T-1} a(n)z^{-n}\left[1 + z^{-T} + z^{-2T} + \cdots\right]$$

$$= \sum_{k=0}^{\infty} \left(\frac{1}{z^T}\right)^k \sum_{n=0}^{T-1} a(n)z^{-n}$$

$$= \frac{1}{1 - \frac{1}{z^T}} \sum_{n=0}^{T-1} a(n)z^{-n}$$

$$= \frac{z^T}{z^T - 1} \sum_{n=0}^{T-1} a(n)z^{-n}.$$

Now, $|z| > \max\{1, \rho\}$ since the geometric series $\sum_{k=0}^{\infty}(\frac{1}{z^T})^k$ converges for $|z| > 1$. This completes the proof. $\square$

Example 3.8. The sequence $a(n) = (-1)^n$, $n = 0, 1, 2, \cdots$ is periodic with period $T = 2$. Hence its z-transform is given by

$$Z[a(n)] = \frac{z^2}{z^2 - 1} \sum_{n=0}^{1} (-1)^n z^{-n} = \frac{z^2}{z^2 - 1}\left[1 - z^{-1}\right]. \quad \square$$

We end the section with the following example.

Example 3.9. Let a be a nonzero constant and consider the sequence

$$u(n) = (a)^{|n|} = \begin{cases} a^n, & \text{for } n \geq 0 \\ a^{-n}, & \text{for } n < 0. \end{cases}$$

Then the two sided z-transform of the sequence $u(n)$ is

$$Z[u(n)] = \sum_{n=-\infty}^{\infty} u(n)z^{-n} = \sum_{n=-\infty}^{-1} (a)^{-n}z^{-n} + \sum_{n=0}^{\infty}(a)^n z^{-n}.$$

Or,

$$Z[u(n)] = \sum_{n=-\infty}^{-1} (a)^{-n}z^{-n} + \sum_{n=0}^{\infty}(a)^n z^{-n}$$

$$= \sum_{n=-\infty}^{-1} (az)^{-n} + \sum_{n=0}^{\infty}\left(\frac{a}{z}\right)^n$$

$$= az \sum_{n=-\infty}^{-1} (az)^{-n} + \sum_{n=0}^{\infty}\left(\frac{a}{z}\right)^n$$

$$= az\frac{1}{1-az} + \frac{1}{1-a/z}, \quad \text{valid for } |az| < 1, \text{ and } |a/z| < 1$$

$$= \frac{z(1-az)}{(1-az)(z-a)}, \quad |a| < |z| < 1/|a|$$

$$= \frac{z}{z-a}, \quad |a| < |z| < 1/|a|.$$

3.3.1 Exercises

Exercise 3.5. Show that

$$Z[a^n n^2] = \frac{az^2 + a^2 z}{(z-a)^3}.$$

Exercise 3.6. Find

$$Z[a^n \cos(n\theta)] \text{ and } Z[a^n \sin(n\theta)].$$

Exercise 3.7. Find

$$Z[n\cos(n\theta)] \text{ and } Z[n\sin(n\theta)].$$

Exercise 3.8. Prove part (a) of Theorem 3.3.4.

Exercise 3.9. Compute

(a) $Z[[n+1]^2 n]$.

(b) $Z[[n+1]^2 2^n \sin(n\theta)]$.

Exercise 3.10. Show that

$$Z[[n]^k] = \frac{k!z}{(z-1)^{k+1}}.$$

Exercise 3.11. Prove part (c) and additionally, compute $a(3)$ and $a(4)$ of Theorem 3.3.7.

Exercise 3.12. Given that $A(z) = \dfrac{1}{z-1} + \dfrac{z}{z^2+1}$, compute

(a) $Z[a(n+1)]$ (b) $Z[a(n+2)]$ (c) $Z[a(n+3)]$.

Exercise 3.13. Given that $A(z) = e^{\frac{1}{z}}$, compute $Z[a(n+1)]$.

Exercise 3.14. Make use of the (Final Value Theorem), Theorem 3.3.8 to compute $a(\infty)$ for:

(a) $A(z) = \dfrac{z^2}{(z-1)(z-2)}$,

(b) $A(z) = \dfrac{z^2}{(z+1)(z-2)}$.

Exercise 3.15. Verify the statement of Theorem 3.3.8 by considering the sequence $a(n) = (\frac{1}{2})^n$, $n = 0, 1, \ldots$.

Exercise 3.16. Use (b) of Theorem 3.3.6 to show that for $0 \le r \le n$, we have

$$Z\left[\binom{n}{r}\right] = (1 + z^{-1})^n.$$

Exercise 3.17. Show that

$$Z\left[\binom{n+r}{r}\right] = (1 - z^{-1})^{-n-1}.$$

Exercise 3.18. Find

$$Z\left[\frac{1}{(n+r)!}\right].$$

Hint: Use the left-shifting property of Theorem 3.3.4.

Exercise 3.19. Find

(a) $Z[\dfrac{1}{(n+1)!}]$,

(b) $Z[\frac{1}{(n+2)!}]$,

(c) $Z[\frac{1}{(n-2)!}]$.

Exercise 3.20. Determine the period T and the z-transform of each of the following sequences:

(a) $\{3, 2, 1, 0, 3, 2, 1, 0, \cdots\}$,

(b) $\{3, 3, 3, \cdots\}$,

(c) $\{3, 1, 4, 1, 5, 9, 3, 1, 4, 1, 5, 9, \cdots\}$.

Exercise 3.21. Show the sequence

$$a(n + 1) = 1 - a^2(n), \quad a(0) = 0, \ n = 1, 2, \ldots$$

is periodic and find its z-transform.

Exercise 3.22. Consider the sequence

$$a(n + 1) = pa(n) + 10, \quad a(1) = 5, \ n = 2, 3, \ldots,$$

which is periodic of period 2. Find p and the z-transform of the sequence.

Exercise 3.23. Show the sequence

$$a(n + 1) = \frac{a(n) - 3}{a(n) - 2}, \quad a(1) = 3, \ n = 2, 3, \cdots$$

is periodic and find its z-transform.
Hint: Compute the first 6 terms, etc.

3.4   Convolution

This section is devoted to the convolution of two or more sequences. The convolution operation plays a crucial role in the context of the z-transform in the field of signal processing and other pertinent subjects in engineering and applied sciences. We begin with the following definition.

Definition 3.4.1. The *convolution* of two sequences $u(n)$ and $v(n)$ is defined as

$$u * v = \sum_{m=0}^{n} u(m)v(n - m). \tag{3.4.1}$$

For example,

$$2^n * 3^n = \sum_{m=0}^{n} 2^m 3^{n-m},$$

and

$$1 * a(n) = \sum_{m=0}^{n} a(m).$$

The next theorem is referred to as the *convolution theorem*

Theorem 3.4.1 (Convolution Theorem). *Let $U(z) = Z[u(n)]$ for $|z| > l_1$ and $V(z) = Z[v(n)]$ for $|z| > l_2$, then for $|z| > \max\{l_1, l_2\}$ we have*

$$Z[u * v] = U(z)V(z).$$

Proof.

$$Z[u * v] = \sum_{n=0}^{\infty} u(n) * v(n)z^{-n} = \sum_{n=0}^{\infty} \left[\sum_{m=0}^{n} u(m)v(n - m) \right] z^{-n}$$

$$= \sum_{n=0}^{\infty} \sum_{m=0}^{n} u(m)v(n - m)z^{-n} \text{ (rearranging the order of summation)}$$

$$= \sum_{m=0}^{\infty} \sum_{n=m}^{\infty} u(m)v(n - m)z^{-n} \text{ (set } n - m = k)$$

$$= \sum_{m=0}^{\infty} \sum_{k=0}^{\infty} u(m)v(k)z^{-m-k}$$

$$= \left(\sum_{m=0}^{\infty} u(m)z^{-m} \right) \left(\sum_{k=0}^{\infty} v(k)z^{-k} \right) = U(z)V(z).$$

This completes the proof. $\qquad\square$

The graph below shows the convolution of two sequences.

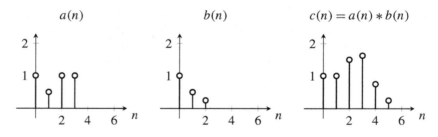

Example 3.10. Compute $Z[\sum_{m=0}^{n} a^m]$. Notice that $1 * a^n = \sum_{m=0}^{n} a^m$ and hence by the convolution theorem we have that

$$Z\left[\sum_{m=0}^{n} a^m \right] = Z[1]Z[a^m] = \frac{z}{z-1} \frac{z}{z-a},$$

for $|z| > \max\{1, |a|\}$. $\qquad\square$

In the next example, we make use of the final theorem and try to recover the sequence from its convolution.

Example 3.11. Consider the recurrence relation

$$a_0 = 1, \quad a(n+1) = \sum_{j=0}^{n} a(j)a(n-j).$$

Let $a(n)$ be the solution of the recurrence relation and denote its z-transform with $A(z)$, that is defined for large $|z|$. From Corollary 3.1 we have

$$Z[a(n+1)] = z[A(z) - a(0)],$$

while the right-hand side is the convolution of $a(n)$ with itself, so has z-transform $A(z)^2$. Thus we have

$$z(A-1) = A^2,$$

yielding

$$A = \frac{z \pm \sqrt{z^2 - 4z}}{2}.$$

Which square root should we take? We want $\lim_{|z| \to \infty} A(z) = a_0 = 1$. If we express A as

$$\frac{z}{2}\left(1 \pm \sqrt{1 - \frac{4}{z}}\right),$$

then plainly we must take

$$A = \frac{z}{2}\left(1 - \left(1 - \frac{4}{z}\right)^{\frac{1}{2}}\right)$$

to have a chance. Now we know that

$$(1+w)^{\frac{1}{2}} = 1 + \frac{1}{2}w + \frac{1}{2!}\left(\frac{1}{2}\right)\left(-\frac{1}{2}\right)w^2 + \frac{1}{3!}\left(\frac{1}{2}\right)\left(-\frac{1}{2}\right)\left(-\frac{3}{2}\right)w^3$$
$$+ \frac{1}{4!}\left(\frac{1}{2}\right)\left(-\frac{1}{2}\right)\left(-\frac{3}{2}\right)\left(-\frac{5}{2}\right)w^4 + \cdots,$$

so

$$\left(1 - \frac{4}{z}\right)^{\frac{1}{2}} = 1 - \frac{2}{z} - \frac{1}{2!}\frac{4}{z^2} - \frac{1}{3!} \cdot 3 \cdot \frac{8}{z^3} - \frac{1}{4!} \cdot 3 \cdot 5 \cdot \frac{16}{z^4} - \frac{1}{5!} \cdot 3 \cdot 5 \cdot 7 \cdot \frac{32}{z^5}$$
$$- \cdots$$
$$= 1 - \frac{2}{z} - \frac{1}{2!}\frac{1! \, 2^2}{1 \, z^2} - \frac{1}{3!}\frac{3! \, 2^3}{2 \, z^3} - \frac{1}{4!}\frac{5! \, 2^4}{2.4 \, z^4} - \frac{1}{5!}\frac{7! \, 2^5}{2.4.6 \, z^5} - \cdots$$

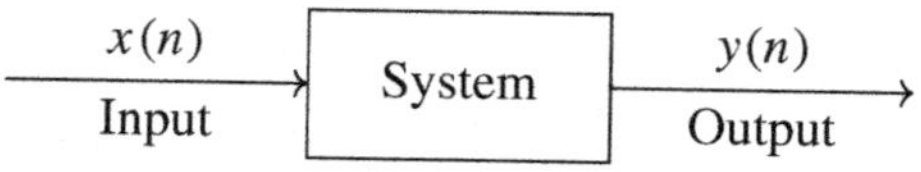

FIGURE 3.1 A generic representation of a system.

$$= 1 - \frac{2}{z} - \frac{1}{2!}\frac{1!}{1}\frac{2^2}{z^2} - \frac{1}{3!}\frac{3!}{2}\frac{2^3}{z^3} - \frac{1}{4!}\frac{5!}{2^2 2!}\frac{2^4}{z^4} - \frac{1}{5!}\frac{7!}{2^3 3!}\frac{2^5}{z^5} - \cdots$$

$$= 1 - \frac{2}{z} - \frac{1}{2!}\frac{1!}{0!}\frac{4}{z^2} - \frac{1}{3!}\frac{3!}{1!}\frac{4}{z^3} - \frac{1}{4!}\frac{5!}{2!}\frac{4}{z^4} - \frac{1}{5!}\frac{7!}{3!}\frac{4}{z^5} - \cdots$$

$$= 1 - \frac{2}{z} - \sum_{n=2}^{\infty} \frac{4(2n-3)!}{n!(n-2)!z^n}.$$

Rearranging the terms we get

$$\frac{z}{2}\left(1 - \left(1 - \frac{4}{z}\right)^{\frac{1}{2}}\right) = 1 + \sum_{n=2}^{\infty} \frac{2(2n-3)!}{n!(n-2)!z^{n-1}}$$

$$= 1 + \sum_{n=1}^{\infty} \frac{2(2n-1)!}{(n+1)!(n-1)!z^n}$$

$$= 1 + \sum_{n=1}^{\infty} \frac{(2n)!}{(n+1)!n!z^n}$$

$$= \sum_{n=0}^{\infty} \frac{(2n)!}{(n+1)!n!z^n}.$$

This means that

$$a(n) = \frac{(2n)!}{(n+1)!n!},$$

for all $n \in \mathbb{N}_0$.

Now we look at some engineering applications. A *system* is an abstract object that accepts input signals and responds by producing output signals. In Fig. 3.1, we display an abstract system of an input-output system.

Examples of such systems:

- Mechanical systems: speeds, displacement, pressure, volume, temperature, ...
- Chemical and biological systems: concentrations of cells and reactants, neuronal activity, cardiac signals, ...
- Electrical circuits: voltages, currents, temperature, ...
- Environmental systems: chemical composition of atmosphere, wind patterns, surface and atmospheric temperatures, pollution levels, ...

- Economic systems: stock prices, unemployment rate, tax rate, interest rate, growth domestic product, ...
- Chemical and biological systems: concentrations of cells and reactants, ...
- Social systems: opinions, gossip, online sentiment, political polls, ...
- Audio/visual systems: music, speech recordings, images, video, ...

Consider a system having input $x(n)$ and output $y(n)$. Assume the input is an impulse function. That is $a(n) = \delta(n)$. We denote this impulsive response with the signal $h(n)$. Recall, from the shifting property of the impulse signal (3.2.3), for an arbitrary signal $x(n)$ we have

$$a(n) = \sum_{k=-\infty}^{\infty} a(k)\delta_k(n).$$

Since our system is time-invariant (constant coefficients), the response of the system corresponding to the input $\delta_k(n)$ is $h(n - k)$. Thus, by linearity, we have the response to $x(n)\delta_k(n)$ is $x(n)h(n - k)$. In addition, the response to $\sum_{k=-\infty}^{\infty} x(n)\delta_k(n)$ is then

$$y(n) = \sum_{k=-\infty}^{\infty} x(n)h(n - k),$$

which is two-sided convolution. Or

$$x(n) * h(n) = \sum_{k=-\infty}^{\infty} x(n)h(n - k).$$

Thus, we have the following crucial characteristic of discrete systems that if an input signal is $x(n)$ and its impulse response is $h(n)$, then the system's output is $y(n) = x(n) * h(n)$. We have the following example.

Example 3.12. Find a formula for the output $y(n)$ of the system with input signal

$$x(n) = \begin{cases} 1, & \text{for } 0 \leq n \leq 4, \\ 0, & \text{otherwise}, \end{cases}$$

and impulse response

$$h(n) = \begin{cases} 1, & \text{for } 0 \leq n \leq 4, \\ 0, & \text{otherwise}. \end{cases}$$

From the above discussion, we have that

$$y(n) = x(n) * h(n) = \sum_{k=-\infty}^{\infty} x(k)h(n - k).$$

Due to the fact that $x(n) = 0$, for $k < 0$ and $h(n - k) = 0$, for $k > n$, the previous formula reduces to the regular convolution. That is

$$y(n) = \sum_{k=0}^{n} x(k)h(n - k).$$

For example,

$$y(0) = \sum_{k=0}^{0} x(k)h(0 - k) = x(0)h(0) = (1)(1) = 1,$$

$$y(1) = \sum_{k=0}^{1} x(k)h(1 - k) = x(0)h(1) + x(1)h(0) = 1 + (1)(1) = 2.$$

Continuing in this fashion, we see that

$$y(2) = 3, \quad y(3) = 4, \quad y(4) = 5, \quad y(5) = 4, \quad y(6) = 3, \quad y(7) = 2,$$
$$y(8) = 1, \quad \text{and} \quad y(n) = 0, \ n \geq 9. \quad \square$$

Let $y(n)$ be the output system. We know that

$$y(n) = \sum_{k=-\infty}^{\infty} x(n)h(n - k).$$

Then the system is *causal* if and only if its impulse response $h(n) = 0$ for all $n < 0$. Suppose the input signal $x(n)$ is bounded. That is, there exists a positive constant D such that $|x(n)| \leq D$. Then,

$$|y(n)| = \left| \sum_{k=-\infty}^{\infty} x(n)h(n - k) \right|$$

$$\leq \sum_{k=-\infty}^{\infty} |x(n)||h(n - k)|$$

$$\leq D \sum_{k=-\infty}^{\infty} |h(n - k)|.$$

Thus, if $\sum_{k=-\infty}^{\infty} |h(n - k)| < \infty$, or *absolutely summable*, then the output $y(n)$ will be bounded for all n. Hence, a discrete system is *stable* if and only if

$$\sum_{k=-\infty}^{\infty} |h(k)| < \infty,$$

since convolution commutes.

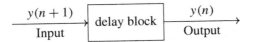

FIGURE 3.2 Delay block.

Example 3.13. Consider the discrete system with impulse response,

$$h(n) = \begin{cases} 0, & \text{for } n < 0, \\ \alpha^n, & \text{otherwise.} \end{cases}$$

Then

$$\sum_{k=-\infty}^{\infty} |h(k)| = \sum_{k=0}^{\infty} |\alpha|^k = \begin{cases} \frac{1}{1-|\alpha|}, & \text{for } |\alpha| < 1 \\ \infty, & \text{for } |\alpha| \ge 1 \end{cases}.$$

Thus, the system is stable if and only if $|\alpha| < 1$. □

In some cases it is useful to represent an Nth-order difference equation using block diagrams. Consider the Nth-order difference equation

$$y(n + N) = -a_{N-1}y(n + N - 1) - \ldots - a_0 y(n) + b_0 d(n). \tag{3.4.2}$$

We will use *delay blocks* to represent (3.4.2). When the signal $y(n + 1)$ is fed into a *delay block*, the output is $y(n)$, as depicted in Fig. 3.2.

To use these blocks to represent (3.4.2), we simply chain a sequence of these blocks in series, and feed the term with the highest order into the first block in the chain, as shown in Fig. 3.3.

This series chain of delay blocks provides us with all of the signals needed to represent (3.4.2). In particular, from (3.4.2) we observe that $y(n + N)$ is a linear combination of the signals $y(n + N - 1), \ldots, y(n), d(n)$. In other words, to generate the signal $y(n + N)$ we multiply the signals from the corresponding delay blocks by their respective coefficients, and add them all together.

Assume the system represented by Fig. 3.3 is causal. That is, $y(-1) = y(-2) = \ldots = 0$. Then $Z[y(n)] = Y(z)$, $Z[y(n-1)] = z^{-1}Y(z)$, $Z[y(n-2)] = z^{-2}Y(z), \ldots$, $Z[y(n-m)] = z^{-m}Y(z)$. We see that D can be replaced with the factor of z^{-1}. In other words, if we replace the input and output signals by their z-transforms, it is evident that in the z-transform domain the shift or delay becomes a multiplication by the factor z^{-1}.

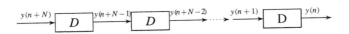

FIGURE 3.3 Chained delay blocks.

3.4.1 Exercises

Exercise 3.24. Find

$$Z\left[\sum_{m=0}^{n} 2^{n-m} \sin(5m)\right].$$

Exercise 3.25. Compute:

(a) $1 * 2$,
(b) $2 * n$,
(c) $n * n$,
(d) $a^n * b^n$, where a, b are constants with $a \neq b$.

Exercise 3.26. For two sequences $\{a(n)\}_{n=1}^{\infty}$ and $\{b(n)\}_{n=1}^{\infty}$, define

$$\{a(n)\}_{n=1}^{\infty} * \{b(n)\}_{n=1}^{\infty} = \left\{\sum_{k=1}^{n} a(k)b(n-k+1)\right\}_{n=1}^{\infty}.$$

Show that

$$\{1, -1, 2, -2, 3, -3, 4, -4, 5, -5 \cdots\} * \{1, 2, 3, 4, 5, 6, 7, 8, 9, 10, \cdots\}$$
$$= \{1, 1, 3, 3, 6, 6, 10, 10, 15, 15, \cdots\}.$$

Exercise 3.27. Let $0 < \alpha < 1$. Find a formula for the output $y(n)$ of the system with input signal

$$x(n) = \begin{cases} \alpha^n, & \text{for } n \geq 0 \\ 0, & \text{otherwise.} \end{cases}$$

and impulse response

$$h(n) = \begin{cases} 1, & \text{for } n \geq 0 \\ 0, & \text{for } n < 0. \end{cases}$$

ANS:

$$y(n) = \begin{cases} \frac{1-\alpha^{n+1}}{1-\alpha}, & \text{for } n \geq 0 \\ 0, & \text{otherwise.} \end{cases}$$

Is the system stable?

Exercise 3.28. Show the system with impulse response, $h(n) = \frac{1}{n^2+1}$ is stable.

3.5 Inverse z-transform and applications

In the previous sections we were concerned with finding the z-transform of a given sequence. Now we turn the problem around, namely; given $A(z)$, we want to find the original sequence $a(n)$. We have the following definition.

Definition 3.5.1. Let $A(z)$ be the z-transform of a given sequence $a(n)$. We denote the z-inverse of $a(n)$ by $Z^{-1}[A(z)]$ such that

$$Z^{-1}[A(z)] = a(n).$$

Given a function $A(z)$, we can find the sequence $a(n)$ by one of the following methods:

- Inspection or direct inversion,
- Direct division,
- Partial fractions,
- Residues (Inverse integral),
- Power-Series,
- Convolution theorem.

We will explain each method by series of examples.

- **Inspection**

Example 3.14. Find $a(n)$ when $A(z) = 1 + \frac{1}{3}z^{-1} + \frac{1}{9}z^{-2} + \frac{1}{27}z^{-3} + \cdots$.

Our aim is to rewrite $A(z)$ as an infinite power series in z^{-1} and compare it with the definition of the z-transform. First, $A(z) = \sum_{n=0}^{\infty} a(n)z^{-n}$. Now, the given $A(z)$ can be easily put in the form

$$A(z) = \sum_{n=0}^{\infty} \left(\frac{1}{3}\right)^n z^{-n}.$$

Comparing both infinite series we see that

$$a(n) = \left(\frac{1}{3}\right)^n. \quad \square$$

Example 3.15. Find $a(n)$ when $A(z) = \dfrac{z^2}{(z-1)^2}$. We will make use of the identity

$$(1-x)^{-n} = 1 + nx + \frac{n(n+1)}{2!}x^2 + \frac{n(n+1)(n+2)}{3!}x^3 + \cdots.$$

Next we consider $A(z)$.

$$A(z) = \left(\frac{(z-1)}{z}\right)^{-2} = \left(1 - \frac{1}{z}\right)^{-2}.$$

Using the previous stated formula, we have

$$A(z) = 1 + 2z^{-1} + 3z^{-2} + 4z^{-3} + \cdots = \sum_{n=0}^{\infty} (n+1)z^{-n}.$$

A quick comparison with $A(z) = \sum_{n=0}^{\infty} a(n)z^{-n}$, leads to

$$a(n) = n + 1.$$

We do a quick check. By Example 3.1 and Example 3.3 we get

$$Z[n+1] = \frac{z}{(z-1)^2} + \frac{z}{z-1} = \frac{z^2}{(z-1)^2}. \quad \Box$$

Example 3.16. Find $a(n)$ when

$$A(z) = 2 + \frac{3z}{z-1} + \frac{z}{5z-1}.$$

We will find the z-inverse term by term. First,

$$Z^{-1}[2] = 2Z^{-1}[1] = 2\delta(n).$$

On the other hand,

$$Z^{-1}\left[\frac{3z}{z-1}\right] = 3Z^{-1}\left[\frac{z}{z-1}\right] = 3\mathcal{U}(n).$$

Finally,

$$Z^{-1}\left[\frac{z}{5z-1}\right] = \frac{1}{5}Z^{-1}\left[\frac{z}{z-\frac{1}{5}}\right] = \frac{1}{5}\left(\frac{1}{5}\right)^n = \left(\frac{1}{5}\right)^{n+1}.$$

Consequently,

$$a(n) = 2\delta(n) + 3\mathcal{U}(n) + \left(\frac{1}{5}\right)^{n+1}. \quad \Box$$

- **Long division**

We give an example to illustrate the method.

Example 3.17 ([45]). Find $a(n)$ when

$$A(z) = \frac{z}{z^2 - 3z + 2}.$$

Here is the long division to rewrite our problem.

$$
\begin{array}{r}
z^{-1} \quad +3z^{-2} \quad +7z^{-3} \quad +\cdots \\
z^2 - 3z + 2 \sqrt{\ z} \\
z \quad -3 \quad +2z^{-1} \\
\hline
3 \quad -2z^{-1} \\
3 \quad -9z^{-1} \quad +6z^{-2} \\
\hline
7z^{-1} \quad -21z^{-2} \quad -6z^{-3} \\
\hline
15z^{-2} \quad -14z^{-3} \\
\vdots
\end{array}
$$

This results into

$$
A(z) = z^{-1} + 3z^{-2} + 7z^{-3} + \cdots = \sum_{n=0}^{\infty} (2^n - 1) z^{-n}.
$$

Again, comparing this with the definition of the z-transform we detect that

$$
a(n) = 2^n - 1, \quad n \in \mathbb{N}_0. \quad \square
$$

• Partial fractions

Partial fractions is widely used method to find the inverse z-transform. We will provide several examples, including solving difference equations when explaining the idea.

Example 3.18. Find $a(n)$ given that

$$
\frac{A(z)}{z} = \frac{2z+3}{(2z-1)(z+1)}, \quad |z| > 1.
$$

Next we use partial fractions. That is

$$
\frac{2z+3}{(2z-1)(z+1)} = \frac{F}{2z-1} + \frac{B}{z+1}.
$$

Take common denominator on the right side.

$$
\frac{2z+3}{(2z-1)(z+1)} = \frac{F(z+1) + B(2z-1)}{(2z-1)(z+1)}.
$$

Since the denominators are identical, the numerators are identical and hence,

$$
2z + 3 = F(z+1) + B(2z-1).
$$

To be able to compare coefficients of power z, we rewrite the previous expression

$$2z + 3 = (F + 2B)z + F - B.$$

Equating coefficients of 1 and z, we arrive at the two equations

$$F + 2B = 2, \quad F - B = 3.$$

After some calculations we obtain $F = \frac{8}{3}$, $B = -\frac{1}{3}$. Taking the z-inverse inverse we arrive at

$$a(n)) = Z^{-1}\big[A(z)\big].$$

Or,

$$
\begin{aligned}
a(n) &= \frac{8}{3}Z^{-1}\left[\frac{z}{2z-1}\right] - \frac{1}{3}Z^{-1}\left[\frac{z}{z+1}\right]\\
&= \frac{4}{3}Z^{-1}\left[\frac{z}{z-\frac{1}{2}}\right] - \frac{1}{3}Z^{-1}\left[\frac{z}{z+1}\right]\\
&= \frac{4}{3}\left(\frac{1}{2}\right)n^n - \frac{1}{3}(-1)n^n. \quad \square
\end{aligned}
$$

One of the major uses of the z-transform is to find solutions of difference equations. If we consider a difference equation in $x(n)$, we proceed by assuming that there is a solution in which the sequence $x(n)$ has a z-transform defined for $|z| > \rho$ for some finite ρ. We have the following example.

Example 3.19 (Delay switch).   Consider the first-order difference equation

$$x(n+1) + x(n) = g(n), \quad x(0) = 0, \ \in \mathbb{N}_0 \tag{3.5.1}$$

where

$$
g(n) = \begin{cases} 2^n, & \text{for } 0 \leq n \leq k-1, \\ 0, & \text{for } n \geq k. \end{cases}
$$

We may express the sequence $g(n)$ by

$$g(n) = 2^n - 2^n \mathcal{U}_n(k),$$

where $\mathcal{U}_n(k)$ is given by Definition 3.3.1. Let $X(z) = Z[x(n)]$ and take the z-transform on both sides of (3.5.1) and have

$$(z+1)X(z) = \frac{z}{z-2} - Z\big[2^n \mathcal{U}_n(k)\big].$$

For the proper use of Theorem 3.3.5, we notice that

$$Z\big[2^n \mathcal{U}_n(k)\big] = Z\big[2^{n-k+k}\mathcal{U}_n(k)\big] = 2^k Z\big[2^{n-k}\mathcal{U}_n(k)\big] = 2^k \frac{z^{1-k}}{z-2}.$$

Therefore,

$$(z+1)X(z) = \frac{z}{z-2} - 2^k \frac{z^{1-k}}{z-2}$$

$$= (1 - 2^k z^{-k})zz - 2\frac{z}{z-2}.$$

Dividing both sides with $z+1$ we arrive at

$$X(z) = \frac{(1-2^k z^{-k})z}{(z+1)(z-2)} = \frac{z}{(z+1)(z-2)} - 2^k z^{-k}\frac{z}{(z+1)(z-2)}.$$

We have to assemble this out of known z-transforms:
Thus we need to find constants α, β, γ, δ such that

$$\frac{\alpha z}{z-2} + \frac{\beta z}{z+1} + \gamma z^{-k}\frac{z}{z-2} + \delta z^{-k}\frac{z}{z+1} = (1 - 2^k z^{-k})\frac{z}{(z+1)(z-2)}.$$

Taking a common denominator and setting numerators equal we arrive at

$$\alpha(z+1) + \beta(z-2) + \gamma z^{-k}(z+1) + \delta z^{-k}(z-2) = 1 - 2^k z^{-k}.$$

Equating coefficients of 1, z, z^{-k+1}, and z^{-k} we get

$$\alpha = \frac{1}{3}, \beta = -\frac{1}{3}, \gamma = -\frac{1}{3}2^k, \quad \text{and} \quad \delta = \frac{1}{3}2^k.$$

Next, we find the inverse z-transform for individual terms and then put them all together to arrive at the answer. Thus,

$$Z^{-1}\left[\frac{\alpha z}{z-2}\right] = \frac{1}{3}2^n, \; Z^{-1}\left[\frac{\beta z}{z+1}\right] = -\frac{1}{3}(-1)^n.$$

Using Theorem 3.3.5, we arrive at

$$Z^{-1}\left[\gamma z^{-k}\frac{z}{z-2}\right] = -\frac{2^k}{3}Z^{-1}\left[\frac{z^{1-k}}{z-2}\right] = -\frac{1}{3}2^k 2^{n-k}.$$

Finally, by similar steps and using Theorem 3.3.5, we have

$$Z^{-1}\left[\delta z^{-k}\frac{z}{z+1}\right] = \frac{1}{3}2^k(-1)^{n-k}.$$

We need to be careful in writing the solution since the delay functions are zeroes for $n \le k-1$. Thus,

$$x(n) = \begin{cases} \frac{1}{3}2^n - \frac{1}{3}(-1)^n, & \text{for } 0 \le n \le k-1, \\ \frac{1}{3}2^n - \frac{1}{3}(-1)^n - \frac{1}{3}2^k 2^{n-k} + \frac{1}{3}2^k(-1)^{n-k}, & \text{for } n \ge k. \end{cases}$$ $\quad\square$

Sometimes we need to do improvisation to be able to carry out with inverse z-transform, as the next example shows.

Example 3.20. Find $a(n)$ when

$$A(z) = \frac{z}{z^2 - 5z + 6}.$$

We write

$$\frac{z}{z^2 - 5z + 6} = \frac{B}{z - 2} + \frac{C}{z - 3},$$

and obtain $B = -2$, and $C = 3$. Thus,

$$a(n) = -2Z^{-1}\left[\frac{1}{z - 2}\right] + 3Z^{-1}\left[\frac{1}{z - 3}\right],$$

which is not obvious how to invert each term. However, the nearest z-transform known to us to either of these two partial fractions is $Z[a^n]$. Based on this observation we redo the problem. We consider, instead

$$\frac{A(z)}{z} = \frac{B}{z - 2} + \frac{C}{z - 3}$$

and obtain, $B = -1$ and $C = 1$. Thus, we have

$$A(z) = -\frac{z}{z - 2} + \frac{z}{z - 3}.$$

Taking inverse z-transform we obtain

$$a(n) = -Z^{-1}\left[\frac{z}{z - 2}\right] + Z^{-1}\left[\frac{z}{z - 3}\right] = -2^n + 3^n. \quad \square$$

One more example.

Example 3.21. Solve the second-order difference equation

$$x(n + 2) + x(n) = 1, \quad x(0) = 0, \ x(1) = 0, \ n \in \mathbb{N}_0.$$

By taking the z-transform on both sides and using the initial conditions we arrive at

$$\frac{X(z)}{z} = \frac{z}{(z - 1)(z^2 + 1)},$$

or

$$\frac{X(z)}{z} = \frac{A}{z - 1} + \frac{Bz + C}{z^2 + 1}.$$

Taking common denominator and equating coefficients from both sides we end up with $A = \frac{1}{2}$, $B = C = -\frac{1}{2}$. Hence,

$$X(z) = \frac{1}{2}\left[\frac{z}{z-1} - \frac{z^2}{z^2+1} - \frac{z}{z^2+1}\right].$$

Hence, using the inverse z-transform we obtain

$$x(n) = \frac{1}{2}\left[1 - \cos\left(\frac{n\pi}{2}\right) - \sin\left(\frac{n\pi}{2}\right)\right]. \quad \square$$

The z-transform is very effective in solving systems of difference equations as we shall see in the next example.

Example 3.22 (Systems). Find $x(n)$ and $y(n)$ that satisfy the system

$$x(n+1) - 10y(n) - 7x(n) = 0, \quad x(0) = 3,$$
$$y(n+1) - 4y(n) - x(n) = 0, \quad y(0) = 2.$$

Let $X(n) = Z[x(n)]$ and $Y(n) = Z[y(n)]$. By taking the z-transform in both equations and using the initial data we arrive at

$$(z-7)X(z) - 10Y(z) - 3z = 0,$$
$$-X(z) + (z-4)Y(z) - 2z = 0.$$

Multiply the second equation with $(z-7)$ and add the resulting equation to the first equation to eliminate $X(z)$ and get

$$\frac{Y(z)}{z} = \frac{2z - 11}{(z-9)(z-2)}.$$

Using partial fraction on $\frac{Y(z)}{z}$ gives

$$\frac{Y(z)}{z} = \frac{A}{z-9} + \frac{B}{z-2}.$$

By similar steps as before, we obtain $A = 1$, and $B = 1$. Solving for $Y(z)$ yields

$$Y(z) = \frac{z}{z-9} + \frac{z}{z-2}.$$

By taking the inverse z-transform we obtain

$$y(n) = 9^n + 2^n.$$

There is no need to redo similar work to solve for $X(z)$ and then use partial fraction; instead, we substitute the obtained $y(n)$ into either original equation and then solve for $x(n)$, directly. Let's make use of

$$y(n+1) - 4y(n) - x(n) = 0.$$

This implies

$$x(n) = y(n+1) - 4y(n) = 9^{n+1} + 2^{n+1} - 4(9^n + 2^n) = 5(9)^n - 2^{n+1}.$$

Note that had we used

$$x(n+1) - 10y(n) - 7x(n) = 0, \quad x(0) = 3,$$

then we would have had to solve a first-order difference equation in $x(n)$. ☐

- **Residues (Inverse integral)**

The *residues method* is deeply rooted in complex integrations. Remember that the z parameter of the z-transform is a complex variable. To generate straightforward inverse z-transforms using so-called residues, however, one need not be an expert in the theory of complex variables. In many situations, inversion with residues is simpler then inversion with partial fractions. We will illustrate the definition with straightforward examples.

Definition 3.5.2. If $G(z)$ is a complex-valued function given by

$$G(z) = \frac{g(z)}{(z - z_0)^k},$$

where $g(z_0) \neq 0$, then $G(z)$ is said to have a *pole* at $z = z_0$ of order k. If $k = 1$, then the pole is said to be *simple*.

The next theorem is concerned with calculating the residues.

Theorem 3.5.1 (Residue). *Let* $Z[a(n)] = A(z)$. *If* $A(z)$ *has a pole at* z_0 *of order* $k = 1, 2, \ldots$ *then the residue* α_k *is given by*

$$\alpha_k = \frac{1}{(k-1)!} \lim_{z \to z_0} \left[\frac{d^{k-1}}{dz^{k-1}} \left(A(z)(z - z_0)^k z^{n-1} \right) \right]. \tag{3.5.2}$$

Theorem 3.5.2. *Let* $Z[a(n)] = A(z)$. *Let* C *be a closed contour that contains all the isolated singularities of* $A(z)$. *Then,*

$$a(n) = \frac{1}{2\pi i} \oint_C A(z) z^{n-1} dz = \text{sum of residues of } A(z).$$

The residue method is a powerful tool for obtaining the inverse z-transform when the method of partial fraction seems tedious.

Example 3.23. Consider

$$A(z) = \frac{2z^2}{(z-2)(z-4)}, \quad |z| > 4.$$

Then $A(z)$ has simple poles at 2 and 4. Hence

$$a(n) = \frac{1}{2\pi i} \oint_C \frac{2z^2}{(z-2)(z-4)} z^{n-1} dz = \text{sum of residues of } A(z).$$

Let α_i, $i = 1, 2$ be the residue at $z = 2$, and $z = 4$, respectively. Then

$$\alpha_1 = \lim_{z \to 2} (z-2) \frac{2z^2}{(z-2)(z-4)} z^{n-1} = -(2)^{n+1},$$

$$\alpha_2 = \lim_{z \to 4} (z-4) \frac{2z^2}{(z-2)(z-4)} z^{n-1} = (4)^{n+1}.$$

Hence,

$$a(n) = \alpha_1 + \alpha_2 = -(2)^{n+1} + (4)^{n+1}. \quad \square$$

Example 3.24. Consider

$$A(z) = \frac{z}{(z-2)^3}, \quad |z| > 2.$$

Then $A(z)$ has a pole at 2 of order 3. By Theorem 3.5.1 the pole α_3 is

$$\alpha_k = \frac{1}{(2)!} \lim_{z \to 2} \left[\frac{d^2}{dz^2} \left(\frac{z}{(z-2)^3} (z-2)^3 z^{n-1} \right) \right]$$

$$= \frac{1}{(2)!} \lim_{z \to 2} \left[\frac{d^2}{dz^2} (z^n) \right]$$

$$= \frac{1}{(2)!} \lim_{z \to 2} \left[n(n-1) z^{n-2} \right]$$

$$= n(n-1) 2^{n-3}.$$

Hence,

$$a(n) = n(n-1) 2^{n-3}. \quad \square$$

Example 3.25. We verify, using the residue method, that if

$$A(z) = \frac{z^2}{z^2 + 1}$$

then $a(n) = \cos(\frac{n\pi}{2})$. We have two simple poles $\pm i$. Then $A(z)$ has simple poles at $\pm i$. Hencen the corresponding residues, α_1 and α_2 are

$$\alpha_1 = \lim_{z \to i} (z-i) \frac{z^{n+1}}{(z-i)(z+i)} = \frac{i^{n+1}}{2i},$$

$$\alpha_2 = \lim_{z \to -i} (z+i) \frac{z^{n+1}}{(z-i)(z+i)} = \frac{(-i)^{n+1}}{(-2i)}.$$

Therefore, using $i = e^{\frac{i\pi}{2}}$ and $-i = e^{\frac{-i\pi}{2}}$ we have

$$a(n) = \frac{1}{2i}\left((i)^{n+1} - (-i)^{n+1}\right)$$

$$= \frac{1}{2i}\left(e^{i(n+1)\frac{\pi}{2}} - e^{-i(n+1)\frac{\pi}{2}}\right)$$

$$= \sin\left((n+1)\frac{\pi}{2}\right)$$

$$= \cos\left(\frac{n\pi}{2}\right). \quad \square$$

• Power series method

The *power series method* is useful when $A(z)$ can be expanded in power series in z. We begin with the following example.

Example 3.26. Let

$$A(z) = \log\left(\frac{z}{z+1}\right),$$

our aim is to find the sequence $a(n)$. We make the substitution $z = \frac{1}{y}$. Then

$$A(z) = \log\left(\frac{\frac{1}{y}}{\frac{1}{y}+1}\right) = \log\left(\frac{1}{1+y}\right) = -\log(1+y).$$

Consequently,

$$A(z) = -\log(1+y)$$

$$= 0 - y + \frac{y^2}{2} - \frac{y^3}{3} + \frac{y^4}{4} + \cdots$$

$$= 0 + -\frac{1}{z} + \frac{1}{2z^2} - \frac{1}{3z^3} + \frac{1}{4z^4} - \cdots$$

$$= 0 + \sum_{n=1}^{\infty} \frac{(-1)^n}{n}.$$

Thus,

$$a(n) = \begin{cases} 0, & \text{for } n = 0, \\ \frac{(-1)^n}{n}, & \text{for } n = 1, 2, 3, \ldots. \end{cases} \quad \square$$

Example 3.27. Let

$$A(z) = e^{-2z^{-1}}.$$

Then using the identity,

$$e^x = \sum_{n=0}^{\infty} \frac{x^n}{n!}$$

we arrive at

$$A(z) = \sum_{n=0}^{\infty} \frac{(-2z^{-1})^n}{n!} = \sum_{n=0}^{\infty} \frac{(-2)^n}{n!} z^{-n}.$$

Thus, by comparison we have

$$a(n) = \frac{(-2)^n}{n!}. \quad \square$$

It is very handy when using power series method to recall the formula

$$\frac{1}{1+x} = \sum_{n=0}^{\infty} (-1)^n x^n, \quad |x| < 1.$$

- **Convolution method**

This is the last method that we discuss. The method of convolution could be messy and not very practical as the next example shows. Recall from the convolution theorem that if $U(z) = Z[u(n)]$ for $|z| > l_1$ and $V(z) = Z[v(n)]$ for $|z| > l_2$, then for $|z| > \max\{l_1, l_2\}$ we have

$$Z[u * v] = U(z)V(z).$$

It follows that

$$Z^{-1}\big[U(z) \cdot V(z)\big] = u(n) * v(n).$$

Example 3.28. Consider

$$A(z) = \frac{z^2}{(z-1)(z-2)}, \quad |z| > 2.$$

Then, we may write $A(z) = U(z) \cdot V(z)$, where

$$U(z) = \frac{z}{z-1} \quad \text{and} \quad V(z) = \frac{z}{z-2}.$$

Consequently,

$$u(n) = (1)^n \quad \text{and} \quad v(n) = (2)^n.$$

Hence,

$$Z^{-1}\left[\frac{z^2}{(z-1)(z-2)}\right] = u(n) * v(n) = \sum_{m=0}^{n} (1)^m (2)^{n-m}.$$

$$\sum_{m=0}^{n} (1)^m (2)^{n-m} = \sum_{m=0}^{n} (2)^{n-m}$$

$$= (2)^n \sum_{m=0}^{n} \left(\frac{1}{2}\right) m^m$$

$$= (2)^n \left[\frac{(\frac{1}{2})m^m}{\frac{1}{2} - 1}\right] ||_{m=0}^{n+1}$$

$$= -(2)^{n+1} \left[\left(\frac{1}{2}\right)n + 1^{n+1} - 1\right]$$

$$= 2^{n+1} - 1. \quad \square$$

For more interesting examples on topics of this section we refer the reader to [45].

3.5.1 Exercises

Exercise 3.29. Find $a(n)$ when $A(z) = 1 + \frac{1}{2}z^{-1} + \frac{1}{4}z^{-2} + \frac{1}{8}z^{-3} + \cdots$.

Exercise 3.30. Find $a(n)$ when

$$A(z) = 2 + \frac{4z^{-10}}{z - 1} + \frac{n!z^6}{(z - 1)^7}.$$

Exercise 3.31. Find $a(n)$ when $A(z) = \dfrac{z^3}{(z - 1)^3}$.

Exercise 3.32. Find $a(n)$ when

$$A(z) = \frac{z^{-4}}{z - 1} + \frac{z^{-3}}{z + 2}.$$

Answer: $a(n) = \mathcal{U}_n(5) + (-2)^{n-4}\mathcal{U}_n(4)$.

Exercise 3.33. Use long term division to find $a(n)$ when

$$A(z) = \frac{z}{z - 1}.$$

Answer: $a(n) = \mathcal{U}(n)$.

Exercise 3.34. Use Long term division to find $a(n)$ when

$$A(z) = \frac{4z^2 + 2z}{2z^2 - 3z + 1}.$$

Answer: $a(n) = 6 - 4(\frac{1}{2})^n$.

Exercise 3.35. Use long division to find $a(n)$ when

$$A(z) = \frac{2z^2 + z}{z^2 - .5z + 0.5}.$$

Answer: $A(z) = 2 + 4z^{-1} + 5z^{-2} + \cdots$

Exercise 3.36 (Delay switch). Use partial fractions to solve the first-order difference equation

$$x(n+1) + x(n) = g(n), \quad x(0) = 0, \in \mathbb{N}_0$$

where

$$g(n) = \begin{cases} 0, & \text{for } 0 \le n \le k - 1, \\ 2^n, & \text{for } n \ge k. \end{cases}$$

Exercise 3.37. Use partial fractions to find $a(n)$ when

$$A(z) = \frac{1}{(1 - z^{-1})(2 - z^{-1})}.$$

Exercise 3.38. Use partial fractions to find $a(n)$ when

$$A(z) = \frac{z}{(z - 3)(z - 4)}.$$

Exercise 3.39. Use partial fractions to find $a(n)$ when

$$A(z) = \frac{5z}{(z^2 - 4z + 4)(z + 2)}.$$

Answer: $a(n) = 5/8(n2^n) - (5/16)2^n + (5/16)(-2)^n$.

Exercise 3.40. Use partial fractions to find $a(n)$ when

$$A(z) = \frac{5z^3 - 23z^2 + 26z}{z^3 - 7z^2 + 15z - 9}.$$

Answer: $a(n) = 2 + 3(3)^n + \frac{1}{3}n(3)^n$.

Exercise 3.41. Find $a(n)$ when

$$A(z) = \frac{z^2(z - 1/2)}{(z - 2/3)(z - 1/3)(z - 1/4)}.$$

Exercise 3.42. Solve the first-order difference equation

$$x(n+1) + x(n) = 1, \quad x(0) = 0, \, n \in \mathbb{N}_0.$$

Exercise 3.43. Solve the second-order difference equation

$$x(n+2) = x(n+1) + x(n), \quad x(0) = x(1) = 1, \ n \in \mathbb{N}_0.$$

Exercise 3.44. Solve the second-order difference equation

$$x(n+2) + 6x(n+1) + 9x(n) = 2^n, \quad x(0) = 0, \ x(1) = 0, \ n \in \mathbb{N}_0.$$

Answer: $x(n) = \frac{1}{25}[2^n - (-3)^n + (5/3)n(-3)^n]$.

Exercise 3.45. Solve the second-order difference equation

$$x(n+2) - 5x(n+1) + 6x(n) = n(n-1), \quad x(0) = 0, \ x(1) = 0, \ n \in \mathbb{N}_0.$$

Exercise 3.46. Solve the third-order difference equation

$$x(n+3) - 6x(n+2) + 12x(n+1) - 8x(n) = 0,$$
$$x(0) = -1, \ x(1) = 0, \ x(2) = 1, \ n \in \mathbb{N}_0.$$

Exercise 3.47. Solve the second-order difference equation

$$x(n+2) + x(n) = 3^n, \quad x(0) = 0, \ x(1) = 0, \ n \in \mathbb{N}_0.$$

Exercise 3.48. Solve for $x(n)$ and $y(n)$ that satisfy the system

$$x(n+1) - y(n+1) + x(n) = 2^n, \quad x(0) = 1,$$
$$y(n+1) - x(n+1) + y(n) = 2^n, \quad y(0) = 1.$$

Answer: $x(n) = y(n) = 2^n$.

Exercise 3.49. Use the residue method to find $a(n)$ where

$$A(z) = \frac{5z}{(z^2 - 4z + 4)(z+2)(2z-1)}.$$

Exercise 3.50. Use the residue method to find $a(n)$ where

$$A(z) = \frac{5z(z+1)}{(z-1)^3}.$$

Exercise 3.51. Use the residue method to show that

$$Z^{-1}\left[\frac{z^2}{z^2+1}\right] = \sin\left(\frac{n\pi}{2}\right).$$

Exercise 3.52. Let

$$A(z) = \frac{1}{1 - 1.5z^{-1} + 0.5z^{-2}}, \quad |z| > 1.$$

Use long division to obtain the first three terms of the power series representation of $A(z)$, and then, with the aid of Example 3.1, find the inverse z-transforms of those terms.

Answer: $a(n) = \delta(n) + 1.5\delta_1(n) + 1.75\delta_2(n)$.

Exercise 3.53. Use the power series method to find $a(n)$ where

$$A(z) = e^{\frac{2}{z}}.$$

Exercise 3.54. Use the power series method to find $a(n)$ where

$$A(z) = \frac{1}{1 + \frac{1}{2}z^{-1}}.$$

Exercise 3.55. Use the power series method to find $a(n)$ where

$$A(z) = \frac{z}{z - 1}.$$

Exercise 3.56. Use convolution to find $a(n)$ when

$$A(z) = \frac{z^2}{(z - 1)(2z - 1)}, \quad |z| > 1.$$

Exercise 3.57. Use convolution to find $a(n)$ when

$$A(z) = \frac{z^2}{(z + 3)^3}, \quad |z| > 3.$$

Exercise 3.58. Solve:

(a) $x(n + 1) = 3 \cdot 5^{n+1} - 4 \sum_{m=0}^{n} 5^{n-m} x(m)$.

(b) $x(n + 1) = 3 + 9 \sum_{m=0}^{n} (n - m) x(m)$.

3.5.2 Frequently used Laplace transforms

Function $a(n)$	Transform $A(z) = \sum_{n=0}^{\infty} a(n)z^{-n}$
$\delta(n)$	1
$\mathcal{U}(n)$	$z/(z-1)$, $\|z\| > 1$
1	$z/(z-1)$, $\|z\| > 1$
b, b is constant	$bz/(z-1)$, $\|z\| > 1$
n	$z/(z-1)^2$, $\|z\| > 1$
n^2	$\frac{z^2+z}{(z-1)^3}$
n^3	$\frac{z(z^2+4z+1)}{(z-1)^4}$

$[n+k-1]^k$	$\frac{n!z^k}{(z-1)^{k+1}}$, $\|z\| > 1$
$[n]^k$	$\frac{n!z}{(z-1)^{k+1}}$, $\|z\| > 1$
a^n	$z/(z-a)$, $\|z\| > \|a\|$
na^n	$az/(z-a)^2$, $\|z\| > \|a\|$
$(-1)^n$,	$z/(z+1)$, $\|z\| > 1$
$\sin n\theta$	$\frac{z\sin(\theta)}{z^2-2z\cos(\theta)+1}$, $\|z\| > 1$
$\cos n\theta$	$\frac{z(z-\cos(\theta))}{z^2-2z\cos(\theta)+1}$, $\|z\| > 1$
$n\sin n\theta$	$\frac{z(z^2-1)\sin(\theta)}{(z^2-2z\cos(\theta)+1)^2}$, $\|z\| > 1$
$n\cos n\theta$	$\frac{z(z^2\cos\theta-2z+\cos(\theta))}{(z^2-2z\cos(\theta)+1)^2}$, $\|z\| > 1$
$\frac{1}{n!}$	$e^{\frac{1}{z}}$
$\frac{1}{(n-r)!}$	$z^{-r}e^{\frac{1}{z}}$
$\mathcal{U}_n(k)$	$\frac{z^{1-k}}{z-1}$, $\|z\| > 1$
$\delta_k(n)$	z^{-k}
$e^{-an}\sin n\theta$	$\frac{ze^{-a}\sin(\theta)}{z^2-2ze^{-a}\cos(\theta)+e^{-2a}}$, $\|z\| > 1$
$e^{-an}\cos n\theta$	$\frac{z^2-ze^{-a}\cos(\theta)}{z^2-2ze^{-a}\cos(\theta)+e^{-2a}}$, $\|z\| > 1$

3.6 Engineering applications

In this section, we continue with engineering applications, building upon what we started in Section 3.4. We will begin with the concept of *transfer function*, but first, we need the following results concerning right-shifting. The next corollary is parallel to Corollary 3.1. From (a) of Theorem 3.3.4, we have that for positive integer k and a sequence $u(n)$ with $Z[u(n)] = U(z)$

$$Z\big[u(n-k)\big] = z^{-k}\left[U(z) + \sum_{m=-k}^{-1} u(m)z^{-m}\right], \quad n \geq k.$$

Corollary 3.2 (Right-shifting). *Let $x(n)$ be a sequence for $n \geq 0$. Then for $Z[x(n)] = X(z)$ we have*

(a) $Z[x(n-1)] = z^{-1}X(z) + x(-1)$,
(b) $Z[x(n-2)] = z^{-2}X(z) + x(-1)z^{-1} + x(-2)$,
(c) $Z[x(n-3)] = x(-3) + z^{-3}X(z) + z^{-2}x(-1) + z^{-1}x(-2)$.

For constants p and q and a given sequence $x(n)$, we consider the first-order difference equation

$$y(n) + py(n-1) - qx(n) = 0, \quad n \in \mathbb{N}. \tag{3.6.1}$$

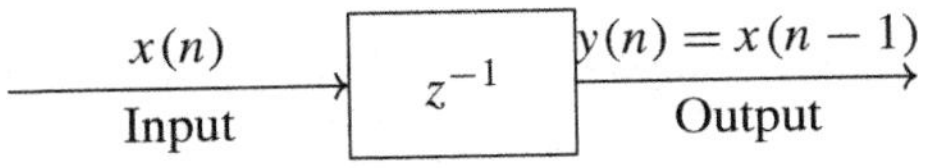

FIGURE 3.4 Input-output.

Letting $Z[x(n)] = X(z)$ and $Z[y(n)] = Y(z)$, and by applying right-shifting (Corollary 3.2) we obtain

$$Y(z) = \frac{qX(z)}{1 + pz^{-1}} - \frac{py(-1)}{1 + pz^{-1}}.$$

If the initial condition $y(-1) = 0$, then we are left with

$$Y(z) = \frac{qX(z)}{1 + pz^{-1}},$$

that we refer to as *zero-state response*. On the other hand, if we have a zero input $(x(n) = 0, \ y(-1) \neq 0)$, then

$$Y(z) = -\frac{py(-1)}{1 + pz^{-1}},$$

that we refer to as *zero-input response*. As noted before, the given sequence $x(n)$ is considered to be the input sequence and $y(n)$, the solution to (3.6.1), is regarded as the output sequence, as seen in Fig. 3.4.

We make the following definition.

Definition 3.6.1. For a causal system with input signal $x(n)$ and output signal $y(n)$, with $x(n) = 0, \ n = -1, -2, -3, \ldots$ the *transfer function* $H(z)$ is defined by

$$H(z) = \frac{Y(z)}{X(z)} = \frac{z\text{-transform of output sequence}}{z\text{-transform of input sequence}}.$$

For emphasis, the transfer function is only defined for all zero initial conditions. Thus, for the system (3.6.1), we have

$$H(z) = \frac{q}{1 + pz^{-1}}.$$

Notice that from Definition 3.6.1 it follows that

$$Y(z) = X(z)H(z), \tag{3.6.2}$$

which can be represented by Fig. 3.5.

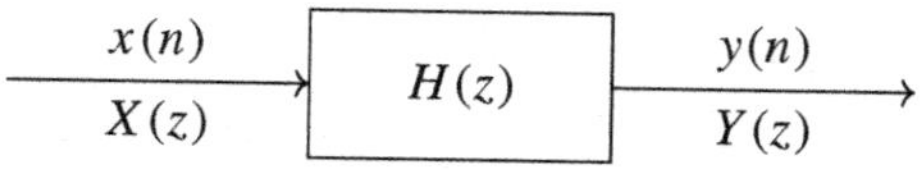

FIGURE 3.5 Block diagram of $H(z)$.

We may easily find that

$$h(n) = Z^{-1}[H(z)] = Z^{-1}\left[\frac{q}{1 + pz^{-1}}\right] = Z^{-1}\left[\frac{qz}{z + p}\right] = q(-p)^n,$$
$$n = 0, 1, 2, \ldots$$

Next we look at the relations between an input signal $x(n)$, the output $y(n)$ and $Z^{-1}[H(z)]$. Let

$$x(n) = \delta(n) = \begin{cases} 1, & \text{for } 0 \le n \le 4, \\ 0, & \text{otherwise.} \end{cases}$$

Then, $X(z) = Z[\delta(n)] = 1$ so $Y(z) = H(z)$ and hence $y(n) = h(n)$. This means that $h(n)$ is the response or output of a system where the input is the unit impulse sequence $\delta(n)$. Therefore, $h(n)$ is referred to as the unit impulse response of the system. In summary, for a linear system with time-invariant, or all of its coefficients are constants

$$H(z) = Z[h(n)] \Leftrightarrow h(n) = Z^{-1}[H(z)].$$

We will leave it as an exercise to show the transfer function for the causal second-order linear system

$$y(n) - a_1 y(n - 1) + a_2 y(n - 2) = bx(n), \tag{3.6.3}$$

where the input sequence $x(n) = 0$, $n = -1, -2, -3, \ldots$ is given by

$$H(z) = \frac{bz^2}{z^2 - a_1 z + a_2}.$$

Suppose we have two coupled linear time-invariant systems with transfer functions $H_1(z)$, and $H_2(z)$, respectively. The output of the first system is the input of the second system. See Fig. 3.6.

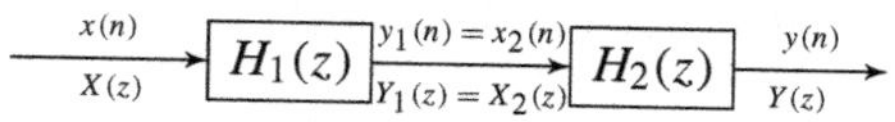

FIGURE 3.6 Coupled two linear systems.

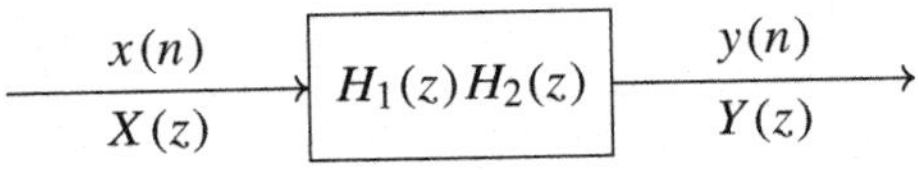

FIGURE 3.7 Series systems with transform function $H_1(z)H_2(z)$.

It follows from the diagram that

$$Y_1(z) = H_1(z)X(z), \text{ and } Y(z) = H_2(z)X_2(z) = H_2(z)Y_1(z).$$

It follows that

$$Y(z) = H_2(z)H_1(z)X(z).$$

Consequently, dividing the final output transform with the input transform we arrive at the formula

$$\frac{Y(z)}{X(z)} = H_2(z)H_1(z).$$

The series system is therefore identical to a single system with the transform function $H_2(z)H_1(z)$, as shown in Fig. 3.7

We have the following example.

Example 3.29. Consider the two time-invariant causal linear systems

$$L_1: \ y(n) - a_1 y(n-1) = a_2 x(n)$$
$$L_2: \ y(n) - b_1 y(n-1) = b_2 x(n)$$

with $x(n) = 0$, $n = -1, -2, -3, \ldots$ Then, by applying Corollary 3.2 we obtain

$$H_1(z) = \frac{a_2}{1 - a_1 z^{-1}}, \quad H_2(z) = \frac{b_2}{1 - b_1 z^{-1}}.$$

Hence the transfer function $H(z)$ for the series is

$$H(z) = H_1(z)H_2(z) = \frac{a_2}{1 - a_1 z^{-1}} \frac{b_2}{1 - b_1 z^{-1}} = \frac{a_2 b_2}{1 - (a_1 + b_1)z^{-1} + a_1 b_1 z^{-2}}.$$

Since $Y(z)$ and $X(z)$ are the output and input transforms, respectively, for the series, we may write

$$Y(z) = H(z)X(z) = \frac{a_2 b_2 X(z)}{1 - (a_1 + b_1)z^{-1} + a_1 b_1 z^{-2}}.$$

By cross multiplying, and distributing the term $Y(z)$ we obtain

$$Y(z) - (a_1 + b_1)z^{-1}Y(z) + a_1 b_1 z^{-2}Y(z) = a_2 b_2 X(z).$$

By contrasting this with Corollary 3.2 we immediately arrive at the

$$y(n) - (a_1 + b_1)y(n-1) + a_1 b_1 y(n-2) = a_2 b_2 x(n),$$

which is a second-order difference equation resulting from connecting two first-order systems in series. $\qquad\square$

Let's look a bit deeper into (3.6.2). By the convolution, we easily observe from (3.6.2) that

$$y(n) = (h * x)(n) = \sum_{m=0}^{n} h(n-m)x(m),$$

is a solution to the corresponding linear time-invariant system that it represents. We are interested in extending the notion of response function to solve linear time-invariant systems with non-zero initial conditions. To better illustrate the concept, we consider the second-order linear time-invariant system given by (3.6.3); namely,

$$y(n) - a_1 y(n-1) + a_2 y(n-2) = bx(n), \quad y(-1) = y_1, \ y(-2) = y_2, \quad (3.6.4)$$

where the input sequence $x(n) = 0$, $n = -1, -2, -3, \dots$ Let $p(n)$ denote the solution to the corresponding homogeneous initial value problem

$$y(n) - a_1 y(n-1) + a_2 y(n-2) = 0, \quad y(-1) = y_1, \ y(-2) = y_2. \quad (3.6.5)$$

We expect the general solution $y(n)$ of (3.6.4) to be given by

$$y(n) = (h * x)(n) + p(n),$$

where $h(n) = Z^{-1}[H(z)]$, is the response function and $H(z)$ is the transfer function and in this case is given by

$$H(z) = \frac{bz^2}{z^2 - a_1 z + a_2}.$$

Indeed, it follows from Theorem 2.3.3 that if $h(n) * x(n)$ is a solution of (3.6.3) and $p(n)$ is a solution to (3.6.5), then $y(n) = (h * x)(n) + p(n)$ is a solution to (3.6.4). Moreover, since $(h * x)(n)$ has initial conditions zero, it readily follows from

$$y(n) = (h * x)(n) + p(n) = \sum_{m=0}^{n} h(n-m)x(m) + p(n),$$

that

$$y(-2) = \sum_{m=0}^{-2} h(-2-m)x(m) + p(-2)$$

$$= \sum_{k=-2}^{0} h(k)x(-2-k) + p(-2)$$

$$= 0 + p(-2) = y_2,$$

where we have made use of the fact that $h(-2) = h(-1) = 0$, and $x(-2) = 0$. In a similar manner, one can show that $y(-1) = p(-1) = y_1$. Thus, we have obtained the following.

Theorem 3.6.1. *The solution of the initial value problem*

$$y(n) - a_1 y(n-1) + a_2 y(n-2) = bx(n), \quad y(-1) = y_1, \ y(-2) = y_2,$$

where the input sequence $x(n) = 0$, $n = -1, -2, -3, \ldots$ *is given by*

$$y(n) = (h * x)(n) + p(n) = \sum_{m=0}^{n} h(n-m)x(m) + p(n),$$

where h is the response of the system and $p(n)$ *is the solution to* (3.6.5).

Sampling:
When a continuous-time signal, denoted as $f(t)$, is sampled at discrete time instants $t = 0, T, 2T, \ldots, nT$, the resulting sequence of sampled values can be expressed as follows:

The continuous-time signal $f(t)$ is sampled at these discrete time instants, resulting in the sequence $f(0), f(T), f(2T), f(3T), \ldots, f(nT), \ldots$. Each value in this sequence corresponds to the value of the continuous-time signal at the respective sampling time instance. The sampling interval T represents the time spacing between samples. It's also known as the sampling period. The index n represents the discrete time instant at which the signal is sampled. The value $n = 0$ corresponds to the initial sample at $t = 0$; $n = 1$ corresponds to the next sample at $t = T$, and so on.

The sequence of values $\{f(0), f(T), f(2T), f(3T), \ldots\}$ represents the discrete-time signal obtained by sampling the continuous-time signal $f(t)$. This discrete-time signal is often denoted as $f(n)$, and its values are given by $f(n) = f(nT)$.

So, the sequence of values $\{f(0), f(1), f(2), \ldots\}$ represents the discrete-time samples of the continuous-time signal $f(t)$ at the time instants nT, where $n \in \mathbb{N}$. We refer to Fig. 3.8. The process of converting a continuous-time signal into a discrete-time sequence through sampling is a fundamental concept in signal processing, and it plays a crucial role in various applications, including digital signal processing and data acquisition systems.

Let $f(t)$ be a continuous signal. We define its z-transform by

$$Z[f(t)] = Z[f(nT)] = \sum_{n=0}^{\infty} f(nT)z^{-n}, \quad |z| > \rho, \qquad (3.6.6)$$

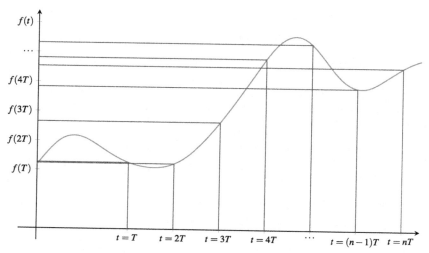

FIGURE 3.8 Sampled function.

where z is a complex number, and ρ is the radius of convergence of $Z[f(nT)]$. Since we have used the notation $f(n) = f(nT)$, we may then write

$$F(z) = \sum_{n=0}^{\infty} f(nT)z^{-n}, \quad |z| > \rho.$$

We furnish the following example.

Example 3.30 (Ramp function). Consider the *ramp function* defined by

$$r(n) = \begin{cases} t, & \text{for } t \geq 0 \\ 0, & \text{for } t < 0 \end{cases}$$

We take the sample values

$$\{r(nT)\} = \{0, T, 2T, \ldots\}.$$

Since $r(n) = r(nT)$, we have the ramp sequence $r(n) = n$, $n = 0, 1, 2, \ldots$ and has the z-transform $\frac{z}{(z-1)^2}$. This leads us to

$$R(z) = TZ[n] = \frac{Tz}{(z-1)^2}.$$

We could have obtained the same answer by aligning the z-transform directly to nT. That is

$$Z[nT] = \sum_{n=0}^{\infty} nTz^{-n} = T \sum_{n=0}^{\infty} nz^{-n} = \frac{Tz}{(z-1)^2}. \quad \square$$

Example 3.31. Let $f(t) = \cos(t\theta)$, $t \geq 0$. Then sampling the continuous sinusoid yields $\{f(nT) = \cos(nT\theta)\}$. Therefore,

$$Z[f(nT)] = Z[\cos(nT\theta)] = \frac{z(z - \cos(\theta T))}{z^2 - 2z\cos(\theta T) + 1}, \quad |z| > 1,$$

where we have replaced θ in the formula for $Z[\cos(n\theta)]$ with θT. $\quad\square$

In sampling, one would have to shift to the right by one sample interval or so. For example, for a signal $y(t)$, we have seen that its sample value is the sequence $\{y(nT)\}$. Thus, $Y(z) = Z[y(nT)]$. If we shift to the right by one sample interval, then

$$Z[y(t - T)]$$
$$= \sum_{n=0}^{\infty} y(nT - T)z^{-n}$$
$$= y(-T) + y(0)z^{-1} + y(1)z^{-2} + y(2)z^{-3} + \ldots + y(k)z^{-k-1} + \ldots$$
$$= y(-T) + z^{-1}[y(0) + y(1)z^{-1} + y(2)z^{-2} + \ldots + y(k)z^{-k} + \ldots]$$
$$= y(-T) + z^{-1}Y(z).$$

In a similar fashion we can show that

$$Z[y(t - 2T)] = \sum_{n=0}^{\infty} y(nT - 2T)z^{-n} = y(-2T) + y(-T)z^{-1} + z^{-2}Y(z).$$
$$(3.6.7)$$

The next theorem gives a formula for shifting to the right a sampling signal by m sample intervals.

Theorem 3.6.2. *Let* $Y(z) = Z[y(nT)]$. *Then*

$$Z[y(t - mT)] = z^{-m}\left[Y(z) + \sum_{j=-m}^{-1} y(jT)z^{-j}\right].$$

Proof.

$$Z[y(t - mT)] = Z[y(nT - mT)] = \sum_{n=0}^{\infty} y(nT - mT)z^{-n} \quad (j = n - m)$$

$$= \sum_{j=-m}^{\infty} y(jT)z^{-j-m}$$

$$= z^{-m}\left[\sum_{j=0}^{\infty} y(jT)z^{-j} - \sum_{j=-m}^{-1} a(jT)z^{-j}\right]$$

$$= z^{-m} \left[Y(z) - \sum_{j=-m}^{-1} y(jT)z^{-j} \right].$$

This completes the proof. $\qquad\qquad\square$

We have the immediate corollary regarding causal sampled signals.

Corollary 3.3. *If the signal* $y(t)$ *is causal, that is* $y(t) = 0$ *for all* $t < 0$, *then*

$$Z[y(t - mT)] = z^{-m}[Y(z).$$

Proof. Since $y(t) = 0$ for all $t < 0$, we have

$$\sum_{j=-m}^{-1} y(jT)z^{-j} = 0,$$

and hence the result. This completes the proof. $\qquad\qquad\square$

The result of the next example follows from Corollary 3.3.

Example 3.32. We know that $\mathcal{U}(t)$ is causal and moreover,

$$Z[\mathcal{U}(t)] = \frac{1}{1 - z^{-1}},$$

and hence

$$Z[\mathcal{U}(t - T)] = z^{-1} \frac{1}{1 - z^{-1}}. \qquad \square$$

We have seen may parallel results between the Laplace transform and the z-transform and in particular, we expect the z-transform of a sampled continuous signal to collapse to the Laplace's transform as the interval of sampling is made smaller and smaller. To see this, assume a continuous signal $y(t)$, $t \geq 0$. Taking the z-transform of the sampled signal we get

$$Z[y(t)] = Z[y(nT)] = \sum_{n=0}^{\infty} y(nT)z^{-n}.$$

Set $z = e^{sT}$ in the previous expression and obtain

$$Y[e^{sT}] = \sum_{n=0}^{\infty} y(nT)e^{-snT}.$$

Let $t_n = nT$, and $\Delta t = t_{n+1} - t_n$. Then taking the limit as $T \to 0$, we have

$$\lim_{T \to 0} Y[e^{sT}]T = \lim_{T \to 0} \sum_{n=0}^{\infty} y(nT)e^{-st_n}T$$

$$= \lim_{T \to 0} \sum_{n=0}^{\infty} y(nT)e^{-st_n} \Delta t$$

$$= \int_{0}^{\infty} y(t)e^{-st}dt := Y(s).$$

In summary, we have shown that, as the sampling interval T approaches zero, the Laplace Transform of the underlying continuous-time signal $y(t)$ approaches the z-transform (times the sampling interval T) of a discrete time signal $y(nT)$.

3.7 Exercises

In Exercises 3.59–3.65, we assume all systems are causal.

Exercise 3.59. For a given input sequence $x(n) = 0$, $n = -1, -2, -3, \ldots$, find the transfer function of the linear system.

$$y(n) + ay(n-1) - bx(n) - cx(n-1) = 0.$$

Exercise 3.60. For a given input sequence $x(n) = 0$, $n = -1, -2, -3, \ldots$, find $h(n)$.

$$y(n) - 2y(n-1) - x(n) = 0.$$

Exercise 3.61. For a given input sequence $x(n) = 0$, $n = -1, -2, -3, \ldots$, find $h(n)$.

$$y(n) + 2y(n-1) - x(n) = 0.$$

Exercise 3.62. Show the transfer function for the second-order linear system

$$y(n) - a_1 y(n-1) + a_2 y(n-2) = bx(n),$$

where the input sequence $x(n) = 0$, $n = -1, -2, -3, \ldots$ is given by

$$H(z) = \frac{bz^2}{z^2 - a_1 z + a_2}.$$

Exercise 3.63. Consider the second-order linear system

$$8y(n) - 6y(n-1) + y(n-2) = 8x(n),$$

where the input sequence $x(n) = 0$, $n = -1, -2, -3, \ldots$

(a) Show the transfer function

$$H(z) = \frac{z^2}{z^2 - \frac{3}{4}z + \frac{1}{8}}.$$

(b) Find $y(n)$ when $x(n) = \delta(n)$, and then show that $y(n) \to 0$, as $n \to \infty$.

(c) Find $y(n)$ when $x(n) = \mathcal{U}(n)$ and then show that $y(n) \to \dfrac{8}{3}$, as $n \to \infty$.
(The terms that decrease to zero as n increases to infinity are called the *transients* and the term $\frac{8}{3}$ is called the *steady state*).

Exercise 3.64. Find the transfer function $H(z)$ for the third-order linear system with $x(n)$ as input and $y(n)$ as output with $x(n) = 0$, $n = -1, -2, -3, \ldots$

$$a_0 y(n) + a_1 y(n-1) + a_2 y(n-2) + a_3 y(n-3)$$
$$= b_0 x(n) + b_1 x(n-1) + b_2 x(n-2).$$

Exercise 3.65. Consider the two time-invariant linear systems with $x(n)$ as input and $y(n)$ as output.

$$L_1: \quad y(n) - 3y(n-1) = x(n)$$
$$L_2: \quad y(n) + 2y(n-1) = 4x(n).$$

(a) Find the transfer function $H(z)$.
(b) Find the difference equation governing the overall system.
(c) Find the output sequence $y(n)$ when $x(n) = 3^{-n}$, $n = 0, 1, 2, \ldots$ and $x(n) = 0$, $n = -1, -2, -3, \ldots$

Exercise 3.66. Use Theorem 3.6.1 to find a formula for the solution of the linear time-invariant system

$$y(n) + 2y(n-1) - 3y(n-2) = x(n), \quad y(-2) = 0, \ y(-1) = 2,$$

where the input sequence $x(n) = 0$, $n = -1, -2, -3, \ldots$

Exercise 3.67. Use Theorem 3.6.1 to find a formula for the solution of the linear time-invariant system

$$y(n) - y(n-1) - 6y(n-2) = x(n), \quad y(-2) = 1, \ y(-1) = 8,$$

where the input sequence $x(n) = 0$, $n = -1, -2, -3, \ldots$

Exercise 3.68. For a continuous signal $f(t)$, $t \geq 0$, let $f(nT)$, $n = 0, 1, 2, \ldots$ be its sampled signal. Show the followings.

(a) If $f(t) = t^2$, then $Z[f(nT)] = \dfrac{T^2 z(z+1)}{(z-1)^3}$,

(b) If $f(t) = e^{-at}$, then $Z[f(nT)] = \dfrac{z}{z - e^{-aT}}$,

(c) If $f(t) = te^{-at}$, then $Z[f(nT)] = \dfrac{Tze^{-aT}}{(z - e^{-aT})^2}$,

(d) If $f(t) = e^{-at}\cos(n\theta)$, then $Z[f(nT)] = \dfrac{1 - e^{-aT}z^{-1}\cos(\theta T)}{1 - 2e^{-aT}z^{-1}\cos(\theta T) + e^{-2aT}z^{-2}}$.

Chapter 4

Systems

This chapter is devoted to the study of systems of difference equations. We will develop the variation of parameters formula in terms of the fundamental matrix solution and study different approaches on how to obtain such a matrix.

4.1 Introduction

In this section we look at homogeneous and non-homogeneous systems of difference equations. We begin by considering the general system of difference equations

$$x_1(n+1) = f_1(n, x_1(n), \ldots, x_k(n))$$
$$x_2(n+1) = f_2(n, x_1(n), \ldots, x_k(n))$$
$$\vdots$$
$$x_k(n+1) = f_k(n, x_1(n), \ldots, x_k(n)).$$

Using the vector notations

$$\mathbf{x} = \begin{pmatrix} x_1 \\ x_2 \\ \vdots \\ x_k \end{pmatrix}$$

and

$$f(n, \mathbf{x}) = \begin{pmatrix} f_1(n, \mathbf{x}) \\ f_2(n, \mathbf{x}) \\ \vdots \\ f_k(n, \mathbf{x}) \end{pmatrix}$$

the above system can be written in the vector form

$$\mathbf{x}(n+1) = f(n, \mathbf{x}). \tag{4.1.1}$$

Consider the kth-order difference equation

$$x(n+k) = g\big(n, x(n), x(n+1), \ldots, x(n+k-1)\big)$$

Difference Equations and Applications. https://doi.org/10.1016/B978-0-44-331492-6.00010-8

and make use of the transformation

$$x_i(n) = x(n+i-1), \quad i = 1, 2, \ldots, k.$$

Then, by varying i we obtain

$$x_1(n) = x(n)$$
$$x_2(n) = x(n+1)$$
$$x_3(n) = x(n+2)$$
$$\vdots$$
$$x_k(n) = x(n+k-1).$$

Using the above calculations we arrive at the system,

$$x_1(n+1) = x(n+1) = x_2(n)$$
$$x_2(n+1) = x(n+2) = x_3(n)$$
$$x_3(n+1) = x(n+3) = x_4(n)$$
$$\vdots$$
$$x_k(n+1) = x(n+k) = g\big(n, x_1(n), x_2(n), \ldots, x_k(n)\big).$$

The right-side of the above system defines a vector function $f(n, \mathbf{x})$.

Example 4.1. Consider the fourth-order difference equation

$$x(n+4) + b_3 x(n+3) + b_2 x(n+2) + b_1 x(n+1) + b_0 x(n).$$

Let

$$x_i(n) = x(n+i-1), \quad i = 1, 2, 3, 4.$$

Then,

$$x_1(n+1) = x_2(n), \quad x_2(n+1) = x_3(n), \quad x_3(n+1) = x_4(n),$$

and

$$x_4(n+1) = x(n+4)$$
$$= -b_3 x(n+3) - b_2 x(n+2) - b_1 x(n+1) - b_0 x(n)$$
$$= -b_3 x_4(n) - b_2 x_3(n) - b_1 x_2(n) - b_0 x_1(n).$$

Using matrix notation the system may take the form

$$x(n+1) = Ax(n),$$

where

$$x(n) = \begin{pmatrix} x_1(n) \\ x_2(n) \\ x_3(n) \\ x_4(n) \end{pmatrix}, \quad A = \begin{pmatrix} 0 & 1 & 0 & 0 \\ 0 & 0 & 1 & 0 \\ 0 & 0 & 0 & 1 \\ -b_0 & -b_1 & -b_2 & -b_3 \end{pmatrix}.$$

Specifically,

$$\begin{pmatrix} x_1(n+1) \\ x_2(n+1) \\ x_3(n+1) \\ x_4(n+1) \end{pmatrix} = \begin{pmatrix} 0 & 1 & 0 & 0 \\ 0 & 0 & 1 & 0 \\ 0 & 0 & 0 & 1 \\ -b_0 & -b_1 & -b_2 & -b_3 \end{pmatrix} \begin{pmatrix} x_1(n) \\ x_2(n) \\ x_3(n) \\ x_4(n) \end{pmatrix}. \quad \square$$

4.1.1 Exercises

Exercise 4.1. Write in matrix notation the fifth-order difference equation

$$x(n+5) - 2x(n+3) + 3x(n+2) + 6x(n+1) - x(n) = 0.$$

4.2 Linear systems

Consider the linear system of difference equations of the form

$$x_1(n+1) = a_{11}(n)x_1(n) + a_{12}(n)x_2(n) + \ldots, a_{1k}(n)x_k(k) + g_1(k)$$
$$x_2(n+1) = a_{21}(n)x_1(n) + a_{22}(n)x_2(t) + \ldots, a_{2k}(n)x_k(n) + g_2(n)$$

$$\ldots$$

$$x_k(n+1) = a_{k1}(n)x_1(n) + a_{k2}(n)x_2(n) + \ldots, a_{kk}(n)x_k(n) + g_k(n)$$

where the functions $a_{ij}, g_i, 1 \leq i \leq k, 1 \leq j \leq k$, are real-valued functions defined on the set $\mathbb{N}_{n_0}$. Using vector and matrix notations, the above system is equivalent to the vector equation

$$x(n+1) = A(n)x(n) + g(n), \tag{4.2.1}$$

where

$$x(n) := \begin{pmatrix} x_1(n) \\ x_2(n) \\ \vdots \\ x_k(n) \end{pmatrix},$$

and

$$A(n) := \begin{pmatrix} a_{11} & a_{12} & \cdots & a_{1k} \\ a_{21} & a_{22} & \cdots & a_{2k} \\ \vdots & \vdots & \ddots & \vdots \\ a_{k1} & a_{k2} & \cdots & a_{kk} \end{pmatrix}, \quad g(n) := \begin{pmatrix} g_1(n) \\ g_2(n) \\ \vdots \\ g_k(n) \end{pmatrix}$$

for $n \in \mathbb{N}_{n_0}$.

Definition 4.2.1. We say the $k \times 1$ vector y is a solution of (4.2.1) on $\mathbb{N}_{n_0}$, provided y is defined on $\mathbb{N}_{n_0}$ and

$$y(n+1)) = A(n)y(n) + g(n),$$

for all $n \in \mathbb{N}_{n_0}$. Moreover, the solution is unique for each $n_0 \in \mathbb{N}_{n_0}$ such that $y(n_0) = y_0$.

Next, we consider the homogeneous system with constant coefficients

$$x(n+1) = Ax(n), \quad x(n_0) = x_0, \tag{4.2.2}$$

where A and x are given in (4.2.1) but with constant coefficients. Along the lines of Section 2.1, we easily find that

$$x(n) = A^{n-n_0}x_0, \quad n \in \mathbb{N}_{n_0}, \tag{4.2.3}$$

is the solution of (4.2.2). The next question is how to compute the matrix A^n. In the case the matrix A is diagonal, in the sense that it is both upper- and lower-triangular, then A^n can be easily computed as shown in the next example.

Example 4.2. Solve

$$x(n+1) = \begin{pmatrix} 2 & 0 \\ 0 & -3 \end{pmatrix} \begin{pmatrix} x_1(n) \\ x_2(n) \end{pmatrix}, \quad x_0 = \begin{pmatrix} x_{01} \\ x_{02} \end{pmatrix}.$$

We have

$$A^2 = A \cdot A = \begin{pmatrix} 2^2 & 0 \\ 0 & (-3)^2 \end{pmatrix},$$

$$A^3 = A \cdot A^2 = \begin{pmatrix} 2^3 & 0 \\ 0 & (-3)^3 \end{pmatrix},$$

$$\vdots$$

$$A^n = A \cdot A^{n-1} = \begin{pmatrix} 2^n & 0 \\ 0 & (-3)^n \end{pmatrix}.$$

Hence, by (4.2.3) the solution is

$$x(n) = \begin{pmatrix} 2^{n-n_0} & 0 \\ 0 & (-3)^{n-n_0} \end{pmatrix} \begin{pmatrix} x_{01} \\ x_{02} \end{pmatrix}. \quad \square$$

We will revisit the diagonalization process a bit later.
Now we start a systematic study of the constant system given by (4.2.2).

Definition 4.2.2. Let A be an $k \times k$ constant matrix, in short "matrix." A number λ is said to be an *eigenvalue* of A if there exists a nonzero vector v of the linear system

$$Av = \lambda v. \tag{4.2.4}$$

The solution vector v is said to be an *eigenvector* corresponding to the eigenvalue λ. We may refer to λ and v as *eigenpair*.

Again, consider the linear and constant homogeneous system,

$$x(n+1) = Ax(n) \tag{4.2.5}$$

where A is an $k \times k$ constant matrix. The fundamental solution of (4.2.5) can be constructed from knowing the eigenvalues and eigenvectors of A.
Now assume a solution x of (4.2.5) in the form

$$x(n) = v\lambda^n,$$

for $v \in \mathbb{R}^k$ and $\lambda \in \mathbb{R}$. Then we have

$$\lambda v \lambda^n = Av\lambda^n.$$

Or

$$(A - \lambda I)v = 0, \text{ if and only if } \det(A - \lambda I) = 0.$$

Thus we have the following straight forward result.

Theorem 4.2.1. *If λ_0, v_0 is an eigenpair of A, then using (4.2.4) we arrive at*

$$x(n) = \lambda_0{}^n v_0 = v_0 \lambda_0{}^n,$$

which is a solution of $x(n+1) = Ax(n)$.

Proof. Let $x(n) = \lambda_0^n v_0$, then

$$x(n+1) = \lambda_0^{n+1} v_0 = \lambda_0 v_0 \lambda_0^n = Av_0\lambda^n = A\lambda_0{}^n v_0 = Ax(n),$$

as desired. This completes the proof. $\square$

Theorem 4.2.2 (Independent eigenvectors). *Let $v_1, v_2, \ldots, v_p$ be the corresponding eigenvectors to the distinct eigenvalues $\lambda_1, \lambda_2, \ldots, \lambda_p$ of a matrix A. Then $v_1, v_2, \ldots, v_p$ are linearly independent.*

Proof. Suppose $v_1, v_2, \ldots, v_j$ are linearly independent for positive integer j. If $j < p$, then v_{j+1} can be written as a linear combination of the vectors, $v_1, v_2, \ldots, v_j$. That is, there are constants $c_1, c_2, \ldots, c_j$ such that

$$v_{j+1} = c_1 v_1 + c_2 v_2 + \ldots + c_j v_j.$$

Multiply from the left by the matrix A and apply the fact that $A v_i = \lambda_i v_i$ for $i = 1, 2, \ldots j$ to arrive at

$$\begin{aligned} A v_{j+1} &= \lambda_{j+1} v_{j+1} \\ &= \lambda_{j+1}\left(c_1 v_1 + c_2 v_2 + \ldots + c_j v_j\right) \\ &= c_1 \lambda_{j+1} v_1 + c_2 \lambda_{j+1} v_2 + \ldots + c_j \lambda_{j+1} v_j. \end{aligned}$$

On the other hand

$$\begin{aligned} A v_{j+1} &= A\left(c_1 v_1 + c_2 v_2 + \ldots + c_j v_j\right) \\ &= c_1 A v_1 + c_2 A v_2 + \ldots + c_j A v_j \\ &= c_1 \lambda_1 v_1 + c_2 \lambda_2 v_2 + \ldots + c_j \lambda_j v_j. \end{aligned}$$

Subtracting the two equations gives

$$c_1(\lambda_{j+1} - \lambda_1)v_1 + c_2(\lambda_{j+1} - \lambda_2)v_2 + \ldots + c_j(\lambda_{j+1} - \lambda_j)v_j = 0.$$

Since $v_1, v_2, \ldots, v_j$ are linearly independent, we must have that

$$c_1(\lambda_{j+1} - \lambda_1) = 0, \; c_2(\lambda_{j+1} - \lambda_2) = 0, \ldots, c_j(\lambda_{j+1} - \lambda_j) = 0.$$

But since

$$\lambda_{j+1} - \lambda_j \neq 0, \;\; \text{for all } j = 1, 2, \ldots, n$$

this could only be possible if

$$c_1 = c_2, \ldots, c_j = 0.$$

This implies that the vector

$$v_{j+1} = 0, \text{ (zero vector)}$$

which is a contradiction, since v_{j+1} is the eigenvector corresponding to λ_{j+1}. This completes the proof. $\qquad\square$

As a result of the above discussion, we have the following theorem concerning systems with constant matrices such as (4.2.5).

Theorem 4.2.3 (Distinct eigenvalues). *Let $\lambda_1, \lambda_2, \ldots, \lambda_k$ be k distinct real eigenvalues of the matrix A of (4.2.5) and let $V_1, V_2, \ldots, V_k$ be the corresponding eigenvectors. Then the general solution of (4.2.5) on the interval $\mathbb{N}_{n_0}$ is given by*

$$x(n) = c_1 V_1 \lambda_1{}^n + c_2 V_2 \lambda_2{}^n, \ldots, c_k V_k \lambda_k{}^n$$

for constants $c_i, i = 1, 2, \ldots, k$.

Remark 4.1. It is important to know that

- If we let

$$\phi_i = V_i \lambda_i{}^n, \quad i = 1, 2, \ldots, k$$

then we may construct the matrix $\phi(n)$ by taking its columns to be the eigenvectors of A and hence

$$\phi(n) = \left[\phi_1(n), \phi_2(n), \ldots, \phi_k(n) \right].$$

- Each column vector ϕ_i solves (4.2.5). That is

$$\phi_i(n+1) = A\phi_i(n).$$

Example 4.3 (Distinct eigenvalues). Consider the linear system

$$x(n+1) = \begin{pmatrix} -4 & 1 & 1 \\ 1 & 5 & -1 \\ 0 & 1 & -3 \end{pmatrix} \begin{pmatrix} x_1 \\ x_2 \\ x_3 \end{pmatrix}.$$

Then the eigenpairs are given by

$$\lambda_1 = -3, V_1 = \begin{pmatrix} 1 \\ 0 \\ 1 \end{pmatrix}, \quad \lambda_2 = -4, V_2 = \begin{pmatrix} 10 \\ -1 \\ 1 \end{pmatrix}, \quad \lambda_3 = 5, V_3 = \begin{pmatrix} 1 \\ 8 \\ 1 \end{pmatrix}$$

and hence the general solution can be written as

$$x(n) = c_1 \begin{pmatrix} 1 \\ 0 \\ 1 \end{pmatrix} (-3)^n + c_2 \begin{pmatrix} 10 \\ -1 \\ 1 \end{pmatrix} (-4)^n + c_3 \begin{pmatrix} 1 \\ 8 \\ 1 \end{pmatrix} (5)^n.$$

If we let

$$\phi(n) = \begin{pmatrix} (-3)^n & 10(-4)^n & (5)^n \\ 0 & -(-4)^n & 8(5)^n \\ (-3)^n & (-4)^n & (5)^n \end{pmatrix}, \tag{4.2.6}$$

then $\det \phi(n) \neq 0$ and A^{n-n_0} is defined with

$$A^{n-n_0} = \phi(n)\phi^{-1}(n_0).$$

Now suppose we impose that every solution satisfies the initial condition

$$x(n_0) := x_0 = \begin{pmatrix} x_{01} \\ x_{02} \\ x_{03} \end{pmatrix},$$

then the unique solution is

$$x(n) = A^{n-n_0} x_0.$$

Let's verify the second item of Remark 4.1. Let

$$\phi_1(n) = \begin{pmatrix} 1 \\ 0 \\ 1 \end{pmatrix} (-3)^n.$$

Then

$$\phi_1(n+1) = \begin{pmatrix} 1 \\ 0 \\ 1 \end{pmatrix} (-3)^{n+1}.$$

On the other hand

$$A\phi_1(n) = \begin{pmatrix} -4 & 1 & 1 \\ 1 & 5 & -1 \\ 0 & 1 & -3 \end{pmatrix} \begin{pmatrix} 1 \\ 0 \\ 1 \end{pmatrix} (-3)^n$$

$$= \begin{pmatrix} -3 \\ 0 \\ -3 \end{pmatrix} (-3)^n = \begin{pmatrix} 1 \\ 0 \\ 1 \end{pmatrix} (-3)^{n+1}.$$

Thus, we have shown that

$$\phi_1(n+1) = A\phi_1(n). \quad \square$$

Some times an eigenevalue with multiplicity m, does not produce m-linearly independent eigenvectors so that the general solution can be found. In general, if m is a positive integer and $(\lambda - \lambda_1)^m$ is a factor of the characteristic equation $\det(A - \lambda I) = 0$, while $(\lambda - \lambda_1)^{m+1}$ is not a factor, then λ_1 is said to be an eigenvalue of multiplicity m. Following, we discuss two such scenarios:

(a) For some $k \times k$ matrices A it may be possible to find m linearly independent eigenvectors $V_1, V_2, \ldots, V_k$ corresponding to an eigenvalue λ_1 of multiplicity $m \leq k$. In this case, the general solution of the system contains the linear combination

$$c_1 V_1 \lambda_1{}^n + c_2 V_2 \lambda_2{}^n + \ldots + c_k V_k \lambda_k{}^n.$$

(b) If there is one eigenvector corresponding to an eigenvalue of multiplicity m, then m linearly independent solutions of the form

$$\phi_1(n) = V_{11} \lambda_1{}^n$$
$$\phi_2(n) = V_{21} n \lambda_1{}^n + V_{22} \lambda_1{}^n$$

$$\vdots$$

$$\phi_m(n) = V_{m1} \frac{n^{m-1}}{(m-1)!} \lambda_1{}^n + V_{m2} \frac{n^{m-2}}{(m-2)!} \lambda_1{}^n + \cdots + K_{mm} \lambda_1{}^n,$$

where V_{ij} are columns vectors that can always be found, and they are known as *generalized eigenvectors*. For an illustration of case (b), we suppose λ_1 is an eigenvalue of multiplicity two with only one corresponding eigenvector. To find the second eigenvector, in general, we assume a second solution of

$$x(n+1) = Ax(n)$$

of the form

$$\phi_2(n) = V n \lambda_1{}^n + P \lambda_1{}^n, \tag{4.2.7}$$

where

$$P = \begin{pmatrix} p_1 \\ p_2 \\ \vdots \\ p_n \end{pmatrix} \quad \text{and} \quad V = \begin{pmatrix} v_1 \\ v_2 \\ \vdots \\ v_k \end{pmatrix}$$

are to be found. Setting $\phi_2(n+1) = A \phi_2(n)$ gives

$$V(n+1) \lambda_1{}^{n+1} + P \lambda_1{}^{n+1} = A \left[V(n) \lambda_1{}^n + P \lambda_1{}^n \right].$$

Combining like wise terms and factoring $n \lambda_1{}^n$, and $\lambda_1{}^n$ we arrive at

$$(AV - \lambda_1 V) n \lambda_1{}^n + (AP - \lambda_1 P - \lambda_1 K) \lambda_1{}^n = 0.$$

Since the above equation must hold for all n, we must have

$$(A - \lambda_1 I) V = 0 \tag{4.2.8}$$

and

$$(A - \lambda_1 I)P = \lambda_1 V. \tag{4.2.9}$$

Eq. (4.2.8) implies V must be an eigenvector of A associated with the eigenvalue λ_1 and hence the solution $\phi_1(n) = V\lambda_1{}^n$ is readily known. By solving for the vector P in (4.2.9), we obtain the second solution

$$\phi_2(n) = Vn\lambda_1{}^n + P\lambda_1{}^n.$$

We have the following example.

Example 4.4 (Repeated roots). The system

$$x(n+1) = \begin{pmatrix} 3 & -18 \\ 2 & -9 \end{pmatrix} \begin{pmatrix} x_1 \\ x_2 \end{pmatrix} \tag{4.2.10}$$

has the repeated eigenvalue $\lambda_1 = \lambda_2 = -3$. If $V = \begin{pmatrix} v_1 \\ v_2 \end{pmatrix}$ is the corresponding eigenvector, then we have the two equations $6v_1 - 18v_2 = 0$, $2v_1 - 6v_2 = 0$, which are both equivalent to $v_1 = 3v_2$. Consequently, we have the only corresponding eigenvector $V = \begin{pmatrix} 3 \\ 1 \end{pmatrix}$ and so one solution is readily known and given by

$$\phi_1(n) = \begin{pmatrix} 3 \\ 1 \end{pmatrix} (-3)^n.$$

We need to use the previous discussion to obtain a second linearly independent eigenvector that corresponds to the eigenvalue $\lambda = -3$. Let $P = \begin{pmatrix} p_1 \\ p_2 \end{pmatrix}$. Then from Eq. (4.2.9) we have $(A + 3I)P = -3V$, which implies that $6p_1 - 18p_2 = -9$, or $2p_1 - 6p_2 = -3$. Since these two equations are equivalent, we may choose $p_2 = 1$ and find $p_1 = \frac{3}{2}$. For simplicity, we shall choose $p_2 = 0$ so that $p_1 = -\frac{3}{2}$. Using (4.2.7) we obtain

$$\phi_2(n) = \begin{pmatrix} 3 \\ 1 \end{pmatrix} n(-3)^n + \begin{pmatrix} -\frac{3}{2} \\ 0 \end{pmatrix} (-3)^n.$$

Thus the general solution is

$$x(n) = c_1\phi_1(n) + c_2\phi_2(n)$$

$$= c_1 \begin{pmatrix} 3 \\ 1 \end{pmatrix} (-3)^n + c_2 \left(\begin{pmatrix} 3 \\ 1 \end{pmatrix} n(-3)^n + \begin{pmatrix} -\frac{3}{2} \\ 0 \end{pmatrix} (-3)^n \right).$$

Note that the columns of $\phi(n)$ are $\phi_1(n)$ and $\phi_2(n)$ and hence

$$\phi(n) = \begin{pmatrix} 3(-3)^n & 3n(-3)^n - \frac{3}{2}(-3)^n \\ (-3)^n & n(-3)^n \end{pmatrix}.$$

Thus, in terms of A^n the solution takes the form

$$x(n) = A^{n-n_0}x(n_0), \text{ for some initial condition } x(n_0) = x_0,$$

where

$$A^{n-n_0} = \phi(n)\phi^{-1}(n_0).$$

This will be made official in Section 4.4. $\qquad\qquad\qquad\square$

4.3 Putzer algorithm

In the previous section, we saw that computing the power matrix A^n depends on the multiplicity of the eigenvalues of the constant matrix A. The next result, called *Putzer algorithm* allows us to find A^n, when the matrix is constant, regardless of the order or multiplicities of the eigenvalues. The proof of the Putzer's algorithm depends on Cayley-Hamilton Theorem that we state next.

Theorem 4.3.1 (Cayley-Hamilton). *Every $k \times k$ constant matrix satisfies its characteristic equation*

$$p_A(\lambda) := \det(A - \lambda I) = 0.$$

In other words, if

$$p_A(\lambda) = (-1)^k \left(\lambda^k + c_1 \lambda^{n-1} + \ldots + c_{k-1}\lambda + c_k \right),$$

then

$$p_A(A) = (-1)^k \left(A^k + c_1 A^{k-1} + \ldots + c_{k-1}A + c_k I \right).$$

Theorem 4.3.2 (Putzer algorithm). *Suppose A is a $k \times k$ constant matrix with eigenvalues $\lambda_1, \lambda_2, \ldots, \lambda_k$, that may come in any order or multiplicities. Suppose $\mu_1(n), \mu_2(n), \ldots, \mu_k(n)$, satisfy the following relations*

$$\mu_1(n+1) = \lambda_1 \mu_1(n), \quad \mu_1(0) = 1,$$
$$\mu_j(n+1) = \lambda_j \mu_j(n) + \mu_{j-1}(n), \quad \mu_j(0) = 0, \ j = 2, 3, \ldots, k. \qquad (4.3.1)$$

Define the matrices $D_0, D_1, \ldots, D_k$, recursively by

$$D_0 = I, \quad D_j = \prod_{k=1}^{j}(A - \lambda_k I), \ j = 1, 2, \ldots, k. \qquad (4.3.2)$$

Then

$$A^n = \sum_{j=0}^{k-1} \mu_{j+1}(n) D_j. \tag{4.3.3}$$

Proof. First, we make sure that if A^n is given with (4.3.3), then $A^0 = I$. Let $\phi(n) = \sum_{j=0}^{k-1} \mu_{j+1}(n) D_j$, then

$$\phi(0) = \sum_{j=0}^{n-1} \mu_{j+1}(0) D_j$$
$$= \mu_1(0) D_0(0) + \mu_2(0) D_1(0) + \ldots + \mu_n(0) D_{k-1}(0)$$
$$= \mu_1(0) D_0(0) + 0 + \ldots + 0 \quad \text{(by (4.3.1))}$$
$$= \mu_1(0) I = I \quad \text{(by(4.3.2)).}$$

Next we evaluate ϕ at $n+1$ and obtain

$$\phi(n+1) = \sum_{j=0}^{k-1} \mu_{j+1}(n+1) D_j$$
$$= \lambda_1 \mu_1(n) D_0 + \sum_{j=1}^{k-1} \mu_{j+1}(n+1) D_j \text{ (by (4.3.1)).}$$

Replacing j with $j+1$ in (4.3.1) the above expression gives

$$\phi(n+1) = \lambda_1 \mu_1(n) D_0 + \sum_{j=1}^{k-1} (\lambda_{j+1} \mu_{j+1} + \mu_j) P_j \text{ (by (4.3.1)).} \tag{4.3.4}$$

On the other hand

$$A\phi(n) = \sum_{j=0}^{k-1} \mu_{j+1}(n) A D_j.$$

From (4.3.2) we have for $j = 1, 2, \ldots k$, that

$$D_j = \prod_{k=1}^{j} (A - \lambda_k I) = (A - \lambda_j I) \prod_{k=1}^{j-1} (A - \lambda_k I) = (A - \lambda_j I) D_{j-1}.$$

Replacing j with $j+1$ and then solving for $A D_j$ gives

$$A D_j = D_{j+1} + \lambda_{j+1} D_j, \quad j = 1, 2, \ldots k-1.$$

Substituting AD_j into $A\phi(n)$ in the above expression yields,

$$A\phi = \sum_{j=0}^{k-1} \mu_{j+1}(n)[D_{j+1} + \lambda_{j+1} D_j],$$

or

$$A\phi(n) = \lambda_1 \mu_1(n) D_0 + (\lambda_2 \mu_2(n) + \mu_1(n)) D_1$$
$$+ \cdots + (\lambda_k \mu_k(n) + \mu_{k-1}(n)) D_{k-1} + \mu_k(n) D_k.$$

But $D_k = p_A(A) = 0$, by Cayley-Hamilton theorem and hence the last expression of $A\phi$ is the same as the right side of (4.3.4) and so we have $\phi(n+1) = A\phi(n)$ with $\phi(0) = I$. Thus by the uniqueness of solution, expression (4.3.3) holds. This completes the proof. □

Example 4.5. Consider the linear system

$$x(n+1) = \begin{pmatrix} 2 & 2 & -1 \\ 2 & -1 & 2 \\ -1 & 2 & 2 \end{pmatrix} \begin{pmatrix} x_1(n) \\ x_2(n) \\ x_3(n) \end{pmatrix}$$

satisfying the initial condition

$$x(0) := x_0 = \begin{pmatrix} 1 \\ -1 \\ 2 \end{pmatrix}.$$

The matrix $A = \begin{pmatrix} 2 & 2 & -1 \\ 2 & -1 & 2 \\ -1 & 2 & 2 \end{pmatrix}$ has the eigenvalues $3, 3, -3$. You may order them in anyway you want. We chose to let $\lambda_1 = 3$, $\lambda_2 = 3$, and $\lambda_3 = -3$. Using (4.3.2) with $j = 1, 2$ we see that

$$D_0 = I, \quad D_1 = A - 3I, \quad D_2 = (A - 3I)^2.$$

From (4.3.1) we must solve

$$\mu_1(n+1) = 3\mu_1(n), \quad \mu_1(0) = 1,$$
$$\mu_2(n+1) = 3\mu_2(n) + \mu_1(n), \quad \mu_2(0) = 0,$$
$$\mu_3(n+1) = -3\mu_3(n) + \mu_2(n), \quad \mu_3(0) = 0.$$

The first equation has the solution

$$\mu_1(n) = (3)^n.$$

To solve for the two others we make use of (2.2.9), which says

$$x(n+1) = ax(n) + g(n), \quad x(n_0) = x_0$$

has the solution

$$x(n) = a^{n-n_0} x_0 + \sum_{r=n_0}^{n-1} a^{n-r-1} g(r).$$

Thus, since $\mu_2(0) = 0$, the solution of

$$\mu_2(n+1) = 3\mu_2(n) + (3)^n, \quad \mu_2(0) = 0,$$

is

$$\mu_2(n) = \sum_{r=0}^{n-1} (3)^{n-r-1}(3)^r = \sum_{r=0}^{n-1} (3)^{n-1} = n(3)^{n-1}.$$

Substituting $\mu_2(n)$ by its value in $\mu_3(n)$ gives

$$\mu_3(n+1) = -3\mu_3(n) + n(3)^{n-1}, \quad \mu_3(0) = 0.$$

Therefore,

$$\mu_3(n) = \sum_{r=0}^{n-1} (-3)^{n-r-1} r(3)^{r-1}$$

$$= \frac{1}{3}(-3)^{n-1} \sum_{r=0}^{n-1} r(-1)^r.$$

We will perform summation by parts using the formula

$$\sum_{k=n_0}^{n-1} x(k)\Delta y(k) = x(k)y(k)\Big|_{k=n_0}^{n} - \sum_{k=n_0}^{n-1} \Delta x(k) Ey(k).$$

Let $x(r) = r$, and $\Delta y(r) = (-1)^r$. Then $\Delta x(r) = 1$ and $y(r) = \frac{(-1)^r}{-1-1} = -\frac{1}{2}(-1)^r$. Therefore,

$$\sum_{r=0}^{n-1} r(-1)^r = -\frac{r}{2}(-1)^r\Big|_{r=0}^{n} - \frac{1}{2}\sum_{r=0}^{n-1}(-1)^r$$

$$= -\frac{r}{2}(-1)^r\Big|_{r=0}^{n} + \frac{1}{4}(-1)^r\Big|_{r=0}^{n}$$

$$= \frac{n}{2}(-1)^{n+1} + \frac{1}{4}(-1)^n - \frac{1}{4}.$$

Finally,

$$\mu_3(n) = \frac{1}{3}(-3)^{n-1}\sum_{r=0}^{n-1} r(-1)^r$$

$$= \frac{1}{3}(-3)^{n-1}\left[\frac{n}{2}(-1)^{n+1} + \frac{1}{4}(-1)^n - \frac{1}{4}\right]$$

$$= \frac{n3^n}{18} - \frac{(3)^n}{36} + \frac{(-3)^n}{36}.$$

Next we compute D_0, D_1 and D_2.

$$D_0 = \begin{pmatrix} 1 & 0 & 0 \\ 0 & 1 & 0 \\ 0 & 0 & 1 \end{pmatrix}, \quad D_1 = \begin{pmatrix} -1 & 2 & -1 \\ 2 & -4 & 2 \\ -1 & 2 & -1 \end{pmatrix}, \quad \text{and}$$

$$D_2 = \begin{pmatrix} 6 & -12 & 6 \\ -12 & 24 & -12 \\ 6 & -12 & 6 \end{pmatrix}.$$

Next we use (4.3.3) with $k = 3$ to compute A^n. Thus,

$$A^n = \sum_{j=0}^{2} \mu_{j+1}(n)D_j = \mu_1 D_0 + \mu_2 D_1 + \mu_3 D_2$$

$$= (3)^n \begin{pmatrix} 1 & 0 & 0 \\ 0 & 1 & 0 \\ 0 & 0 & 1 \end{pmatrix} + n(-3)^n \begin{pmatrix} -1 & 2 & -1 \\ 2 & -4 & 2 \\ -1 & 2 & -1 \end{pmatrix}$$

$$+ \left[\frac{n3^n}{18} - \frac{(3)^n}{36} + \frac{(-3)^n}{36}\right] \begin{pmatrix} 6 & -12 & 6 \\ -12 & 24 & -12 \\ 6 & -12 & 6 \end{pmatrix}$$

$$= \begin{pmatrix} \frac{5+2n}{6}3^n + \frac{1-6n}{6}(-3)^n & \frac{1-2n}{3}3^n + \frac{6n-1}{3}(-3)^n & \frac{2n-1}{6}3^n + \frac{1-6n}{6}(-3)^n \\ \frac{1-2n}{3}3^n + \frac{6n-1}{3}(-3)^n & \frac{1+4n}{3}3^n + \frac{2-12n}{3}(-3)^n & \frac{1-2n}{3}3^n + \frac{6n-1}{3}(-3)^n \\ \frac{2n-1}{6}3^n + \frac{1-6n}{6}(-3)^n & \frac{1-2n}{3}3^n + \frac{6n-1}{3}(-3)^n & \frac{5+2n}{6}3^n + \frac{1-6n}{6}(-3)^n \end{pmatrix}.$$

Notice that $A^0 = I$, and A^n is symmetric. Finally, the unique solution is given by

$$x(n) = A^n x_0. \quad \square$$

The next theorem is a special case of Theorem 4.3.2, in which the matrix A is 2×2.

Theorem 4.3.3 (Special Putzer algorithm). *Assume A is a 2 × 2 matrix.*

1. *If the two eigenvalues λ_1 and λ_2 of A are distinct, then*

$$A^n = \lambda_1^n I + \frac{\lambda_1^n - \lambda_2^n}{\lambda_1 - \lambda_2}(A - \lambda_1 I), \quad n \geq 1.$$

2. *If $\lambda_1 = \lambda_2 = \lambda$, then*

$$A^n = \lambda^n I + n\lambda^{n-1}(A - \lambda I), \quad n \geq 1.$$

- **Diagonalization**

Briefly, we discuss the concept of diagonalization, which is the process of transformation on a matrix in order to recover a similar matrix that is diagonal. Once a matrix is diagonalized, it becomes very easy to raise it to integer powers. We begin with the following definition.

Definition 4.3.1. Let A be an $k \times k$ matrix. We say that A is *diagonalizable* if there exists an invertible matrix P such that

$$D = P^{-1}AP$$

where D is a diagonal matrix.

Theorem 4.3.4. *Let A be an $k \times k$ matrix. The following are equivalent.*
(a) *The matrix A is diagonalizable.*
(b) *The matrix A has n linearly independent eigenvectors.*

Note that not every matrix is diagonalizable. To see this we consider the matrix $A = \begin{pmatrix} 0 & 1 \\ 0 & 0 \end{pmatrix}$. Then 0 is the only eigenvalue but A is not the zero matrix.

In summary, to diagonalize a matrix, one should perform the following steps:
(1) Compute the eigenvalues of A and the corresponding n linearly independent eigenvectors.
(2) Form the matrix P by taking its columns to be the eigenvectors found in step (1).
(3) The diagonalization is done and given by $D = P^{-1}AP$.

Example 4.6. Consider the matrix

$$A = \begin{pmatrix} 5 & 2 & 3 \\ 0 & 8 & 3 \\ 0 & 0 & 4 \end{pmatrix}.$$

Expanding the determinant along the first row we obtain the third degree equation

$$(5 - \lambda)(8 - \lambda)(4 - \lambda) = 0,$$

which has the three distinct eigenvalues

$$\lambda_1 = 5, \quad \lambda_2 = 8, \quad \text{and} \quad \lambda_3 = 4.$$

The corresponding eigenvectors are

$$\mathbf{p_1} = \begin{pmatrix} 1 \\ 0 \\ 0 \end{pmatrix}, \quad \mathbf{p_2} = \begin{pmatrix} 2 \\ 3 \\ 0 \end{pmatrix}, \quad \text{and} \quad \mathbf{p_3} = \begin{pmatrix} 6 \\ 3 \\ -4 \end{pmatrix}.$$

Thus,

$$P = \begin{pmatrix} 1 & 2 & 6 \\ 0 & 3 & 3 \\ 0 & 0 & -4 \end{pmatrix}.$$

One can easily check that

$$P^{-1}AP = \begin{pmatrix} 5 & 0 & 0 \\ 0 & 8 & 0 \\ 0 & 0 & 4 \end{pmatrix},$$

where

$$P^{-1} = \begin{pmatrix} 1 & -\frac{2}{3} & 1 \\ 0 & \frac{1}{3} & \frac{1}{4} \\ 0 & 0 & -\frac{1}{4} \end{pmatrix}. \quad \square$$

Note that the diagonalization of a matrix is not unique, since you may rename the eigenvalues or remix the columns of the matrix P.

As we have said before, one of the most important applications to diagonalization is the computation of matrix powers. Suppose the matrix A is diagonalizable. Then there exists a matrix P such that $D = P^{-1}AP$, or $PDP^{-1} = A$, where

$$D = \begin{pmatrix} d_{11} & 0 & \cdots & 0 \\ 0 & d_{22} & \ddots & \vdots \\ \vdots & \ddots & \ddots & \vdots \\ 0 & \cdots & 0 & d_{kk} \end{pmatrix}.$$

Consequently, for a positive integer n we have

$$A^n = \underbrace{PDP^{-1} \cdots PDP^{-1} \cdots PDP^{-1}}_{n-\text{times}} = PD^n P^{-1},$$

where

$$D^n = \begin{pmatrix} d_{11}^n & 0 & \cdots & 0 \\ 0 & d_{22}^n & \ddots & \vdots \\ \vdots & \ddots & \ddots & \vdots \\ 0 & \cdots & 0 & d_{kk}^n \end{pmatrix}.$$

Another advantage is that once a matrix is diagonalized, then it is easy to find its inverse if it has one. To see this, let $PDP^{-1} = A$. Then $A^{-1} = \left(PDP^{-1}\right)^{-1} = PD^{-1}P^{-1}$, where

$$D^{-1} = \begin{pmatrix} \frac{1}{d_{11}} & 0 & \cdots & 0 \\ 0 & \frac{1}{d_{22}} & \ddots & \vdots \\ \vdots & \ddots & \ddots & \vdots \\ 0 & \cdots & 0 & \frac{1}{d_{nn}} \end{pmatrix}.$$

Example 4.7. Consider the 2×2 matrix

$$A = \begin{pmatrix} 1 & 2 \\ 4 & 3 \end{pmatrix}.$$

Then the eigenpairs are

$$\lambda_1 = -1, \quad \mathbf{p_1} = \begin{pmatrix} 1 \\ -1 \end{pmatrix}; \quad \lambda_2 = 5, \quad \mathbf{p_2} = \begin{pmatrix} 1 \\ 2 \end{pmatrix}.$$

Hence,

$$P = \begin{pmatrix} 1 & 1 \\ -1 & 2 \end{pmatrix}, \quad \text{and} \quad P^{-1} = \frac{1}{3} \begin{pmatrix} 2 & -1 \\ 1 & 1 \end{pmatrix}.$$

Finally, for positive integer n, we have

$$A^n = PD^nP^{-1} = \begin{pmatrix} 1 & 1 \\ -1 & 2 \end{pmatrix} \begin{pmatrix} -1 & 0 \\ 0 & 5 \end{pmatrix}^n \frac{1}{3} \begin{pmatrix} 2 & -1 \\ 1 & 1 \end{pmatrix}$$

$$= \frac{1}{3} \begin{pmatrix} 2(-1)^n + 5^n & (-1)^{n+1} + 5^n \\ 2(-1)^{n+1} + 2 \cdot 5^n & (-1)^n + 2 \cdot 5^n \end{pmatrix}.$$

In particular, for $n = 100$, we have

$$A^{100} = \frac{1}{3} \begin{pmatrix} 2 + 5^{100} & -1 + 5^{100} \\ -2 + 2 \cdot 5^{100} & 1 + 2 \cdot 5^{100} \end{pmatrix}. \quad \square$$

4.3.1 Exercises

Exercise 4.2. For $n \in \mathbb{Z}$ verify that

$$\phi_1(n) = \begin{pmatrix} (-2)^n \\ -(-2)^n \end{pmatrix} \text{ and } \phi_2(n) = \begin{pmatrix} 3(6)^n \\ 5(6)^n \end{pmatrix}$$

are solutions of

$$x(n+1) = \begin{pmatrix} 1 & 3 \\ 5 & 3 \end{pmatrix} x(n).$$

Exercise 4.3. Use Theorem 4.2.3 to solve the linear system with initial condition

$$x(n+1) = \begin{pmatrix} 1 & -12 & -14 \\ 1 & 2 & -3 \\ 1 & 1 & -2 \end{pmatrix} \begin{pmatrix} x_1(n) \\ x_2(n) \\ x_3(n) \end{pmatrix}, \quad x(0) := x_0 = \begin{pmatrix} 4 \\ 6 \\ -7 \end{pmatrix},$$

for all $n \in \mathbb{N}_0$.

Exercise 4.4. Use Theorem 4.2.3 to solve the linear system

$$x(n+1) = \begin{pmatrix} 1 & 2 & -1 \\ 1 & 0 & 1 \\ 4 & -4 & 5 \end{pmatrix} \begin{pmatrix} x_1(n) \\ x_2(n) \\ x_3(n) \end{pmatrix}$$

satisfying the initial condition

$$x(n_0) := x_0 = \begin{pmatrix} 1 \\ 0 \\ 2 \end{pmatrix},$$

for all $n \in \mathbb{N}_{n_0}$.

Exercise 4.5. Use the method of Example 4.4 to find A^n of the given system.

(a) $x(n+1) = \begin{pmatrix} 3 & -1 \\ 9 & -3 \end{pmatrix} \begin{pmatrix} x_1(n) \\ x_2(n) \end{pmatrix}$

(b) $x(n+1) = \begin{pmatrix} -1 & 3 \\ -3 & 5 \end{pmatrix} \begin{pmatrix} x_1(n) \\ x_2(n) \end{pmatrix}.$

Exercise 4.6. Use Theorem 4.3.2 to compute A^n for $x(n+1) = Ax(n)$, where

(a)

$$A = \begin{pmatrix} 2 & 0 & 0 \\ 1 & 2 & 0 \\ 1 & 0 & 3 \end{pmatrix}.$$

(b)

$$A = \begin{pmatrix} 2 & -4 & 2 \\ -2 & 1 & 2 \\ 4 & 2 & 5 \end{pmatrix}.$$

Exercise 4.7. Use Theorem 4.3.2 to solve

$$x(n+1) = \begin{pmatrix} -1 & -2 & 2 \\ 3 & 5 & -3 \\ 3 & 4 & -2 \end{pmatrix} \begin{pmatrix} x_1(n) \\ x_2(n) \\ x_3(n) \end{pmatrix}$$

satisfying the initial condition

$$x(0) := x_0 = \begin{pmatrix} 1 \\ -1 \\ 1 \end{pmatrix}.$$

Exercise 4.8. Use Theorem 4.3.3 to solve

$$x(n+1) = \begin{pmatrix} 2 & 0 \\ 4 & -2 \end{pmatrix} \begin{pmatrix} x_1(n) \\ x_2(n) \end{pmatrix}$$

satisfying the initial condition

$$x(0) := x_0 = \begin{pmatrix} 1 \\ -1 \end{pmatrix}.$$

Exercise 4.9. Use Theorem 4.3.3 to solve

$$x(n+1) = \begin{pmatrix} 3 & 0 \\ 7 & 3 \end{pmatrix} \begin{pmatrix} x_1(n) \\ x_2(n) \end{pmatrix}$$

satisfying the initial condition

$$x(0) := x_0 = \begin{pmatrix} -1 \\ 1 \end{pmatrix}.$$

Exercise 4.10. Use Theorem 4.3.2 to solve

$$x(n+1) = \begin{pmatrix} 8 & -2 & 2 \\ -2 & 5 & 4 \\ 2 & 4 & 5 \end{pmatrix} \begin{pmatrix} x_1(n) \\ x_2(n) \\ x_3(n) \end{pmatrix}$$

satisfying the initial condition

$$x(0) := x_0 = \begin{pmatrix} 1 \\ -1 \\ 0 \end{pmatrix}.$$

Exercise 4.11. Let

$$A = \begin{pmatrix} 0 & a \\ -a & 0 \end{pmatrix}.$$

(a) Compute

$$A^2, \ A^3, \ A^4.$$

(b) Show that

$$A^{2n} = \begin{pmatrix} (-1)^n a^{2n} & 0 \\ 0 & (-1)^n a^{2n} \end{pmatrix}, \quad n = 0, 1, 2, \cdots$$

(c) Show that

$$A^{2n+1} = \begin{pmatrix} 0 & (-1)^n a^{2n+1} \\ (-1)^{n+1} a^{2n+1} & 0 \end{pmatrix}, \quad n = 0, 1, 2, \cdots$$

4.4 Time-varying systems

Consider the homogeneous time-varying system

$$x(n+1) = A(n)x(n) \tag{4.4.1}$$

where

$$x := \begin{pmatrix} x_1(n) \\ x_2(n) \\ \vdots \\ x_k(n) \end{pmatrix}, \quad x(n+1) := \begin{pmatrix} x_1(n+1) \\ x_2(n+1) \\ \vdots \\ x_k(n+1) \end{pmatrix}$$

and

$$A(n) := \begin{pmatrix} a_{11}(n) & a_{12}(n) & \cdots & a_{1k}(n) \\ a_{21}(n) & a_{22}(n) & \cdots & a_{2k}(n) \\ \vdots & \vdots & \ddots & \vdots \\ a_{k1}(n) & a_{k2}(n) & \cdots & a_{kk}(n) \end{pmatrix},$$

for $n \in \mathbb{N}_{n_0}$.

Definition 4.4.1 (Fundamental set of solutions). A set of k solutions of the linear difference system (4.4.1) all defined on the same set $\mathbb{N}_{n_0}$, is called a *fundamental set of solutions* on $\mathbb{N}_{n_0}$ if the solutions are linearly independent functions on $\mathbb{N}_{n_0}$.

We have the following corollary.

Corollary 4.1. *Let $A(n)$ be an $k \times k$ matrix of coefficients $a_{ij}(n)$ that are defined on $\mathbb{N}_{n_0}$. If $\{\phi_1(n), \phi_2(n), \ldots, \phi_k(n)\}$ form a fundamental set of solutions on $\mathbb{N}_{n_0}$, then the general solution of (4.4.1) is given by*

$$x(n) = c_1\phi_1(n) + c_2\phi_2(n) + \ldots + c_n\phi_k(n),$$

for constants $c_i, i = 1, 2, \ldots, k$.

Definition 4.4.2 (Fundamental matrix). A matrix solution Φ is called a *fundamental matrix solution* (or in short, fundamental matrix) of (4.4.1) on an interval $\mathbb{N}_{n_0}$ if its columns form a fundamental set of solutions. If in addition $\Phi(n_0) = I$, a fundamental matrix solution is called the *principal fundamental matrix solution* (or in short, principal matrix).

Theorem 4.4.1. *The matrix Φ is a fundamental matrix of $x(n + 1) = A(n)x(n)$ at n_0 if and only if*

(a) Φ *is a solution of $x(n + 1) = A(n)x(n)$ and*
(b) $\det\Phi(n_0) \neq 0$.

Proof. We know that $\Phi(n + 1) = A(n)\Phi(n)$. Let

$$\Phi = \left[\phi_1(n), \phi_2(n), \ldots, \phi_n(n)\right],$$

where the columns of Φ consist of $\phi_1(n), \phi_2(n), \ldots, \phi_k(n)$. Then

$$\Phi(n + 1) = \left[\phi_1(n + 1), \phi_2(n + 1), \ldots, \phi_k(n + 1)\right],$$

and

$$A\Phi = \left[A\phi_1(n), A\phi_2(n), \ldots, A\phi_k(n)\right].$$

But two matrices are equal if and only if their corresponding columns are the same. Thus

$$\Phi(n + 1) = A\Phi(n)$$

if and only if

$$x_j(n+1) = Ax_j$$

or, if and only if each column of Φ is a solution.

The condition (b) is equivalent to

$$\phi_1(n_0), \phi_2(n_0), \ldots, \phi_k(n_0)$$

being linearly independent. For a complete proof we refer to Theorem 2.3.2. This completes the proof.   □

Example 4.8. Consider

$$x(n+1) = Ax(n),$$

where

$$A = \begin{pmatrix} 10 & -24 \\ 4 & -10 \end{pmatrix}.$$

Let

$$\Phi(n) = 2^n \begin{pmatrix} 3 - 2(-1)^n & -6 + 6(-1)^n \\ 1 - (-1)^n & -2 + 3(-1)^n \end{pmatrix}.$$

One can easily verify, by computing $\Phi(n+1)$ and $A\Phi(n)$, that $\Phi(n+1) = A\Phi(n)$. Thus, $\Phi(n)$ is the fundamental matrix for the system. Note that since the matrix A is constant and $\Phi(0) = I$, $\Phi(n)$ is actually A^n.   □

For simplicity, in the next two theorems and in long expressions, we suppress the argument n.

Theorem 4.4.2. If Φ is a fundamental matrix of (4.4.1), then ΦC, where C is an arbitrary $n \times n$ nonsingular constant matrix is a fundamental matrix of (4.4.1). Conversely, if Ψ is another fundamental matrix to (4.4.1), then there exists constant nonsingular $k \times k$ matrix C such that $\Psi = \Phi C$ for all $n \in \mathbb{N}_{n_0}$.

Proof. Let Φ be a fundamental matrix of (4.4.1) on $\mathbb{N}_{n_0}$. Then

$$(\Phi C)(n+1) = \Phi(n+1)C = (A(n)\Phi)C = A(n)(\Phi C).$$

So, ΦC solves (4.4.1). Next

$$\det(\Phi C) = \det\Phi \, \det C \neq 0$$

and by Theorem 4.4.1 ΦC is a fundamental matrix of (4.4.1). Conversely, let Φ and Ψ be two fundamental matrix solutions of (4.4.1). Since $\Phi\Phi^{-1} = I$, using the product rule

$$\triangle(x(n)y(n)) = \triangle x(n)Ey(n) + x(n)\triangle y(n)$$

we get

$$\triangle\left(\Phi(n)\Phi^{-1}(n)\right) = \triangle\Phi(n)E\Phi^{-1}(n) + \Phi(n)\triangle\Phi^{-1}(n) = 0.$$

Solving for $\triangle\Phi^{-1}(n)$, we obtain

$$\triangle\Phi^{-1}(n) = -\Phi^{-1}(n)\triangle\Phi(n)E\Phi^{-1}(n).$$

Additionally,

$$\triangle(\Phi^{-1}\Psi)(n) = \triangle\Phi^{-1}E\Psi + \Phi^{-1}\triangle\Psi$$
$$= -\Phi^{-1}\triangle\Phi E\Phi^{-1}E\Psi + \Phi^{-1}\triangle\Psi. \qquad (4.4.2)$$

We perform some algebra so we can further reduce (4.4.2).

$$-\Phi^{-1}\triangle\Phi E\Phi^{-1}E\Psi + \Phi^{-1}\triangle\Psi = -\Phi^{-1}[E\Phi - \Phi]E\Phi^{-1}E\Psi + \Phi^{-1}\triangle\Psi$$
$$= -\Phi^{-1}E\Phi E\Phi^{-1}E\Psi + \Phi^{-1}\Phi E\Phi^{-1}E\Psi + \Phi^{-1}\triangle\Psi$$
$$= -\Phi^{-1}E\Psi + E\Phi^{-1}E\Psi + \Phi^{-1}\triangle\Psi$$
$$= -\Phi^{-1}A(n)\Psi(n) + \Phi^{-1}(n)A^{-1}(n)A(n)\Psi(n) + \Phi^{-1}[A(n)\Psi(n) - \Psi(n)]$$
$$= -\Phi^{-1}A(n)\Psi(n) + \Phi^{-1}(n)\Psi(n) + \Phi^{-1}A(n)\Psi(n) - \Phi^{-1}(n)\Psi(n)$$
$$= 0.$$

Thus, we have shown that

$$\triangle(\Phi^{-1}\Psi)(n) = 0,$$

which implies that $\Phi^{-1}\Psi = C$, for some $k \times k$ constant matrix C. Therefore, $\Psi = \Phi C$. Furthermore, as Ψ and Φ are fundamental matrix solutions, we have $\det\Phi \neq 0$, and $\det\Psi \neq 0$ and hence $\det C \neq 0$. This completes the proof. $\square$

We remark that, it is not true in general that $C\Phi$ is a fundamental matrix of (4.4.1).

Theorem 4.4.3. *If Φ is a fundamental matrix of (4.4.1) on $\mathbb{N}_{n_0}$, then Φc solves (4.4.1) on $\mathbb{N}_{n_0}$ for every $k \times 1$ constant matrix c.*

Proof. Let Φ be a fundamental matrix of (4.4.1) on $\mathbb{N}_{n_0}$. Then

$$(\Phi c)(n+1) = \Phi(n+1)c = (A(n)\Phi(n))c = A(n)(\Phi(n)c).$$

So, $\Phi(n)c$ solves (4.4.1). $\square$

Theorem 4.4.4. *If $\Phi(n)$ is a fundamental matrix of (4.4.1) on $\mathbb{N}_{n_0}$, then $\Phi(n)c$, with $c = \Phi^{-1}(n_0)x_0$ is a solution of (4.4.1) with $x(n_0) = x_0$.*

Proof. First, $\Phi^{-1}(n_0)$ exists by the definition of fundamental matrix. Hence, by Theorem 4.4.3 $x(n) = \Phi(n)\Phi^{-1}(n_0)x_0$ solves (4.4.1). Moreover, $x(n_0) = \Phi(n_0)\Phi^{-1}(n_0)x_0 = Ix_0 = x_0$. This completes the proof. $\square$

We remark that $\Phi(n)\Phi^{-1}(n_0)$ is the principal matrix solution of (4.4.1).

Remark 4.2. When $A(n) = A$ is a constant matrix, then given any fundamental matrix $\Phi(n)$, we know that $A^n = \Phi(n)\Phi^{-1}(0)$ or $\Phi(n) = A^n\Phi(0)$. Thus, $\Phi^{-1}(s) = \Phi^{-1}(0)(A^s)^{-1} = \Phi^{-1}(0)A^{-s}$. Thus,

$$\Phi(n)(\Phi(n_0))^{-1} = [A^n\Phi(0)][\Phi^{-1}(0)A^{-n_0}] = A^{n-n_0},$$

and

$$\Phi(n)(\Phi(s))^{-1} = [A^n\Phi(0)][\Phi^{-1}(0)A^{-s}] = A^{n-s}.$$

4.5   Non-homogeneous systems

Now we consider the non-homogeneous system

$$x(n+1) = A(n)x(n) + g(n) \tag{4.5.1}$$

where $A(n)$ is an $k \times k$ matrix and $g(n)$ is an $k \times 1$ vector that are defined on $\mathbb{N}_{n_0}$. The system (4.5.1) has a unique solution for every $n_0 \in \mathbb{N}_{n_0}$, if an initial condition is specified. Our aim is to obtain the *general solution* of the non-homogeneous system (4.5.1). As before, we associate with the non-homogeneous system the complementary homogeneous system

$$x(n+1) = A(n)x(n). \tag{4.5.2}$$

It is enough to find the general solution of (4.5.1) knowing the solution of (4.5.2), and such solution will be called the *variation of parameters formula*.

Definition 4.5.1. Let x_h be the homogeneous solution of (4.5.2) and x_p be the particular solution of (4.5.1) on $\mathbb{N}_{n_0}$. Then the general solution $x : \mathbb{N}_{n_0} \to \mathbb{R}^k$ of (4.5.1) on $\mathbb{N}_{n_0}$ is given by

$$x(n) = x_h(n) + x_p(n).$$

In the previous section, we learnt that if Φ is the fundamental matrix of the homogeneous system (4.5.2), then $x_h = \Phi(n)c$ for some constant $k \times 1$ vector c.

Theorem 4.5.1. *Let $A(n)$ be an $k \times k$ matrix and $g(n)$ be an $k \times 1$ vector that are defined on $\mathbb{N}_{n_0}$. Suppose Φ is the fundamental matrix of the homogeneous system (4.5.2). Then $x(n)$ is a solution of*

$$x(n+1) = A(n)x(n) + g(n), \quad x(n_0) = x_0, \ n \geq n_0 \tag{4.5.3}$$

on $\mathbb{N}_{n_0}$ *if and only if* $x(n)$ *satisfies*

$$x(n) = \Phi(n)\Phi^{-1}(n_0)x_0 + \sum_{s=n_0}^{n-1} \Phi(n)\Phi^{-1}(s+1)g(s) \qquad (4.5.4)$$

where $n_0 \in \mathbb{N}_{n_0}$ *and* $x_0 \in \mathbb{R}^k$.

Proof. First we note that the uniqueness of the solution of (4.5.3) follows from Theorem 5.8.5. Let Φ be the fundamental matrix of the homogeneous system (4.5.2). Suppose x satisfies (4.5.4). Evaluating x at $n + 1$ we arrive at

$$x(n+1) = \Phi(n+1)\Phi^{-1}(n_0)x_0 + \sum_{s=n_0}^{n} \Phi(n+1)\Phi^{-1}(s+1)g(s)$$

$$= A(n)\Phi(n)\Phi^{-1}(n_0)x_0 + A(n)\Phi(n)\sum_{s=n_0}^{n} \Phi^{-1}(s+1)g(s) + g(n)$$

$$= A(n)\left[\Phi(n)\Phi^{-1}(n_0)x_0 + \sum_{s=n_0}^{n} \Phi(n)\Phi^{-1}(s+1)g(s)\right] + g(n)$$

$$= A(n)x(n) + g(n).$$

Evaluating $x(n)$ given by (4.5.4) at n_0 gives $x(n_0) = Ix_0 = x_0$.

Left to show if $x(n)$ is a solution of (4.5.3), then it satisfies (4.5.4). We know that, $x_h = \Phi(n)c$ for some constant $k \times 1$ vector c. We assume a particular solution x_p of the form

$$x_p(n) = \Phi(n)u(n)$$

for some function u defined on $\mathbb{N}_{n_0}$, that is to be found. Substituting x_p into (4.5.3) gives

$$\Phi(n+1)u(n+1) = A(n)\Phi(n)u(n) + g(n).$$

Multiply from the left by $\Phi^{-1}(n+1)$ to get

$$u(n+1) = \Phi^{-1}(n+1)A(n)\Phi(n)u(n) + \Phi^{-1}(n+1)g(n)$$
$$= \Phi^{-1}(n+1)\Phi(n+1)u(n) + \Phi^{-1}(n+1)g(n) \qquad (4.5.5)$$
$$= u(n) + \Phi^{-1}(n+1)g(n),$$

where we have used

$$A(n)\Phi(n) = \Phi(n+1).$$

Rearranging the terms in (4.5.5) we arrive at the first-order difference equation in $u(n)$

$$\Delta u(n) = \Phi^{-1}(n+1)g(n).$$

Summing both sides from $s = n_0$ to $s = n - 1$ and taking the constant of summation to be zero we arrive at

$$u(n) = \sum_{s=n_0}^{n-1} \Phi^{-1}(s+1)g(s).$$

Since by assumption $x_p(n) = \Phi(n)u(n)$, we have

$$x_p(n) = \Phi(n) \sum_{s=n_0}^{n-1} \Phi^{-1}(s+1)g(s).$$

Hence the general solution is

$$x(n) = x_h(n) + x_p(n) = \Phi(n)c + \sum_{s=n_0}^{n-1} \Phi(n)\Phi^{-1}(s+1)g(s). \qquad (4.5.6)$$

Using the initial condition in (4.5.6) we get

$$x_0 = \Phi(n_0)c + 0,$$

or

$$c = \Phi^{-1}(n_0)x_0.$$

A substitution of c into (4.5.6) gives (4.5.4). This completes the proof. $\qquad \Box$

Eq. (4.5.4) is called *variation of parameters formula*.

Remark 4.3. By Remark 4.2, if $A(n) = A$, then (4.5.4) takes the form

$$x(n) = A^{n-n_0}x_0 + \sum_{s=n_0}^{n-1} A^{n-s-1}g(s), \quad n \geq n_0. \qquad (4.5.7)$$

Example 4.9. Solve the non-homogeneous system

$$x(n+1) = \begin{pmatrix} -1 & 0 & 4 \\ 0 & -1 & 2 \\ 0 & 0 & 1 \end{pmatrix} x(n) + \begin{pmatrix} n \\ 2 \\ 0 \end{pmatrix}, \quad x(0) = \begin{pmatrix} 7 \\ 3 \\ 1 \end{pmatrix}.$$

We leave it as an exercise to verify that A^n is given by

$$A^n = \begin{pmatrix} (-1)^n & 0 & 2(1-(-1)^n) \\ 0 & (-1)^n & 1-(-1)^n \\ 0 & 0 & 1 \end{pmatrix}.$$

We make use of (4.5.7) to find the solution.

$$\sum_{s=n_0}^{n-1} A^{n-s-1} g(s)$$

$$= \sum_{s=n_0}^{n-1} \begin{pmatrix} (-1)^{n-s-1} & 0 & 2\left(1-(-1)^{n-s-1}\right) \\ 0 & (-1)^{n-s-1} & 1-(-1)^{n-s-1} \\ 0 & 0 & 1 \end{pmatrix} \begin{pmatrix} s \\ 2 \\ 0 \end{pmatrix}. \quad \square$$

4.5.1 Exercises

Exercise 4.12. Solve the non-homogeneous system

$$x(n+1) = \begin{pmatrix} 1 & 4 \\ 4 & 3 \end{pmatrix} x(n) + \begin{pmatrix} n \\ 2 \end{pmatrix}, \quad x(0) = \begin{pmatrix} 7 \\ 3 \end{pmatrix}.$$

Exercise 4.13. Solve the non-homogeneous system

$$x_1(n+1) = 2x_1(n) + x_2(n) + n$$
$$x_2(n+1) = 2x_2(n) + 1$$
$$x_1(0) = 1, \quad x_2(0) = 0.$$

Answer: $x_1(n) = 2^n + n2^{n-1} - \frac{3}{4}n$, $x_2(n) = 2^n - 1$.

Exercise 4.14. Solve the non-homogeneous system

$$x(n+1) = \begin{pmatrix} -1 & 0 & 4 \\ 0 & -1 & 2 \\ 0 & 0 & 1 \end{pmatrix} x(n) + \begin{pmatrix} n \\ 2 \\ 0 \end{pmatrix}, \quad x(0) = \begin{pmatrix} 7 \\ 3 \\ 1 \end{pmatrix}.$$

Exercise 4.15. Solve the non-homogeneous system

$$x(n+1) = \begin{pmatrix} 2 & 2 & 1 \\ 1 & 3 & 1 \\ 1 & 2 & 2 \end{pmatrix} x(n) + \begin{pmatrix} 1 \\ 2 \\ 0 \end{pmatrix}, \quad x(0) = \begin{pmatrix} 1 \\ 1 \\ -1 \end{pmatrix}.$$

Exercise 4.16. Finish the calculations to find the solution of the system in Example 4.9.

Chapter 5

Stability

In this chapter we explore different types of stability. We begin with the study of stability of scalar equations and then move on to the stability of systems.

5.1 Scalar equations

Consider the scalar non-linear difference equation

$$x(n+1) = f(x(n)) \tag{5.1.1}$$

where $f : \mathbb{R} \to \mathbb{R}$ that is continuous. In many cases, it is impossible to find all solutions to (5.1.1) when it is non-linear. Our ultimate goal is to study the behavior of its solutions for large n and determine whether they approach a certain value or shoot out to infinity. It is possible that the value that solutions approach at infinity could be a solution of the non-linear difference equation. In this case, we say solutions approach *constant solution* or an *equilibrium solution* or a *fixed point*. We have the following definition.

Definition 5.1.1. A real-valued number x^* in the domain of f is said to be an *equilibrium solution,* or *constant solution*, or a *fixed point* of (5.1.1) if

$$f(x^*) = x^*.$$

Consider (5.1.1) and rewrite it in the form

$$\triangle x(n) = -x(n) + f(x(n)).$$

In order for this equation to have a fixed point, say $x(n)$, then we must have $\triangle x(n) = 0$, which implies that $-x(n) + f(x(n)) = 0$, or $x = f(x)$, and hence the name fixed point. Constant solutions or fixed points in difference equations are essential because they provide valuable information about stability and long-term behavior of the system. They serve as a starting point for analyzing more complex solutions and understanding the dynamics of the system in question. The implication of fixed point or equilibrium solution is that if we choose the initial condition equal to the equilibrium solution, we will stay there forever. That is, substituting the equilibrium solution back into the equation gives the same value for all future iterations. Here is an example.

Difference Equations and Applications. https://doi.org/10.1016/B978-0-44-331492-6.00011-X

Example 5.1. Consider the difference equation

$$x(n+1) = 2 + \frac{1}{2}x(n), \quad x(0) = x_0.$$

Then setting the right-hand side equal to x yields the equilibrium solution $x^* = 4$. Based on what we just said, set $x(0) = x_0 = x^* = 4$, then

$$x(1) = 2 + \frac{1}{2}x(0) = 2 + \frac{1}{2}(4) = 4,$$

$$x(2) = 2 + \frac{1}{2}x(1) = 2 + \frac{1}{2}(4) = 4,$$

$$x(3) = 2 + \frac{1}{2}x(2) = 2 + \frac{1}{2}(4) = 4,$$

$$\vdots \quad \square$$

In practice, it would be easier to talk about *zero solution*, or $x = 0$. To do so, we translate every nonzero equilibrium point to zero by the change of variable

$$\tilde{x}(n) = x(n) - x^*,$$

and have from (5.1.1) that $x(n+1) = \tilde{x}(n+1) + x^* = f(\tilde{x}(n) + x^*)$, or

$$\tilde{x}(n+1) = f(\tilde{x}(n) + x^*) - x^* = g(\tilde{x}(n)).$$

We remark that $\tilde{x}(n) = 0$, corresponds to $x = x^*$. In this case, $\tilde{x}(n) \equiv 0$ is an equilibrium solution of (5.1.1), that we might call *the zero solution* of (5.1.1). Thus, in most cases we may require $f(0) = 0$ when we talk about the zero solution. Now we examine the dynamic of (5.1.1) in the neighborhood of its equilibrium solution x^*. If (5.1.1) is in an equilibrium state x^* and $\eta(n)$ represents a small perturbation from that state, then $x(n) = x^* + \eta(n)$. It follows that $x(n+1) = x^* + \eta(n+1)$. Thus, using Taylor series expansion about x^* we have

$$x(n+1) = \eta(n+1) + x^* = f(x^* + \eta) = f(x^*) + \frac{f'(x^*)}{1!}(x^* + \eta - x^*)$$

$$+ \frac{f''(x^*)}{2!}(x^* + \eta - x^*)^2 + \dots$$

$$= f(x^*) + f'(x^*)\eta + \frac{f''(x^*)}{2!}\eta^2 + \dots$$

$$= f(x^*) + f'(x^*)\eta + O(\eta^2) + \dots.$$

Since $f(x^*) = x^*$, we arrive at

$$\eta(n+1) = -x^* + f(x^*) + f'(x^*)\eta + O(\eta^2) + \dots$$

$$= f'(x^*)\eta + O(\eta^2) + \dots.$$

If $f'(x^*) \neq 0$, then the term $|f'(x^*)\eta| \gg |\frac{f''(x^*)}{2!}\eta^2|$. Thus we may neglect the term $O(\eta^2)$ and higher order terms for that same reason. Then we arrive at the linearization of the system about the equilibrium state x^* given by

$$\eta(n+1) = s\eta(n), \qquad\qquad (5.1.2)$$

where

$$s = f'(x^*)$$

is the slope of $f(x)$ at x^*. Eq. (5.1.2) has the solution

$$\eta(n) = c(s)^n \qquad\qquad (5.1.3)$$

for some constant c. It is clear from (5.1.3) that if $|s| > 1$, then solutions grow exponentially and the equilibrium solution x^* is unstable. On the other hand, if $|s| < 1$ then the solutions decay exponentially to zero and the equilibrium solution x^* is asymptotically stable and hence it is stable. If $|s| = 1$ then nothing can be said about the stability. Note that different stability maybe obtained for different systems when $|s| = 1$. Thus, we may refer to the term s as the *stability indicator*. Formal stability definitions will be stated very soon.

Example 5.2. Consider

$$x(n+1) = x^2(n).$$

Setting $x^2 = x$ we obtain two equilibrium points $x^* = 0, 1$. Now $s = f'(x) = 2x$. Hence, for $x^* = 0$, we have $s = f'(x^*) = f'(0) = 0$. Therefore, $x^* = 0$ is asymptotically stable. On the other hand, for $x^* = 1$, we have $s = f'(x^*) = f'(1) = 2$. Therefore, $x^* = 1$ is unstable. $\qquad\square$

Before we state the stability definition we observe the following. Consider Eq. (5.1.1) with $x(0) = x_0$. Then $x(1) = f(x(0)) = f(x_0)$, and $x(2) = f(x(1)) = f(f(x_0)) = f^2(x_0)$. This scheme will generate the sequence

$$x_0, f(x_0), f^2(x_0), \cdots$$

the value $f(x_0)$ is called the first iterate of x_0 and so forth. Set $f^0(x_0) = x_0$, then the set of all iterate

$$\{f^n(x_0) : n \in \mathbb{N}_{n_0}\}$$

is called the *orbit* of x_0. Setting $x(n) = f^n(x_0)$, yields

$$x(n+1) = f^{n+1}(x_0) = f\left(f^n(x_0)\right) = f(x(n)).$$

For the next definition and theorem, we assign $x(0) = x_0$ to Eq. (5.1.1).

Definition 5.1.2. The constant solution, or equilibrium point of (5.1.1), x^* is

(i) *stable* (S) if for each $\varepsilon > 0$, there is a $\delta = \delta(\varepsilon) > 0$ such that $|x_0 - x^*| < \delta$ implies $|f^n(x) - x^*| < \varepsilon$, for all $n \geq 0$.

(ii) It is *unstable* if it is not stable.

(iii) It is *asymptotically stable* (AS) if it is (S) and $x(n) \to x^*$, as $n \to \infty$.

We are ready for our first theorem regarding stability of the equilibrium solution.

Theorem 5.1.1. *Assume f is continuously differentiable in some open interval containing its equilibrium point x^*.*

- *If $|f'(x^*)| < 1$, then x^* is (AS).*
- *If $|f'(x^*)| > 1$, then x^* is unstable.*

Proof. Assume $|f'(x^*)| < 1$. Then by the continuity of f', we have for $u \in (x^* - \delta, x^* + \delta) = I$ and $\alpha \in (0, 1)$, that

$$|f'(u)| \leq \alpha.$$

By the Mean Value Theorem, for $u, z \in I$ we have that

$$|f(u) - f(z)| = |f'(c)||u - z| \leq \alpha|u - z|.$$

Since the above inequality holds for each $u, z \in I$, we conclude

$$|f(u) - x^*| \leq \alpha|u - x^*| < \delta,$$

which implies that $f(u)$ is in I. Hence if we set $\delta = \varepsilon$, we conclude that x^* is stable. Furthermore, for $n \geq 0$, and $u \in I$ we have

$$|f^{n+1}(u) - x^*| \leq \alpha|f^n(u) - x^*|.$$

By an induction argument, we see that

$$
\begin{aligned}
|f^n(u) - x^*| &\leq \alpha|f^{n-1}(u) - x^*| \\
&\leq \alpha^2|f^{n-2}(u) - x^*| \\
&\vdots \\
&\leq \alpha^{n-1}|f(u) - x^*| \\
&\leq \alpha^n|u - x^*|.
\end{aligned}
$$

Thus,

$$|f^{n+1}(u) - x^*| \leq \alpha^n|u - x^*|,$$

from which we conclude that

$$|f^{n+1}(u) - x^*| \to 0, \text{ as } n \to \infty,$$

and hence x^* is (AS).

For the proof of the second part of the theorem, we have for $d > 1$ and $u, z \in I = (x^* - \varepsilon, x^* - \varepsilon)$, for some $\varepsilon > 0$ that

$$|f(u) - f(z)| = |f'(c)||u - z|$$
$$\geq d|u - z|,$$

for all $u, z \in I$. Again, by an induction argument it can be shown that

$$|f^n(u) - x^*| \geq d^n|u - x^*|,$$

from which we conclude that

$$|f^n(u) - x^*| \to \infty, \text{ as } n \to \infty.$$

In other words, every solution that originates in I, except the equilibrium solution, will leave I for sufficiently large n. Hence, the equilibrium solution x^* is unstable. This completes the proof. $\qquad\square$

Example 5.3 (Population growth). In Section 2.1.2, we considered the discrete Verhulst process given by

$$N(n + 1) = rN(n)\left(1 - \frac{N(n)}{K}\right), \quad K > 0,$$

which is kind the discrete analogue of the continuous logistic growth model. The model incorporates a *carrying capacity* K of the environment due to limited resources. In order to have a positive population we must restrict N to the interval $0 < N < K$. Setting the right-side equal to N gives the two equilibrium solutions $N^* = 0, \frac{(r-1)K}{r}$. For biological purpose, we require $r > 1$ so that $N^* = \frac{(r-1)K}{r}$ is a feasible equilibrium solution. Setting

$$f(N) = rN\left(1 - \frac{N}{K}\right)$$

we obtain

$$f'(N) = r - \frac{2r}{K}N = r\left(1 - \frac{2}{K}N\right).$$

For the equilibrium solution $N^* = 0$, we have $f'(0) = r > 1$, and we conclude $N^* = 0$ is unstable. As for equilibrium solution $N^* = \frac{(r-1)K}{r}$, we have

$$\left|f'(N^*)\right| = \left|r\left(1 - \frac{2}{K}\frac{(r-1)K}{r}\right)\right| = |2 - r| < 1$$

for $1 < r < 3$. We conclude that $N^* = \frac{(r-1)K}{r}$ is (AS) for $1 < r < 3$, and unstable for $r > 3$. Note that for $1 < r < 3$ we have $N^* < K$, which is in line with our requirement that $0 < N < K$. This is vindictive of the bifurcation diagram in Fig. 2.4. $\qquad\qquad\qquad\qquad\qquad\qquad\qquad\qquad\qquad\qquad\qquad\qquad$ □

- **Staircase diagram**

In Subsection 2.1.2 we considered the general non-linear first-order difference equation

$$N(n+1) = f(N(n)), \quad n > 0, \ n \in \mathbb{N} \qquad (5.1.4)$$

where f is continuous. Equations of the form (5.1.4) are rich in information regarding the growth or decay of the population. Now, we consider a graphical technique to obtain some insight to the equilibrium solutions. We proceed in the following manner. Select an initial value N_0 to have hold of the next iterate $N_1 = f(N_0)$ which is a point on the graph $f(N)$. We continue in this way to obtain $N_i = f(N_{i-1})$, $i = 1, 2, \cdots$ We are able to achieve this by using the line $y = N$ to reflect each value of N_{n+1} back to the N_n-axis. We see that by performing this procedure, we are bouncing between the curves $y = N$ and $y = f(N)$, which results in a recursive graphical method. This method is referred to as a *staircase diagram*, or *cobweb diagram*. We illustrate the method with several examples.

Example 5.4. Consider the difference equation

$$x(n+1) = \cos(x(n)) = f(x)$$

and set $\cos(x) = x$ to find the equilibrium solutions. There is only one equilibrium solution, which is the intersection of the two graphs $y = \cos(x)$, and $y = x$, and it is labeled with a dark bullet. See Fig. 5.3. Start with an initial point $x(1)$ that is to the left of x^*. Then move upward to the point $(x(1), x(2)) = (x(1), f(x(1)))$. From the point $(x(1), x(2))$ moves horizontally until you intersect with the line $y = x$, which we labeled $(x(2), x(2))$. From $(x(2), x(2))$ move vertically and downward and intersect the graph $y = \cos(x)$, and the point of intersection is $(x(2), x(3)) = (x(2), f(x(2)))$. From that point, move horizontally toward the line $y = x$, and your intersection point is then $(x(3), x(3))$. Continuing with this process we obtain the sequence $x(n) \to x^*$ as $n \to \infty$. Similar results are obtained if we start with an initial point that is bigger than x^*. We conclude that the equilibrium point x^* is stable. $\qquad\qquad\qquad\qquad\qquad$ □

Example 5.5. Consider the non-linear first-order difference equation

$$x(n+1) = 2x(1-x),$$

which has the two equilibrium points $x^* = 0, \frac{1}{2}$.

According to Theorem 5.1.1 the equilibrium solution $x^* = 0$ is unstable and $x^* = \frac{1}{2}$ is stable. As in the previous example, we will try to verify this using the staircase diagram. Pick a starting point $x(1)$ to the left of the equilibrium point x^* and then move up to the graph $f(x) = 2x(1-x)$. From that point,

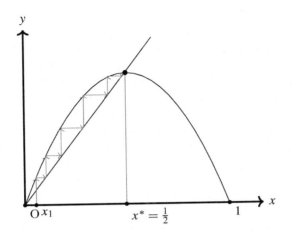

FIGURE 5.1   Staircase for $f(x) = 2x(1 - x)$.

move horizontally until you touch the line $y = x$. A continuation of this process shows the convergence of the solution, or sequence $x(n)$ to the equilibrium point $x^* = \frac{1}{2}$. See Fig. 5.1. In Fig. 5.2, we establish the staircase methods for different values of the growth rate r.   □

According to Fig. 5.2, chaotic behavior occurs for $x(n + 1) = rx(n)(1 - x(n))$ as r slightly increases toward 3.9.

5.1.1   Exercises

Exercise 5.1. Let $f(x) = x^2(n)$. Compute $f^k(x_0)$, for $k = 1, 2, 3, 4, 5$, where $x_0 = 0.7$. From this calculation is it true that

$$\lim_{k \to \infty} f^k(x_0) = 0?$$

Exercise 5.2 (Population growth).   Consider the logistic model

$$y(n + 1) = y(n)(r - dy(n)),$$

where r and d are positive constants. Make the change of variable $y(n) = \frac{r}{d}x(n)$ to rewrite the model as

$$x(n + 1) = rx(n)(1 - x(n)),$$

and analyze it based on the analysis of Example 5.3.

Exercise 5.3 (Population growth).   Consider the non-linear difference equation for population growth

$$x(n + 1) = \frac{dx(n)}{b + x(n)},$$

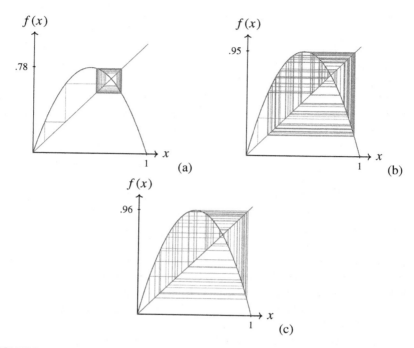

FIGURE 5.2 Cobweb diagrams for $x(n + 1) = rx(n)(1 - x(n))$ for different values of r. In (a), we have $r = 3.2$, while in (b), $r = 3.8$, and $r = 3.9$ in (c).

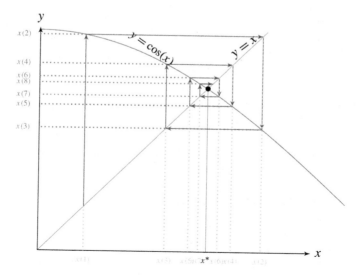

FIGURE 5.3 Staircase method for $f(x) = \cos(x)$.

where b and d are positive constants with $d > b$. Show that the two equilibrium solutions are

$$x^* = 0, \ d - b.$$

Put a mutual condition on d and b so that the equilibrium solution $x^* = d - b$ is stable.

Exercise 5.4. Redo the staircase of Example 5.5 starting from an initial point x_1, where $x^* < x_1 < 1$.

Exercise 5.5. Find the equilibrium solutions and determine the stability situation of those who are positive for the difference equations:

(a) $x(n + 1) = x(n)(1 + r) - b$, where r and b are positive constant.
(b) $x(n + 1) = rx(n) - b$, where r and b are positive constants with $r > 1$.
(c) $x(n + 1) = -x^2(n)(1 - x(n))$.
(d) $x(n + 1) = x(n) \ln(x^2(n))$.

Exercise 5.6. Consider the difference equation $x(n + 1) = \mu x(n) + x^3(n)$, where $0 < \mu < 1$.

(a) Show that

$$x_1^* = 0, \ x_2^* = \sqrt{1 - \mu}, \ x_3^* = -\sqrt{1 - \mu}$$

are equilibrium solutions.
(b) Show that x_1^* is stable for all $0 < \mu < 1$.
(c) Show that x_2^* is unstable for all $0 < \mu < 1$.
(d) Use the staircase diagram to validate your answers with respect to the stability of x_1^* and x_2^*.

Exercise 5.7. Consider the *Ricker equation* that models fish populations

$$x(n + 1) = \mu x(n) e^{-\beta x(n)}$$

where μ denotes the maximal growth rate of the population and β is the inhibition of growth caused by overpopulation.

(a) Show that the equation has an equilibrium solution given by

$$x^* = \frac{\ln(\mu)}{\beta}.$$

(b) Show that, if

$$|1 - \ln(\mu)| < 1,$$

then the equilibrium solution x^* is stable.

Exercise 5.8. Analyze the difference equation

$$x(n + 1) = \alpha x^2(n)(1 - x(n))$$

for positive α.

Exercise 5.9. Find all positive equilibria for the difference equation

$$x(n+1) = \frac{\alpha x^2(n)}{d + x^2(n)}$$

and determine their stability in terms of the positive constants α and d. Compare with cobwebbing.

Exercise 5.10. Discuss the stability of the equilibrium solutions in terms of the positive constants r, d and h for the difference equation

$$x(n+1) = \frac{\alpha x(n)}{d + x(n)} - h.$$

Exercise 5.11. Find all positive equilibria for the modified logistic equation

$$x(n+1) = \alpha x(n)\left(1 - x^2(n)\right)$$

for positive α, and for each equilibrium point determine the values of α, which makes that point stable.

5.2   Stability of systems

Consider the non-autonomous system of difference equations

$$x(n+1) = f(n, x(n)), \quad x(n_0) = x_0 \tag{5.2.1}$$

where $f : [0, \infty) \times D \to \mathbb{R}^n$, $D \subset \mathbb{R}^n$ that is open. We say a vector $x^* \in \mathbb{R}^n$ is an *equilibrium*, or *constant solution*, or *equilibrium solution* of (5.2.1) if

$$f(n, x^*) = x^*.$$

In practice, it would be easier to talk about *zero solution*, or $x = 0$. To do so, we translate every nonzero equilibrium point to zero by the change of variables

$$\tilde{x}(n) = x(n) - x^*,$$

and have from (5.2.1) that

$$x(n+1) = \tilde{x}(n+1) + x^* = f(n, \tilde{x}(n) + x^*),$$

or

$$\tilde{x}(n+1) = -x^* + f(n, \tilde{x}(n) + x^*) \equiv g(n, \tilde{x}(n)).$$

In this case, $\tilde{x}(n) \equiv 0$ is an equilibrium point of (5.2.1), that we might call "*the zero solution*" of (5.2.1). Thus, in most cases we may require $f(n, 0) = 0$, when we talk about the zero solution. Before we embark on formal definitions, we must be precise when talking about maximum or minimum in the sense of distances. This brings us to the notion of a *norm*.

Definition 5.2.1 (Norm). A real-valued function on a vector space V is called a *norm,* and denoted by $\|x\|$, if it has the following properties:

(i) $\|x\| > 0$ for all $x \neq 0$, $x \in V$.
(ii) $\|x\| = 0$ if $x = 0$.
(iii) $\|\alpha x\| = |\alpha| \|x\|$ for all $x \in V$, $\alpha \in \mathbb{R}$.
(iv) $\|x + y\| \leq \|x\| + \|y\|$ (triangle inequality).

Example 5.6. The space $(\mathbb{R}^n, +, \cdot)$ over the field $\mathbb{R}$ is a vector space (with the usual vector addition, $+$ and scalar multiplication, $\cdot$) and there are many suitable norms on it. For example, if $x = (x_1, x_2, \ldots, x_n)$ then

(i) $\|x\| = \max\limits_{1 \leq i \leq n} |x_i|$,

(ii) $\|x\| = \sqrt{\sum\limits_{i=1}^{n} x_i^2}$,

(iii) $\|x\| = \sum\limits_{i=1}^{n} |x_i|$,

(iv) $\|x\|_p = \left(\sum\limits_{i=1}^{n} |x_i|^p \right)^{1/p}$, $\quad p \geq 1$

are all suitable norms. Norm (ii) is the Euclidean norm: the norm of a vector is its Euclidean distance to the zero vector and the metric defined from this norm is the usual Euclidean metric. Norm (iii) generates the "taxi-cab" metric on $\mathbb{R}^2$ and Norm (iv) is the l_p norm. $\qquad \square$

Definition 5.2.2. The zero solution ($x = 0$) of (5.2.1);

(a) is *stable* (S) if for each $\epsilon > 0$ and $n_0 \geq 0$, there is a $\delta = \delta(n_0, \epsilon) > 0$ such that $|x(n_0)| < \delta$ implies $|x(n, n_0, x_0)| < \varepsilon$,
(b) is *uniformly stable* (US) if δ independent of n_0,
(c) is *unstable* if it is not stable,
(d) is *asymptotically stable* (AS) if it is stable and $\lim\limits_{n \to \infty} |x(n, n_0, x_0)| = 0$,
(e) is *uniformly asymptotically stable* (UAS) if it is (US) and there exists a $\gamma > 0$ with the property that for each $\mu > 0$ there exists $T = T(\mu) > 0$ such that $|x(n_0)| < \gamma$, $n \geq n_0 + T$ implies $|x(n, n_0, x_0)| < \mu$,
(f) is *uniformly exponentially stable* (UES) if there exists a $\delta > 0$ and an $\eta \in (0, 1)$ such that for $|x(n_0)| < \delta$ we have $|x(n, n_0, x_0)| < C(|x_0|)\eta^{n-n_0}$, where the constant C is positive and depends on the initial condition,
(f) is l_p-*stable* if it is (S) and for some $p > 0$, we have

$$\sum_{n=n_0}^{\infty} |x(n, n_0, x_0)|^p < \infty,$$

(g) is *Uniformly l_p-stable* if the previous summation converges uniformly with respect to n_0,

(h) is *globally asymptotically stable* (GAS) if it is (AS) for all initial value x_0. (No restriction on the initial condition).

There are various stability definitions that can be found in many of the referenced books. As far as this book is concerned, the stability definitions that we have listed are the most used and practical ones. We furnish a few examples to illustrate the previous definitions. In the next example, we consider a simple difference equation to illustrate that a zero solution can be uniformly stable and asymptotically stable but not uniformly asymptotically stable.

Example 5.7. Consider the difference equation

$$x(n+1) = \frac{n}{n+1} x(n), \quad x(n_0) = x_0 \neq 0, \quad n \geq n_0 \geq 1. \tag{5.2.2}$$

Let $z(n) = nx(n)$. Then $x(n+1) = \frac{z(n+1)}{n+1}$, and hence (5.2.2) becomes

$$z(n+1) = z(n), \quad z(n_0) = x_0 n_0,$$

which has the solution $z(n) = n_0 x_0$. Consequently, the solution to (5.2.2) is then found to be

$$x(n) := x(n, n_0, x_0) = \frac{x_0 n_0}{n}.$$

We claim the zero solution is (US). To see this we let $\epsilon > 0$ and set $\delta = \epsilon$. Then for $|x_0| < \delta$, we have for $n \geq n_0$ that $|x(n, n_0, x_0)| = \left| \frac{x_0 n_0}{n} \right| \leq |x_0| < \delta = \epsilon$. So the zero solution is (US). Clearly the zero solution is (AS), since

$\lim\limits_{n \to \infty} |x(n, n_0, x_0)| = \lim\limits_{n \to \infty} \left| \frac{x_0 n_0}{n} \right| = 0$. To show that the zero solution is not (UAS), we set $n_0 = n$, and then notice that

$$x(2n, n, x_0) = \frac{x_0 n}{2n} \to \frac{x_0}{2} \neq 0$$

which implies that the zero solution is not (UAS). □

Example 5.8. The difference equation

$$x(n+1) = x(n), \quad x(n_0) = x_0$$

has the constant solution $x(n) = x_0$, and hence the zero solution is (US) but not (AS). On the other hand the first-order difference equation

$$x(n+1) = 2x(n), \quad x(n_0) = x_0$$

has the solution $x(n) = x_0 2^{n-n_0}$ and consequently, its zero solution is unstable. □

Example 5.9. Let $|a| < 1$ and consider the difference equation

$$x(n+1) = ax(n), \quad x(n_0) = x_0.$$

Then its solution is given by

$$x(n, n_0, x_0) = x_0 a^{n-n_0}, \quad n \ge n_0 \ge 0.$$

We claim this solution is (US) and (UAS). To see this, for any $\epsilon > 0$ we set $\delta = \epsilon$. Then for $|x_0| < \delta$, we have that $|x(n, n_0, x_0)| = |x_0 a^{n-n_0}| \le |x_0| \le \epsilon = \delta$. So the zero solution is (US). For the (UAS) we must show that there exists a $\gamma > 0$ with the property that for each $\mu > 0$ there exists $T = T(\mu) > 0$ such that $|x(n_0)| < \gamma$, $n \ge n_0 + T$ implies $|x(t, t_0, x_0)| < \mu$. Set $\delta = 1$ of the uniform stability and let $\gamma = \delta = 1$. Let $T(\mu) = \dfrac{\ln(\mu)}{\ln(|a|)}$, then for $n \ge n_0 + T(\mu)$,

$$|x(n, n_0, x_0)| = |x_0 a^{n-n_0}| \le |a|^{T(\mu)} \le |a|^{\frac{\ln(\mu)}{\ln|a|}} = \mu.$$

This shows the zero solution is (UAS). In fact, the zero solution is exponentially stable and l_1 stable. $\qquad\square$

This brings us to the following easy to prove theorem.

Theorem 5.2.1. *If the zero solution of (5.2.1) is (UES), then it is l_p-stable.*

Proof. If the zero solution is (UES), then for $\eta \in (0, 1)$ its solutions satisfy $|x(n, n_0, x_0)| \le C(|x_0|)\eta^{n-n_0}$, and hence for $p > 1$

$$\sum_{n=n_0}^{\infty} |x(n, n_0, x_0)|^p | \le (C(|x_0|))^p \sum_{n=n_0}^{\infty} (\eta^{n-n_0})^p \le (C(|x_0|))^p \frac{1}{1 - \eta^p}.$$

This completes the proof. $\qquad\square$

We end this section by discussing the stability of the *constant solution*. For $n \in \mathbb{N}_{n_0}$, $n_0 \ge 0$, define the operator

$$\mathcal{L}x(n) = a_k x(n+k) + a_{k-1} x(n+k-1) + a_{k-2} x(n+k-2) + \cdots + a_0 x(n)$$

$$= \sum_{i=0}^{k} a_{k-i} x(n+k-i). \tag{5.2.3}$$

Then, (5.2.3) is equivalent to

$$\mathcal{L}x(n) = 0. \tag{5.2.4}$$

Eq. (5.2.4) is the homogeneous equation that corresponds to the non-homogeneous equation

$$\mathcal{L}x(n) = g(n). \tag{5.2.5}$$

We state the following definitions.

Definition 5.2.3. Let $n \in \mathbb{N}_{n_0}$, $n_0 \geq 0$, and $\overline{x}(n)$ be a solution of (5.2.5). We say $\overline{x}(n)$

(a) is *stable* (S), if for any other solution $x(n)$ of (5.2.5), the difference between the two solutions remains bounded. That is, $|\overline{x}(n) - x(n)| \leq M$, for positive constant M,
(b) is *asymptotically stable* (AS) if $|\overline{x}(n) - x(n)| \to 0$, as $n \to \infty$,
(c) is *unstable* if it is *not stable*.

Note that if the function $g(n) = $ constant, then $\overline{x}(n)$ is constant and we refer to it as a *constant solution*.

Example 5.10. Consider the first-order difference equation

$$x(n+1) - \frac{1}{2}x(n) = \frac{1}{2}, \quad n \geq 0.$$

Then $\overline{x}(n) = 1$ is a constant solution and for constant c_1, $x(n) = c_1 \left(\frac{1}{2}\right)^n + 1$ is another solution. Hence,

$$|\overline{x}(n) - x(n)| = |1 - c_1 \left(\frac{1}{2}\right)^n - 1| \to 0$$

as $n \to \infty$, and therefore, the constant solution $\overline{x}(n) = 1$ is (AS). Notice, the constant solution $\overline{x}(n)$ is nothing but the particular solution $x_p(n)$ that we discussed in Section 2.5. □

Example 5.11. Consider the second-order difference equation

$$x(n+2) - x(n+1) + \frac{1}{4}x(n) = 1, \quad n \geq 0.$$

Then $\overline{x}(n) = 4$ is a constant solution and for constant c_1, c_2, $x(n) = c_1 \left(\frac{1}{2}\right)^n + c_2 n \left(\frac{1}{2}\right)^n + 4$ is another solution. Hence,

$$|\overline{x}(n) - x(n)| = |4 - c_1 \left(\frac{1}{2}\right)^n - c_2 n \left(\frac{1}{2}\right)^n - 4| \to 0$$

as $n \to \infty$, and therefore, the constant solution $\overline{x}(n) = 4$ is (AS). □

Example 5.12. Consider the second-order difference equation

$$x(n+2) - x(n) = 2, \quad n \geq 0.$$

Then $\overline{x}(n) = n$ is a solution and for constant c_1, c_2,

$$x(n) = c_1 + c_2(-1)^n + n$$

is another solution. Hence,

$$|\bar{x}(n) - x(n)| = |n - c_1 - c_2(-1)^n - n|$$
$$= |-c_1 - c_2(-1)^n| \le |c_1| + |c_2|$$

and therefore, the solution $\bar{x}(n) = n$ is (S) and not (AS). Note that the constant solution is not unique. For example $\bar{x}(n) = 3$ is another constant solution but it is unstable. □

5.2.1 Exercises

Exercise 5.12. Discuss the stability as we did in Example 5.7 for the following (IVPs):

(a)

$$nx(n+1) = (n-1)x(n), \quad x(n_0) = x_0 \ne 0, \quad n \ge n_0 \ge 2.$$

(b)

$$nx(n+1) = \frac{(n-1)}{2}x(n), \quad x(n_0) = x_0 \ne 0, \quad n \ge n_0 \ge 2.$$

Exercise 5.13. Show that the zero solution of the difference equation

$$x(n+1) = a(n)x(n), \quad x(n_0) = x_0 \ne 0, \quad n \ge n_0 \ge 1$$

is (UAS) provided that

$$|a(n)| \le \alpha, \quad \alpha \in (0, 1), \quad n \in \mathbb{N}_{n_0}.$$

Exercise 5.14. Let $n \in \mathbb{N}_{n_0}$, $n_0 \ge 0$. Based on Definition 5.2.3, discuss the stability of the following difference equations:

(a) $x(n+1) - \frac{1}{4}x(n) = 3.$

(b) $x(n+2) + 9x(n) = \sin(3n).$

(c) $4x(n+2) + 4x(n+1) + x(n) = n^2.$

(d) $x(n+2) - \frac{1}{4}x(n) = \sin(n).$

(e) $3x(n+2) - 6x(n+1) + 3x(n) = n(n-1).$

5.3 $x(n+1) = A(n)x(n)$

We are interested in analyzing the stability of linear time-varying systems of difference equations of the form

$$x(n+1) = A(n)x(n), \quad x(n_0) = x_0 \tag{5.3.1}$$

where $A(n)$ is an $k \times k$ matrix with entries defined on the set $\mathbb{N}_{n_0}$. In our analysis we rely on the concept of the fundamental matrix.

Definition 5.3.1 (Norm). A norm of a square matrix A is a non-negative real number denoted by $\|A\|$. There are several different ways of defining a matrix norm, but they are all share the following properties.

   (i) $\|A\| \geq 0$ for any square matrix
   (ii) $\|A\| = 0$ if and only if $A = 0$ (0 matrix)
  (iii) $\|kA\| = |c|\|A\|$, for any constant c
  (iv) $\|A + B\| \leq \|A\| + \|B\|$, where B is a square matrix
   (v) $\|AB\| \leq \|A\|\|B\|$.

Example 5.13. Let A be a matrix with entries a_{ij}, $1 \leq i \leq k$, $1 \leq j \leq k$; that is $A = (a_{ij})$, then we define the following possible norms on the matrix A.

  (i) The 1-norm:

$$\|A\|_1 = \max_{1 \leq j \leq n} \left(\sum_{i=1}^{n} |a_{ij}| \right)$$

    (the maximum absolute columns sum).
 (ii) The infinity norm:

$$\|A\|_\infty = \max_{1 \leq i \leq n} \left(\sum_{j=1}^{n} |a_{ij}| \right)$$

    (sum the absolute values along each row and then take the biggest value).
(iii) The Euclidean norm:

$$\|A\|_E = \sqrt{\sum_{i=1}^{n} \sum_{j=1}^{n} (a_{ij})^2}$$

(square root of sum of each entry).

We begin with the following theorem.

Theorem 5.3.1. *Let $\Phi(n)$ be the fundamental matrix of (5.3.1). Then the zero solution of (5.3.1) is*

(a) *stable if and only if there exists a positive constant M such that*

$$|\Phi(n)| \leq M, \quad n \geq 0.$$

(b) *Asymptotically stable if and only if*

$$|\Phi(n)| \to 0, \quad as\ n \to \infty.$$

Proof. (a) ($\Leftarrow$) Let $\Phi(n)$ be the fundamental matrix of (5.3.1). Then we have seen that the unique solution $x(n)$ of (5.3.1) is given by $x(n) = \Phi(n)\Phi^{-1}(n_0)x_0$.

Thus for any $\epsilon > 0$ such that $|x_0| < \epsilon$ set $\delta = \frac{\epsilon}{|\Phi^{-1}(n_0)|M}$. Then

$$|x(n)| = |\Phi(n)\Phi^{-1}(n_0)x_0| \leq |\Phi(n)||\Phi^{-1}(n_0)||x_0| \leq M|\Phi^{-1}(n_0)|\delta = \epsilon.$$

This proves the zero solution is stable.

($\Rightarrow$) Set $\epsilon = 1$ from the stability proof. Then

$$|x(n)| = |\Phi(n)\Phi^{-1}(n_0)x_0| < 1, \text{ for } n \geq n_0 \text{ if } |x_0| < \delta(1, n_0),$$

which implies that

$$|\Phi(n)\Phi^{-1}(n_0)| < \frac{1}{\delta(1, n_0)}.$$

Therefore,

$$|\Phi(n)| = |\Phi(n)\Phi^{-1}(n_0)\Phi(n_0)| \leq |\Phi(n)\Phi^{-1}(n_0)||\Phi(n_0)| \leq \frac{1}{\delta(1, n_0)}|\Phi(n_0)|$$

$$:= M.$$

This completes the proof of (a).

Next we prove (b). We already know the zero solution is stable. Now,

$$|x(n)| = |\Phi(n)\Phi^{-1}(n_0)x_0| \to 0, \text{ as } t \to \infty$$

if and only if

$$|\Phi(n)| \to 0, \text{ as } n \to \infty.$$

This completes the proof. $\qquad\square$

Example 5.14. Consider the system

$$x(n+1) = \begin{pmatrix} 0 & \frac{2+(-1)^n}{2} \\ \frac{2+(-1)^n}{2} & 0 \end{pmatrix} x(n), \; x(n_0) = x_0, \; n \geq n \geq n_0.$$

Set

$$\Phi(n) = 2^{-n-1} \begin{pmatrix} (\sqrt{3})^n + (-\sqrt{3})^n & (\sqrt{3})^{n+1} + (-\sqrt{3})^{n+1} \\ (\sqrt{3})^{n+1} + (-\sqrt{3})^{n+1} & (\sqrt{3})^n + (-\sqrt{3})^n \end{pmatrix}.$$

One can verify that

$$\Phi(n+1) = A(n)\Phi(n) \text{ and } \det\Phi(n) \neq 0.$$

Hence $\Phi(n)$ is a fundamental matrix for the system. Then, by Theorem 5.3.1, the zero solution is stable since $|\Phi(n)| \leq \sqrt{3} + 6$, for all $n \geq 0$, where we have summed the absolute values all the entries in the matrix. In fact, the solution is given by

$$x(n) = \Phi(n)\Phi^{-1}(n_0)x_0. \quad \square$$

Theorem 5.3.2. *Let $\Phi(n)$ be the fundamental matrix of (5.3.1). Then the zero solution of (5.3.1) is uniformly stable if and only if there exists a positive constant M such that*

$$|\Phi(n)\Phi^{-1}(s)| \leq M, \quad n \geq s \geq 0. \tag{5.3.2}$$

Proof. ($\Leftarrow$) Let $\Phi(n)$ be the fundamental matrix of (5.3.1). Then the unique solution $x(n)$ of (5.3.1) is given by $x(n) = \Phi(n)\Phi^{-1}(s)x(s)$. Since (5.3.2) holds for all s such that $n \geq s \geq 0$, we have that

$$|x(n)| = |\Phi(n)\Phi^{-1}(s)x(s)| \leq |\Phi(n)\Phi^{-1}(s)||x_0|.$$

Thus, for any $\epsilon > 0$ chose $\delta = \frac{\epsilon}{M}$, such that for $|x_0| < \delta$ imply

$$|x(n)| \leq |\Phi(n)||\Phi^{-1}(s)||x_0| \leq M\delta = \epsilon.$$

($\Rightarrow$) Set $\epsilon = 1$ from the uniform stability proof. Then

$$|x(n)| = |\Phi(n)\Phi^{-1}(s)x(s)| < 1, \text{ for } n \geq s \text{ if } |x(s)| < \delta(1),$$

which implies that

$$|\Phi(n)\Phi^{-1}(s)| \leq \frac{1}{\delta(1)} := M.$$

This completes the proof. $\qquad\qquad\qquad\qquad\qquad\qquad\qquad \square$

The next theorem provides necessary and sufficient conditions for the (UAS).

Theorem 5.3.3. *Let $\Phi(n)$ be the fundamental matrix of (5.3.1). Then the zero solution of (5.3.1) is uniformly asymptotically stable if and only if there exist positive constants M and $\beta \in (0, 1)$ such that*

$$|\Phi(n)\Phi^{-1}(s)| \leq M\beta^{n-s}, \quad n \geq s \geq n_0. \tag{5.3.3}$$

Proof. ($\Leftarrow$) Let $\Phi(n)$ be the fundamental matrix of (5.3.1). Then the unique solution x of (5.3.1) is given by $x(n) = \Phi(n)\Phi^{-1}(s)x(s)$. It is evident from (5.3.3) that the zero solution is (US). Thus, for all s such that $n \geq s \geq n_0$, we have that

$$|x(n)| = |\Phi(n)\Phi^{-1}(s)x(s)| \leq |\Phi(n)\Phi^{-1}(s)||x_0|.$$

As a result, we may set $\gamma = 1$. Now we take $T(\mu) = \dfrac{1}{\ln(\beta)} \text{Ln}(\dfrac{\mu}{M}), 0 < \mu < M$. Then for $|x_0| < 1, n \geq n_0 + T(\mu)$ we have

$$|x(n)| = |\Phi(n)\Phi^{-1}(n_0)x_0| < M\beta^{n-n_0} = M\beta^{T(\mu)} \leq \mu.$$

($\Rightarrow$) Since the zero solution is (UAS) there exists a $\gamma > 0$ with the property that for each $\mu > 0$ there exists $T = T(\mu) > 0$ such that $|x(n_0)| < \gamma, n \geq n_0 + T$ implies $|x(n, n_0, x_0)| < \mu$. Let $\alpha = \frac{\mu}{\gamma}$. Then $|x_0| \leq \gamma$, and $n \geq n_0 + T(\mu)$ yield

$$|\Phi(n)\Phi^{-1}(n_0)x_0| \leq \mu. \tag{5.3.4}$$

Therefore,

$$|\Phi(n + T(\mu))\Phi^{-1}(n)| \leq \alpha < 1, \ n \geq 0. \tag{5.3.5}$$

For $n \geq n_0$ there is a positive integer k such that

$$n_0 + nT(\mu) \leq n \leq n_0 + (k + 1)T(\mu).$$

Now let $n_j = n_0 + (k + 1)T(\mu), j = 1, 2, \ldots, k$. Then by (5.3.4) and (5.3.5)

$$\begin{aligned} |\Phi(n)\Phi^{-1}(n_0)| &= |\Phi(n)\Phi^{-1}(n_k)\Phi(n_k)\Phi^{-1}(n_{k-1}) \ldots \Phi(n_1)\Phi^{-1}(n_0)| \\ &\leq |\Phi(n)\Phi^{-1}(n_k)| \left| \prod_{n=0}^{k-1} \Phi(n_{j+1})\Phi^{-1}(n_j) \right| \\ &\leq M\alpha^k = \frac{M}{\alpha}\left(\alpha^{\frac{1}{T(\mu)}} \right)^{(k+1)T(\mu)} \\ &= D\beta^{(k+1)T(\mu)}, \end{aligned}$$

where $D = \frac{M}{\alpha}$, and $\beta = \alpha^{\frac{1}{T(\mu)}}$. Notice that $n_0 + nT(\mu) \leq n \leq n_0 + (k+1)T(\mu)$, which implies that

$$nT(\mu) \leq n - n_0 \leq (k + 1)T(\mu).$$

Thus,

$$\begin{aligned} |\Phi(n)\Phi^{-1}(n_0)| &\leq D\beta^{(k+1)T(\mu)} \\ &\leq D\beta^{(n-n_0)}, \end{aligned}$$

since $\beta \in (0, 1)$. This completes the proof. $\square$

Remark 5.1. Due to Theorem 5.3.3, for linear systems such as (5.3.1), (UAS) is equivalent (UES).

Theorems 5.3.1–5.3.3 are not practical since one would have to know the fundamental matrix of a time-varying system. However, the results of the next theorem are simple but very effective in determining uniform stability and uniform asymptotic stability for time-varying systems. We ask the reader to look up reference [14] in Elaydi's book.

Theorem 5.3.4. *Consider system (5.3.1).*

(a) *If*

$$||A||_1 = \max_{1 \le j \le k} \left(\sum_{i=1}^{k} |a_{ij}| \right) \le 1, \quad j = 1, 2, \cdots, k$$

then the zero solution of (5.3.1) is (US).

(b) *If*

$$||A||_1 = \max_{1 \le j \le k} \left(\sum_{i=1}^{k} |a_{ij}| \right) \le \alpha, \quad j = 1, 2, \cdots, k$$

where $\alpha \in (0, 1)$, then the zero solution of (5.3.1) is (UAS).

5.3.1 Exercises

Exercise 5.15. In Example 5.14 verify that

$$\Phi(n + 1) = A(n)\Phi(n).$$

Exercise 5.16. Verify that

$$\Phi(n) = \frac{1}{2} \begin{pmatrix} (2)^{1-n} - (-2)^{1-n} & (\frac{3}{2})^n - (-\frac{3}{2})^n \\ (2)^{-n} - (-2)^{-n} & (\frac{3}{2})^n - (-\frac{3}{2})^n \end{pmatrix}$$

is a fundamental matrix for the system

$$x(n + 1) = \begin{pmatrix} 0 & \frac{2+(-1)^n}{2} \\ \frac{2-(-1)^n}{2} & 0 \end{pmatrix} x(n),$$

and then determine the stability of the zero solution. Additionally, write down the solution to the system for $x(0) = \begin{pmatrix} 1 \\ 2 \end{pmatrix}$.

Exercise 5.17. Use Theorem 5.3.4 to determine whether or not the zero solution of $x(n + 1) = A(n)x(n)$ is (US) or (UAS), where $A(n)$ is the matrix:

(a) $\begin{pmatrix} 0 & \frac{\cos(n\pi)}{3} \\ -\frac{1}{2} & \frac{\sin(n)}{3} \end{pmatrix}$ (b) $\begin{pmatrix} 0 & \frac{n+1}{n+2} \\ -1 & -1 \end{pmatrix}$.

5.4 $x(n+1) = Ax(n)$

In the previous section, we characterized the stability of time-varying systems with respect to their corresponding fundamental matrices by requiring growth conditions on them. However, for constant systems, that is, A is a constant matrix, those growth conditions can be determined based on the magnitude of the eigenvalues of the matrix A. We begin with the following definition.

Definition 5.4.1. Let A be a square matrix. We denote the *spectrum* of A by

$$\sigma(A) = \{\lambda : \det(A - \lambda I) = 0\}.$$

Thus, the notation

$$\mathcal{M}(\sigma(A)) < \zeta,$$

for some real number ζ is to indicate that all the eigenvalues of A have magnitude less than ζ.

For example, if $\sigma(A) = 2 \pm i$, then

$$\mathcal{M}(\sigma(A)) = \sqrt{4+1}.$$

As we have said before, we are interested in providing conditions with respect to the magnitude of the eigenvalues of A so that we deduce stability results regarding the zero solution of

$$x(n+1) = Ax(n), \quad x(n_0) = x_0, \quad n \geq n_0 \geq 0, \tag{5.4.1}$$

where A is an $k \times k$ constant matrix.

Theorem 5.4.1 (Distinct eigenvalues). *Let $\lambda_1, \lambda_2, \ldots, \lambda_k$ be k distinct eigenvalues of the matrix A of (5.4.1) with $\mathcal{M}(\sigma(A)) < 1$. Then there are positive constants L and $\eta \in (0, 1)$ such that*

$$|A|^{n-n_0} \leq L\eta^{n-n_0}.$$

Moreover, the zero solution of (5.4.1) is (UAS).

Proof. The proof is a direct consequence of Theorem 4.3.2. Thus,

$$A^n = \sum_{j=0}^{k-1} \mu_{j+1}(n) D_j,$$

where

$$\mu_1(n+1) = \lambda_1 \mu_1(n), \quad \mu_1(0) = 1,$$
$$\mu_j(n+1) = \lambda_j \mu_j(n) + \mu_{j-1}(n), \quad \mu_j(0) = 0, \quad j = 2, 3, \ldots, k.$$

Solving for $\mu_1(n)$ we get

$$\mu_1(n) = \lambda_1^n.$$

Similarly, we have

$$\mu_2(n+1) = \lambda_2\mu_2(n) + \lambda_1^n, \quad \mu_2(0) = 0,$$

and

$$\mu_2(n) = \frac{1}{\lambda_1 - \lambda_2}\left(\lambda_1^n - \lambda_2^n\right).$$

We see that, in the case of distinct eigenvalues, Putzer's algorithm collapses to Sylvester's formula, in the sense that

$$A^n = \sum_{i=1}^{k} \lambda_i^n Z_i(A),$$

where

$$Z_i(A) = \frac{\prod_{r\neq i}(A - \lambda_r I)}{\prod_{r\neq i}(\lambda_i - \lambda_r)}.$$

Setting

$$\eta = \max\{|\lambda_i| : i = 1, 2, \cdots k\} < 1$$

and since $\mathcal{M}(\sigma(A)) < 1$ we have

$$\left|\prod_{r=i}^{k}(A - \lambda_r I)\right| \leq D,$$

for some positive constant D. Thus,

$$
\begin{aligned}
|A|^n = \left|\sum_{i=1}^{k}\lambda_i^n Z_i(A)\right| \\
\leq \sum_{i=1}^{k}|\lambda_i^n||Z_i(A)| \\
\leq \frac{kD}{\left|\prod_{r\neq i}(\lambda_i - \lambda_r)\right|}\eta^n.
\end{aligned}
$$

Thus,

$$|A|^{n-n_0} \leq L\eta^{n-n_0}.$$

Now the stability results follow from Theorem 5.3.3. $\qquad\square$

Theorem 5.4.2. *Let $\lambda_1, \lambda_2, \ldots, \lambda_k$ be eigenvalues of the matrix A of (5.4.1) with $\mathcal{M}(\sigma(A)) \leq 1$, where the eigenvalues with $\mathcal{M}(\sigma(A)) = 1$ are simple (multiplicity one). Then all solutions of (5.4.1) are bounded and its zero solution is (US).*

The theorems say the following: some eigenvalues with $\mathcal{M}(\sigma(A)) < 1$, could be repeated. However, those with $\mathcal{M}(\sigma(A)) = 1$, cannot. For example, $\lambda_1 = 1$, $\lambda_2 = \frac{1}{\sqrt{2}} \pm i \frac{1}{\sqrt{2}}$ are fine; however, the list $\lambda_1 = 1$, $\lambda_2 = \frac{1}{\sqrt{2}} \pm i \frac{1}{\sqrt{2}}$, $\lambda_3 = 1$ is not allowable.

Proof. The proof is done with the aid of Theorem 4.3.2. Without loss of generality we assume $n_0 = 0$, and then the solution is

$$A^n x_0 = \sum_{j=0}^{k-1} \mu_{j+1}(n) D_j x_0,$$

where

$$\mu_1(n+1) = \lambda_1 \mu_1(n), \quad \mu_1(0) = 1,$$
$$\mu_j(n+1) = \lambda_j \mu_j(n) + \mu_{j-1}(n), \quad \mu_j(0) = 0, \quad j = 2, 3, \ldots, k.$$

Since we have mixed eigenvalues we rearrange them such that $|\lambda_i| = 1$, for $i = 1, 2, \cdots j - 1$, and $|\lambda_i| < 1$, for $i = j, j + 1, \cdots k$. First we pick up the case when $|\lambda_i| = 1$. Solving for $\mu_1(n)$ we get

$$\mu_1(n) = \lambda_1^n.$$

Similarly, we have

$$\mu_2(n+1) = \lambda_2 \mu_2(n) + \lambda_1^n, \quad \mu_2(0) = 0,$$

and

$$\mu_2(n) = \frac{1}{\lambda_1 - \lambda_2} \left(\lambda_1^n - \lambda_2^n \right) = C_{12} \lambda_1^n + C_{22} \lambda_2^n.$$

A continuation of this process leads to

$$\mu_i(n) = \sum_{i=1}^{j-1} C_{1i} \lambda_i^n.$$

Since $|\lambda_i| = 1$, for $i = 1, 2, \cdots j - 1$, we have

$$|\mu_i(n)| \leq \sum_{i=1}^{j-1} |C_{1i} \lambda_i^n| \leq \sum_{i=1}^{j-1} |C_{1i}| \equiv B,$$

for positive constant B. Now we compute $\mu_i(n)$ from $i = j$. Using the equation,

$$\mu_j(n+1) = \lambda_j \mu_j(n) + \mu_{j-1}(n),$$

we obtain

$$|\mu_j(n+1)| = |\lambda_j||\mu_j(n)| + B.$$

Let

$$\beta = \max\{|\lambda_j|, \cdots, |\lambda_{j+1}|\} < 1.$$

Then

$$|\mu_j(n+1)| \leq \beta|\mu_j(n)| + B.$$

With the aid of formula (2.1.6) we arrive at

$$|\mu_j(n)| \leq \beta^n|\mu_0| + B\frac{\beta^n - 1}{\beta - 1}.$$

Applying $0 = \mu_j(0)$, we obtain $\mu_0 = 0$ and the solution reduces to

$$|\mu_j(n)| \leq B\frac{\beta^n - 1}{\beta - 1}, \quad j, j+1, \ldots, k.$$

Then

$$A^n x_0 = \sum_{i=0}^{k-1} \mu_{j+1}(n) D_j x_0$$

$$= \sum_{i=0}^{j-1} \mu_{j+1}(n) D_j x_0 + \sum_{i=j}^{k-1} \mu_{j+1}(n) D_j x_0.$$

Using the fact that for any constant matrix D, there is a positive constant Q so that

$$|Dv| \leq Q|v|,$$

for all $v \in \mathbb{R}^k$. Thus,

$$|A^n x_0| \leq \sum_{i=0}^{j-1} |\mu_{j+1}(n)||D_j x_0| + \sum_{i=j}^{k-1} |\mu_{j+1}(n)||D_j x_0|$$

$$\leq B \sum_{i=0}^{j-1} |D_j x_0| + B \sum_{i=0}^{j-1} \frac{\beta^n - 1}{\beta - 1} |D_j x_0|$$

$$\leq B \sum_{i=0}^{j-1} |D_j x_0| + B \sum_{i=0}^{j-1} |D_j x_0|$$

$$= 2B \left(\sum_{i=0}^{j-1} Q_i \right) |x_0|$$

$$\leq L|x_0|,$$

for $n \geq 0$, and some positive constant L. Now, we know the solution for (5.4.1) with $x(n_0) = x_0$ is $x(n) = A^{n-n_0}x_0$. Thus,

$$|x(n)| = |A^{n-n_0}x_0| \leq L|x_0|$$

which shows all solutions of (5.4.1) are bounded. As for the (US), for any positive $\varepsilon > 0$, set $\delta = \frac{\varepsilon}{L}$. Then for $|x_0| < \delta$, we have that

$$|x(n, n_0, x_0)| \leq L|x_0| \leq L\delta = L\left(\frac{\varepsilon}{L}\right) = \varepsilon.$$

This completes the proof. $\qquad\qquad\qquad\qquad\qquad\qquad\qquad\square$

The next theorem allows multiplicity of any order of the eigenvalues as long as their magnitudes are less than one.

Theorem 5.4.3. *Let $\lambda_1, \lambda_2, \ldots, \lambda_k$ be eigenvalues of the matrix A with any order of multiplicity with $\mathcal{M}(\sigma(A)) \leq \beta < 1$. Then all solutions of (5.4.1) are bounded and its zero solution is (AS).*

Proof. The proof is done by appealing to Theorem 4.3.2. Without loss of generality we assume $n_0 = 0$, and then the solution is

$$A^n x_0 = \sum_{j=0}^{k-1} \mu_{j+1}(n) D_j x_0.$$

As before, solving for $\mu_1(n)$ we get

$$|\mu_1(n)| \leq \beta^n.$$

Then form

$$\mu_2(n+1) = \lambda_2 \mu_2(n) + \mu_1(n), \quad \mu_2(0) = 0,$$

we arrive at

$$|\mu_2(n+1)| \leq \beta|\mu_2(n)| + \beta^n, \quad \mu_2(0) = 0,$$

with solution

$$|\mu_2(n)| \leq n \left(\mathcal{M}(\sigma(A))\right)^{n-1} \leq n \left(\frac{\mathcal{M}(\sigma(A))}{\delta}\right)^{n-1} \delta^{n-1}.$$

It is clear by L'Hospital's rule that

$$\lim_{n \to \infty} n \left(\frac{\mathcal{M}(\sigma(A))}{\delta} \right)^{n-1} = 0,$$

and so there is a positive constant B_1 so that

$$|\mu_2(n)| \leq B_1 \delta^n.$$

In a similar fashion it can be shown that

$$|\mu_3(n)| \leq \frac{n(n-1)}{2} n \left(\mathcal{M}(\sigma(A)) \right)^{n-1},$$

and for $n \geq 0$, we have that

$$|\mu_3(n)| \leq B_2 \delta^n,$$

for a positive constant B_2. Then, inductively, we can show that

$$|\mu_k(n)| \leq G \delta^n, \quad n \geq 0$$

for a positive constant G. Then, using

$$|Dv| \leq Q|v|,$$

for all $v \in \mathbb{R}^k$, we obtain

$$|A^n x_0| \leq G \delta^n |x_0| \sum_{i=0}^{k-1} |D_i|$$
$$\leq G^* \delta^n |x_0|,$$

for $n \geq 0$, and some positive constant G^*. As the solution of (5.4.1) with $x(0) = x_0$ is $x(n) = A^n x_0$, we arrive at

$$|x(n)| = |A^n x_0| \leq G^* |x_0|$$

which shows all solutions of (5.4.1) are bounded. As for the (S), for any positive $\varepsilon > 0$, set $\delta = \frac{\varepsilon}{L}$. Then for $|x_0|| < \delta$, we have that

$$|x(n, 0, x_0)| \leq G^* |x_0| \leq G^* \delta = G^* \left(\frac{\varepsilon}{G^*} \right) = \varepsilon.$$

Hence the zero solution is stable. Now, as $\delta < 1$, we automatically have

$$\lim_{n \to \infty} |x(n, 0, x_0)| \leq \lim_{n \to \infty} (G^* \delta^n |x_0|) = 0.$$

This completes the proof. $\square$

We have the following corollary.

Corollary 5.1. *If any of the eigenvalues λ_i, $i = 1, 2, \cdots, k$ of A has a magnitude $|\lambda_i| > 1$, then solutions of (5.4.1) become unbounded and its zero solution is unstable.*

Proof. Suppose $|\lambda_1| > 1$ has a corresponding eigenvector V_1. Then $x(n) = \lambda_1^n V_1$ is a solution of (5.4.1) with

$$\lim_{n \to \infty} |x(n)| = \lim_{n \to \infty} |\lambda_1^n| |V_1| \to \infty.$$

This completes the proof. $\qquad\qquad\qquad\qquad\qquad\qquad\qquad\qquad\qquad\square$

We end this section with the following remark.

Remark 5.2. Theorems 5.4.1–5.4.3 do not hold for time-varying systems of the forms $x(n + 1) = A(n)x(n)$. To see this, consider the system in Exercise 5.16. It has the eigenvalues

$$\lambda = \pm \frac{\sqrt{3}}{2},$$

and hence both eigenvalues have the magnitude of $|\lambda| = \frac{\sqrt{3}}{2} < 1$. However, the fundamental matrix of the system is unbounded and therefore by Theorem 5.3.1 the zero solution of the system is unstable.

5.4.1   Exercises

Exercise 5.18. Give the details for obtaining the estimate on $|\mu_3|$ of Theorem 5.4.3.

Exercise 5.19. Determine whether or not the zero solution of $x(n + 1) = A(n)x(n)$ is (S), unstable, (US), (UAS), or (ES) where $A(n)$ is the matrix:

(a) $\begin{pmatrix} 0 & -1 \\ 1 & 0 \end{pmatrix}$ (b) $\begin{pmatrix} 0 & 1 \\ -1 & -2 \end{pmatrix}$ (c) $\begin{pmatrix} 1/2 & 3/2 & 0 \\ 3/2 & 1/2 & 0 \\ 0 & 0 & 1 \end{pmatrix}$

(d) $\begin{pmatrix} 1/3 & 0 & 0 \\ -1 & -1/2 & 5/4 \\ 5/(12) & 0 & 1/2 \end{pmatrix}$ (e) $\begin{pmatrix} 1/4 & 1 \\ 1/4 & 1/4 \end{pmatrix}$.

5.5   Perturbed linear systems

We begin this section by introducing Gronwall's inequality, which plays an important role in having estimate on solutions. We state and prove variant forms of discrete Gronwall's Inequality.

Theorem 5.5.1 (Discrete Gronwall's inequality). *As before, let* $\mathbb{N}_{n_0} = \{n_0, n_0 + 1, n_0 + 2, \cdots\}$ *where n_0 is a fixed nonnegative integer. Let $u(n)$, $\alpha(n)$, $\beta(n)$ and $\gamma(n)$ be nonnegative scalar sequences for all $n \geq n_0$. Let $n \in \mathbb{N}_{n_0}$ and assume, for all $n \geq n_0$ the inequality*

$$u(n) \leq \alpha(n) + \beta(n) \sum_{s=n_0}^{n-1} \gamma(s)u(s) \tag{5.5.1}$$

holds. Then,

$$u(n) \leq \alpha(n) + \beta(n) \sum_{s=n_0}^{n-1} \alpha(s)\gamma(s) \prod_{r=s+1}^{n-1} \big(1 + \beta(r)\big)\gamma(r) \tag{5.5.2}$$

holds for all $n \geq n_0$.

Proof. Define $\varphi : \mathbb{N}_{n_0} \to \mathbb{R}$ by $\varphi(n) = \sum_{s=n_0}^{n-1} \gamma(s)u(s)$. Then,

$$\Delta\varphi(n) = \gamma(n)u(n), \quad \varphi(n_0) = 0.$$

Substituting $u(n) \leq \alpha(n) + \beta(n)\varphi(n)$, in the above equality yields,

$$\Delta\varphi(n) = \gamma(n)u(n) \leq \gamma(n)\big(\alpha(n) + \beta(n)\varphi(n)\big),$$

from which we conclude

$$\varphi(n+1) - (1 + \beta(n)\gamma(n))\varphi(n) \leq \alpha(n)\gamma(n). \tag{5.5.3}$$

Since all sequences are nonnegative, we have that $1 + \beta(n)\gamma(n) > 0$, for all $n \geq n_0$.

Thus, (5.5.3) is equivalent to

$$\Delta\Big[\prod_{s=n_0}^{n-1} \big(1 + \beta(s)\gamma(s)\big)^{-1}\varphi(s)\Big] \leq \alpha(n)\gamma(n) \prod_{s=n_0}^{n} \big(1 + \beta(s)\gamma(s)\big)^{-1}.$$

Summing the above inequality from n_0 to $n-1$ yields

$$\prod_{s=n_0}^{n-1} \big(1 + \beta(s)\gamma(s)\big)^{-1}\varphi(n) \leq \sum_{s=n_0}^{n-1} \alpha(s)\gamma(s) \prod_{r=n_0}^{n} \big(1 + \beta(r)\gamma(r)\big)^{-1}$$

or

$$\varphi(n) \leq \sum_{s=n_0}^{n-1} \alpha(s)\gamma(s) \prod_{r=s+1}^{n-1} \big(1 + \beta(r)\gamma(r)\big).$$

As a consequence

$$\sum_{s=n_0}^{n-1} \gamma(s)u(s) \le \sum_{s=n_0}^{n-1} \alpha(s)\gamma(s) \prod_{r=s+1}^{n-1} \left(1 + \beta(r)\gamma(r)\right).$$

Hence

$$u(n) \le \alpha(n) + \beta(n) \sum_{s=n_0}^{n-1} \gamma(s)u(s)$$

$$\le \alpha(n) + \beta(n) \sum_{s=n_0}^{n-1} \alpha(s)\gamma(s) \prod_{r=s+1}^{n-1} \left(1 + \beta(r)\gamma(r)\right).$$

This completes the proof.   □

We have the following special cases of Gronwall's inequality. Also noting that $1 + L\beta\gamma(s) \le e^{L\beta\gamma(s)}$, where $L > 0$ and constant, yields the following.

Corollary 5.2. *Consider $\alpha(n)$, $\beta(n)$ of Theorem 5.5.1 so that $\alpha(n) = \alpha$, $\beta(n) = \beta$, for all $n \in \mathbb{N}_{n_0}$. Then we have*

$$u(n) \le \alpha \prod_{s=n_0}^{n-1} \left(1 + \beta\gamma(s)\right),$$

or

$$u(n) \le \alpha e^{\sum_{s=n_0}^{n-1} \beta\gamma(s)}.$$

Corollary 5.3. *Assume the hypothesis of Corollary 5.2. If*

$$u(n) \le m\left[\alpha + \beta \sum_{s=n_0}^{n-1} \gamma(s)u(s)\right]$$

for all $n \in \mathbb{N}_{n_0}$, $n \ge n_0$ and $m > 0$, then we have

$$u(n) \le \alpha \prod_{s=n_0}^{n-1} \left(1 + m\beta(s)\gamma(s)\right),$$

or

$$u(n) \le \alpha e^{\sum_{s=n_0}^{n-1} m\beta\gamma(s)}.$$

Proof. Define $\varphi : \mathbb{N}_{n_0} \to \mathbb{R}$ by $\varphi(n) = m\beta \sum_{s=n_0}^{n-1} \gamma(s)u(s)$. The rest of the proof follows along the lines of the proof of Theorem 5.5.1.   □

Consider the time-varying system

$$x(n+1) = A(n)x(n), \quad x(n_0) = x_0, \ n \geq 0 \tag{5.5.4}$$

and assume its being perturbed by adding a perturbed term denoted by the matrix $B(n)$. Then we have the linear perturbed system

$$x(n+1) = A(n)x(n) + B(n)x(n), \quad x(n_0) = x_0, \ n \geq 0 \tag{5.5.5}$$

where A and B are $k \times k$ matrices that are defined on the set $\mathbb{N}_{n_0}$. Under what conditions of $B(n)$, is the perturbed system (5.5.5) stable, knowing that the zero solution of (5.5.4) enjoys some type of stability? It turns out that if, in some sense, the norm of matrix B is small, then a type of stability of (5.5.5) can be deduced.

Theorem 5.5.2. *Assume the zero solution of (5.5.4) is uniformly stable and that*

$$\sum_{s=0}^{\infty} ||B(s)|| \leq E \tag{5.5.6}$$

for some positive constant E. Then all solutions of (5.5.5) are bounded and the zero solution of (5.5.5) is uniformly stable.

Proof. Let $\Phi(n)$ be the fundamental matrix of (5.5.4). Since the zero solution of (5.5.4) is (US), there is a positive constant M,

$$||\Phi(n)\Phi^{-1}(s)|| \leq M.$$

By the variation of parameters formula (4.5.4) we have that

$$x(n) = \Phi(n)\Phi^{-1}(n_0)x_0 + \sum_{s=n_0}^{n-1} \Phi(n)\Phi^{-1}(s+1)B(s)x(s).$$

By taking norms on both sides and making use of Gronwall's inequality

$$|x(n)| \leq ||\Phi(n)\Phi^{-1}(n_0)|||x_0| + \sum_{s=n_0}^{n-1} ||\Phi(n)\Phi^{-1}(s+1)|| \ ||B(s)|| \ |x(s)|$$

$$\leq M \left[|x_0| + \sum_{s=n_0}^{n-1} ||B(s)|| \ |x(s)| \right]$$

$$\leq |x_0| \prod_{s=n_0}^{n-1} (1 + M||B(s)||), \quad \text{(by Corollary 5.3)}$$

$$\leq |x_0| e^{M \sum_{s=0}^{n-1} ||B(s)||}$$

$$\leq |x_0| e^{ME}.$$

This shows all solutions are bounded. For the uniform stability set $\delta = \frac{\varepsilon}{e^{ME}}$.
This completes the proof. □

The next example shows that the summability condition (5.5.6) for $||B(n)||$ is necessary.

Example 5.15. Consider the second-order difference equation

$$y(n+2) + \frac{1}{2} y(n) = 0, \quad x(n_0) = x_0 \neq 0.$$

Let

$$y_i(n) = y(n+i-1), \quad i = 1, 2.$$

Then we arrive at the system

$$y(n+1) = \begin{pmatrix} 0 & 1 \\ -\frac{1}{2} & 0 \end{pmatrix} \begin{pmatrix} y_1(n) \\ y_2(n) \end{pmatrix},$$

with

$$y = \begin{pmatrix} y_1 \\ y_2 \end{pmatrix}.$$

Let $A = \begin{pmatrix} 0 & 1 \\ -\frac{1}{2} & 0 \end{pmatrix}$. Then the matrix A has the two eigenvalues $\lambda = \pm \frac{1}{\sqrt{2}} i$, with magnitude

$$|\lambda| = \frac{1}{2} < 1.$$

Thus, the zero solution is (US). Actually it is (UAS), but for the purpose of our illustration all we ask for is (US). Now let us perturb the system by adding a small perturbation term. Thus, we are looking at

$$y(n+2) - \frac{2}{n+1} y(n+1) + \frac{1}{2} y(n) = 0.$$

Then in matrix form we have the perturbed system

$$y(n+1) = \begin{pmatrix} 0 & 1 \\ -\frac{1}{2} & 0 \end{pmatrix} \begin{pmatrix} y_1 \\ y_2 \end{pmatrix} + \begin{pmatrix} 0 & 0 \\ 0 & \frac{2}{n+1} \end{pmatrix} \begin{pmatrix} y_1 \\ y_2 \end{pmatrix}.$$

Then $B = \begin{pmatrix} 0 & 0 \\ 0 & \frac{2}{n+1} \end{pmatrix}$ and

$$\sum_{s=0}^{\infty} ||B(s)|| = \sum_{s=0}^{\infty} \frac{2}{s+1},$$

which diverges. Hence $B(n)$ does not satisfy condition (5.5.6) and we conclude that Theorem 5.5.2 says nothing about the perturbed system. In addition, it is justified to say B is the perturbation matrix since $||B(n)|| = \dfrac{2}{n+1} \to 0$, as $n \to \infty$. $\qquad\qquad\square$

Theorem 5.5.3. *Assume the zero solution of (5.5.4) is uniformly asymptotically stable. That is, if $\Phi(n)$ is the fundamental matrix of (5.5.4), then by (5.3.3) there are positive constants M and $\xi \in (0, 1)$ such that*

$$||\Phi(n)\Phi^{-1}(s)|| \le M\xi^{n-s}, \; n \ge s \ge 0.$$

Suppose $B(n)$ satisfies condition (5.5.6). Then the zero solution of (5.5.5) is uniformly asymptotically stable.

Proof. The proof is similar to the proof of Theorem 5.5.2 but with small changes. Thus,

$$|x(n)| \le ||\Phi(n)\Phi^{-1}(n_0)|||x_0| + \sum_{s=n_0}^{n-1} ||\Phi(n)\Phi^{-1}(s+1)|| \; ||B(s)|| \; |x(s)|$$

$$\le M\xi^{n-n_0}|x_0| + \sum_{s=n_0}^{n-1} M\xi^{n-s-1}||B(s)|| \; |x(s)|.$$

Factor M and multiply both sides with ξ^{-n} and get

$$\xi^{-n}|x(n)| \le M \left[\xi^{-n_0}|x_0| + \sum_{s=n_0}^{n-1} \xi^{-1}\xi^{-s}||B(s)|| \; |x(s)| \right].$$

Set $y(n) = \xi^{-n}|x(n)|$. Then the above expression reduces to

$$y(n) \le M \left[\xi^{-n_0}|x_0| + \sum_{s=n_0}^{n-1} \xi^{-1}||B(s)|| \; y(s) \right]$$

$$\le \xi^{-n_0}|x_0| \prod_{s=n_0}^{n-1} \left(1 + M\xi^{-1}||B(s)|| \right)$$

$$\le \xi^{-n_0}|x_0| e^{M \sum_{s=n_0}^{n-1} \xi^{-1}||B(s)||}$$

$$\leq \xi^{-n_0}|x_0|e^M \sum_{s=0}^{\infty} \xi^{-1}||B(s)||$$

$$\leq \xi^{-n_0}|x_0|e^{M\xi^{-1}E}.$$

In terms of $|x(n)|$ this implies that

$$|x(n)| \leq \xi^{n-n_0}|x_0|e^{M\xi^{-1}E},$$

from which we arrive at the (UAS) by imitating the proof of Theorem 5.3.3. This completes the proof. □

5.5.1 Exercises

Exercise 5.20. Show all solutions of

$$x(n+2) - \frac{2}{n^2+1}x(n+1) + \frac{1}{2}x(n) = 0, \quad x(n_0) = x_0 \neq 0, \ n \geq n_0 \geq 0$$

are bounded and the zero solution is (US).

Exercise 5.21. Show that the zero solution of the non-autonomous system

$$x(n+1) = \begin{pmatrix} 1 + ne^{-n} & -5 + e^{-n} \\ \frac{1}{4} + \frac{1}{(1+n)^2} & -1 \end{pmatrix} x(n), \quad x(n_0) = x_0, \ n \geq n_0 \geq 0$$

is (UAS).

Exercise 5.22. Show that the zero solution of the non-autonomous system

$$x(n+1) = \begin{pmatrix} \frac{1}{(1+n)^2} & 0 & \frac{1}{2} \\ -1 & -\frac{1}{2} & \frac{5}{4} \\ \frac{1}{3} & \frac{1}{(1+n)^2} & 0 \end{pmatrix} x(n), \quad x(n_0) = x_0, \ n \geq n_0 \geq 0$$

is (UAS).

Exercise 5.23. Assume the zero solution of (5.5.4) is uniformly stable. Let $f(n, x(n))$ be an $k \times 1$ vector functions that are continuous in x and defined on $\mathbb{N}$. Show that if

$$\sum_{s=n_0}^{\infty} g(n) < \infty$$

then the zero solution of the non-homogeneous non-linear system

$$x(n+1) = A(n)x(n) + f(n, x(n))$$

is uniformly stable provided that

$$|f(n, x(n)| \le g(n)|x(n)|,$$

where $g(n)$ are positive for every $n \in \mathbb{N}_{n_0}$.

Exercise 5.24. Assume the zero solution of (5.5.4) is uniformly asymptotically stable. Let $f(n, x(n))$ be a $k \times 1$ vector functions that are continuous in x and defined on $\mathbb{N}$. Show that if

$$\sum_{s=n_0}^{\infty} g(n) < \infty$$

then the zero solution of the non-homogeneous non-linear system

$$x(n + 1) = A(n)x(n) + f(n, x(n))$$

is uniformly asymptotically stable provided that

$$|f(n, x(n)| \le g(n)|x(n)|,$$

where $g(n)$ are positive for every $n \in \mathbb{N}_{n_0}$.

Exercise 5.25. For $\alpha, \beta \in (0, 1)$, show that the zero solution of

$$x(n + 1) = (\alpha + \beta^n)x(n), \quad x(n_0) = x_0 \ne 0, \ n \ge n_0 \ge 0$$

is (ES).

5.6 Phase plane analysis

We begin a detailed study of the linear planar autonomous system

$$X(n + 1) = AX(n) \tag{5.6.1}$$

where

$$A = \begin{pmatrix} a & b \\ c & d \end{pmatrix}$$

is a nonsingular matrix and

$$X = \begin{pmatrix} x \\ y \end{pmatrix}.$$

If we associate an initial point

$$X(0) = X_0$$

with system (5.6.1), then its solution takes the form

$$X(n) = A^n X_0.$$

A solution $X(n)$ of (5.6.1) is a point in the Euclidean space of the form $(x(n), y(n))$. The union

$$\cup_{n=-\infty}^{n=\infty}(x(n), y(n))$$

traces out a path and its direction is determined by increasing n. The solution $(x(n), y(n))$ of (5.6.1) as n varies describes parametrically a curve in the plane that we call a *trajectory* or *orbit*. Different initial conditions produce different trajectories.

It is clear that the origin $(0, 0)$ is the only equilibrium solution of system (5.6.1).

Recall that two matrices A and J are similar if there exists an invertible matrix P such that $J = P^{-1}AP$. The matrix J is called the *real Jordan canonical form* of A. The outcome, or nature of the matrix J, is depicted in the following three cases:

(1) If the eigenvalues λ_1 and λ_2 of A are real, but not necessarily distinct with independent eigenvectors, then

$$J = \begin{pmatrix} \lambda_1 & 0 \\ 0 & \lambda_2 \end{pmatrix}.$$

(2) If A has a single eigenvalue λ with a single independent eigenvector, then

$$J = \begin{pmatrix} \lambda & 1 \\ 0 & \lambda \end{pmatrix}.$$

(3) If A has complex conjugate eigenvalues of the form $\alpha \pm i\beta$, then

$$J = \begin{pmatrix} \alpha & \beta \\ -\beta & \alpha \end{pmatrix}.$$

It is readily seen from the above cases that the nature of the eigenvalues of the matrix A play a decisive role in determining the stability of the fixed point $(0, 0)$. To begin our analysis of the phase portrait, we let

$$X(n) = PW(n),$$

then system (5.6.1) becomes $X(n + 1) = PW(n + 1)$, or $AX(n) = PW(n + 1)$, from which we have, $P^{-1}AX(n) = W(n + 1)$. Using $X(n) = PW(n)$ yields

$$W(n + 1) = P^{-1}APW(n) = JW(n).$$

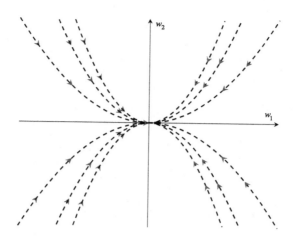

FIGURE 5.4   Sink; $0 < \lambda_1 < \lambda_2 < 1$.

Real Eigenvalues λ_1, λ_2: In this case we have $J = \begin{pmatrix} \lambda_1 & 0 \\ 0 & \lambda_2 \end{pmatrix}$, and hence $W(n + 1) = JW(n)$ transforms the system into two decoupled first-order difference equations,

$$w_1(n + 1) = \lambda_1 w_1(n), \quad w_2(n + 1) = \lambda_2 w_1(n).$$

If we impose the initial condition $W(0) = (w_{10}, w_{20})$ then we have the solution

$$w_1(n) = w_{10}(\lambda_1)^n, \quad w_2(n) = w_{20}(\lambda_2)^n, \quad n \geq 0.$$

• $0 < \lambda_1 < \lambda_2 < 1$. If $w_{10} = 0$, then $|W(n)| \to 0$, as $n \to \infty$, along the w_2-axis. On the other hand, if $w_{20} = 0$, then $|W(n)| \to 0$, as $n \to \infty$, along the w_1-axis. Solving for n in $w_1(n)$ and substituting it in the equation $w_2(n) = w_{20}(\lambda_2)^n$ gives

$$w_2(n) = w_{20}\lambda_2^{\frac{\ln(\frac{w_1}{w_{10}})}{\ln(\lambda_1)}}. \tag{5.6.2}$$

Thus, when both w_{10}, w_{20} are not zero, the points lie along the line (5.6.2). In this case, trajectories will approach the origin $(0, 0)$, and hence the origin is asymptotically stable. In such a case, it is common to say the origin is a *sink* or *stable node*. The trajectories are depicted in Fig. 5.4.

• $\lambda_1 > \lambda_2 > 1$. If $w_{10} = 0$, then $|W(n)| \to \infty$, as $n \to \infty$, along the w_2-axis. On the other hand, if $w_{20} = 0$, then $|W(n)| \to \infty$, as $n \to \infty$, along the w_1-axis. When both w_{10}, w_{20} are not zero, then the points lie along the line (5.6.2). In this case, will diverge to infinity and hence the origin is unstable. In such a case, it is common to say the origin is a *source* or *unstable node*. The trajectories are identical to those in Fig. 5.5(e), in which all arrows are point away from the origin.

- $0 < \lambda_1 < 1$, $\lambda_2 > 1$. If $w_{10} = 0$, then $|W(n)| \to \infty$, as $n \to \infty$, along the w_2-axis. On the other hand, if $w_{20} = 0$, then $|W(n)| \to \infty$, as $n \to \infty$, along the w_1-axis. When both w_{10}, w_{20} are not zero, then the points lie along the line (5.6.2). In this case, trajectories will diverge to infinity and hence the origin is unstable. In such a case, it is common to say the origin is a *source* or *unstable node* or *degenerate node*. The trajectories are depicted in Fig. 5.6(b), with all arrows pointing away from the origin.

- $-1 < \lambda_1 < 0 < 1 < \lambda_2$. If $w_{10} = 0$, then $|W(n)| \to \infty$, as $n \to \infty$, along the w_2-axis. On the other hand, if $w_{20} = 0$, then $|W(n)| \to \infty$, as $n \to \infty$, along the w_1-axis. When both w_{10}, w_{20} are not zero, then the points, or solutions are positioned in a symmetrical way with respect to an axis. For example, when n is odd, the points lie along the trajectories in the first and fourth quadrants; on the other hand, when n is even, the points lie along the trajectories in the second and third quadrants. Thus, we have a reflection over the w_1 axis, for n being even and a reflection over the w_2 axis, for n being odd. In such a case, we have a *saddle node* with *reflection* and the trajectories resemble those of Fig. 5.5(f).

- $\lambda_1 < -1$, $\lambda_2 > 1$. If $w_{10} = 0$, then $|W(n)| \to \infty$, as $n \to \infty$, along the w_2-axis. On the other hand, if $w_{20} = 0$, then $|W(n)| \to \infty$, as $n \to \infty$, along the w_1-axis. When both w_{10}, w_{20} are not zero, then the points, or solutions in a symmetrical way with respect an axis. For example, when n is odd, the points lie along the trajectories in the second and third quadrants; on the other hand, when n is even, the points lie along the trajectories in the first and fourth quadrants. In such a case, all trajectories diverge to infinity and the origin is called a *source with reflection*.

- $\lambda_1 = 1$, $\lambda_2 < \lambda_1$. In this case the solution takes the form

$$W(n) = w_{10}u^1 + w_{20}\lambda_2^n v^1,$$

where u^1 and u^2 are the corresponding eigenvectors.

If $w_{10} = 0$, then $|W(n)| \to |w_{10}|$, as $n \to \infty$. That is the points on the trajectories will trace the vertical line $|w_{10}|$ marching toward the axis $w_2 = 0$. The equilibrium solutions will be located throughout the w_1-axis. This is called a *degenerate node* and the trajectories are depicted in Fig. 5.7(a).

- $\lambda_1 = 1$, $\lambda_2 > \lambda_1$. This is similar to the previous case and the equilibrium solutions will be located throughout the w_1-axis but diverging to infinity. Notice that $|W(n)| \to \infty$, as $n \to \infty$. That is, the points on the trajectories will trace the vertical line $|w_{10}|$ marching toward infinity. This is called a *unstable degenerate node* and the trajectories are depicted in Fig. 5.7(b).

- $0 < \lambda_1 = \lambda_2 < 1$. If $w_{10} = 0$, then $|W(n)| \to 0$, as $n \to \infty$, along the w_2-axis. On the other hand, if $w_{20} = 0$, then $|W(n)| \to 0$, as $n \to \infty$, along the w_1-axis. When both w_{10}, w_{20} are not zero, then the points lie on $w_2(n) = \frac{w_{20}}{w_{10}} w_1(n)$ and the slope of each trajectory depends on the signs of the initial conditions

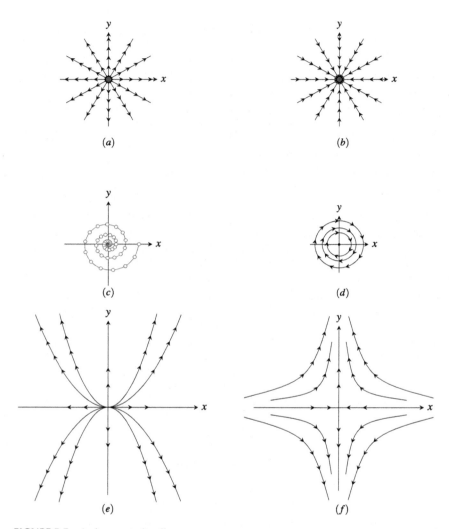

FIGURE 5.5 A phase portrait gallery.

w_{10}, w_{20}. In this case the trajectories are depicted in Fig. 5.5(b) and the origin is called *asymptotically stable node*, or *stable star*.

• $\lambda_1 = \lambda_2 > 1$. If $w_{10} = 0$, then $|W(n)| \to \infty$, as $n \to \infty$, along the w_2-axis. On the other hand, if $w_{20} = 0$, then $|W(n)| \to \infty$, as $n \to \infty$, along the w_1-axis. When both w_{10}, w_{20} are not zero, then the points lie on $w_2(n) = \frac{w_{20}}{w_{10}} w_1(n)$ and the slope of each trajectory depends on the signs of the initial conditions w_{10}, w_{20}. In this case the trajectories are depicted in Fig. 5.5(a) and the origin is called *unstable node*, or *unstable star*.

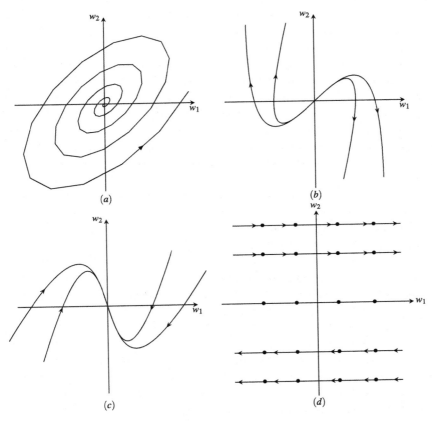

FIGURE 5.6 Unstable spiral; degenerate node; stable node.

Single Eigenvalue λ: In this case we have $J = \begin{pmatrix} \lambda & 1 \\ 0 & \lambda \end{pmatrix}$, and hence $W(n + 1) = JW(n)$ transforms the system into two decoupled first-order difference equations,

$$w_1(n + 1) = \lambda_1 w_1(n) + w_2(n), \quad w_2(n + 1) = \lambda_2 w_1(n).$$

If we impose the initial condition $W(0) = (w_{10}, w_{20})$ then we have the solution

$$w_1(n) = w_{10}\lambda^n + w_{20}\lambda^{n-1}, \quad w_2(n) = w_{20}\lambda^n, \quad n \geq 0.$$

Notice, in such a case we have the ratio

$$\lim_{n \to \infty} \frac{w_2(n)}{w_1(n)} = 0.$$

We will look at several cases.

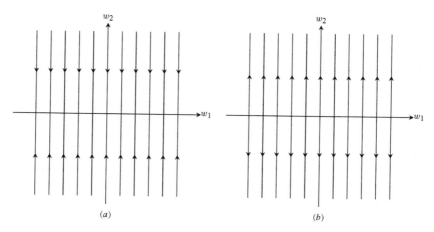

FIGURE 5.7 Line of stable points (a); Line of unstable points (b).

- $0 < |\lambda| < 1$. If $w_{10} = 0$, then $W(n) \to 0$, as $n \to \infty$, along the negative w_2-axis for $w_{20} < 0$, and $W(n)$ approaches zero along the positive w_2-axis for $w_{20} > 0$. On the other hand, if $w_{10} \neq 0$, then $|W(n)| \to 0$, as $n \to \infty$, and the trajectories can be graphed from the curve

$$w_2(n) = \frac{w_1(n)}{|\lambda|} \log_{|\lambda|}(\frac{w_1(n)}{w_{10}}) + w_1(n)\frac{w_{20}}{w_{10}}.$$

It is observed from the solution that for $-1 < \lambda < 0$, the points on the trajectories oscillate about the w_2-axis. The trajectories of such a case are displayed in Fig. 5.6(c) and the origin is referred to as *stable node*.

- $|\lambda| > 1$. This is similar to the previous except points on the trajectories diverge to infinity. The trajectories of such a case are displayed in Fig. 5.6(b) and the origin is referred to as *unstable node*.

- $|\lambda| = 1$. The points on the trajectories are depicted in Fig. 5.6(d) and all points on the w_1-axis are equilibrium points. This case is referred to as *degenerate case (unstable)*.

Complex Eigenvalues: In this case we have $J = \begin{pmatrix} \alpha & \beta \\ -\beta & \alpha \end{pmatrix}$. By the aid of subsection 2.4.3 the solution may take the form

$$W(n) = |\lambda|^n \begin{pmatrix} \cos(n\theta) & \sin(n\theta) \\ -\sin(n\theta) & \cos(n\theta) \end{pmatrix} \begin{pmatrix} w_{10} \\ w_{20} \end{pmatrix},$$

where

$$|\lambda| = \sqrt{\alpha^2 + \beta^2}, \quad \theta = \tan^{-1}(\frac{\beta}{\alpha}).$$

Again, we will consider several cases.

- $|\lambda| = 1 = \alpha^2 + \beta^2$. In this case, the Jordan matrix J is called the *rotation matrix* since all points lie on circular trajectories as shown in Fig. 5.5(d). The circular trajectories are centered at the origin with radii $\sqrt{w_{10}^2 + w_{20}^2}$. In this case we have a *stable center*, or just a *center*.
- $|\lambda| < 1; \alpha^2 + \beta^2 < 1$. In this case, all points on the trajectories will spiral toward the origin. The trajectories are shown in Fig. 5.5(c). Note that as seen in Fig. 5.5(c), nodes in the resulting spiral get closer and closer as they approach the center. In this case, the origin is called *asymptotically stable focus*.
- $|\lambda| > 1; \alpha^2 + \beta^2 > 1$. In this case all points on the trajectories will spiral away from the origin. The trajectories are shown in Fig. 5.6(a). In this case, the origin is called *unstable focus*.

5.6.1 Exercises

Exercise 5.26. Sketch the phase plane diagram and determine the stability of the system $x(n+1) = Ax(n)$ where A is the matrix:

(a) $\begin{pmatrix} 6 & 3 \\ 4 & 7 \end{pmatrix}$ **(b)** $\begin{pmatrix} 0 & 2 \\ 3 & 0 \end{pmatrix}$ **(c)** $\begin{pmatrix} -\frac{1}{6} & -1 \\ \frac{1}{3} & 1 \end{pmatrix}$ **(d)** $\begin{pmatrix} 2 & 5 \\ -1 & -2 \end{pmatrix}$

(e) $\begin{pmatrix} -3 & -2 \\ 13 & 7 \end{pmatrix}$ **(f)** $\begin{pmatrix} -1 & -1 \\ 2 & 1 \end{pmatrix}$ **(g)** $\begin{pmatrix} 3 & -7 \\ 1 & -2 \end{pmatrix}$.

Exercise 5.27. Show that if A is a 2×2 matrix with real entries and two real eigenvalues λ_1, λ_2, then the Jordan canonical form J of A is

$$J = \begin{pmatrix} \lambda_1 & 0 \\ 0 & \lambda_2 \end{pmatrix}.$$

Exercise 5.28. Show that if A is a 2×2 matrix with real entries and has one eigenvalue λ with a single independent eigenvector, then the Jordan canonical form J of A is

$$J = \begin{pmatrix} \lambda & 1 \\ 0 & \lambda \end{pmatrix}.$$

Exercise 5.29. Show that if A is a 2×2 matrix with complex conjugate eigenvalues of the form $\alpha \pm i\beta$, then the Jordan canonical form J of A is

$$J = \begin{pmatrix} \alpha & \beta \\ -\beta & \alpha \end{pmatrix}.$$

5.7 Linearization

Consider the non-linear system of difference equations

$$x_1(n+1) = f_1(x_1(n), \ldots, x_k(n))$$
$$x_2(n+1) = f_2(x_1(n), \ldots, x_k(n))$$
$$\vdots$$
$$x_k(n+1) = f_k(x_1(n), \ldots, x_k(n)).$$

Using the vector notations

$$x = \begin{pmatrix} x_1 \\ x_2 \\ \vdots \\ x_k \end{pmatrix}$$

and

$$f(x) = \begin{pmatrix} f_1(x) \\ f_2(x) \\ \vdots \\ f_k(x), \end{pmatrix}$$

then above system can be written in the vector form

$$x(n+1) = f(x(n)), \tag{5.7.1}$$

where $f : \mathbb{R}^k \to \mathbb{R}^k$ with each component being continuously differentiable. We assume the vector $x^* = (x_1^*, x_2^*, \ldots, x_k^*)$ is a fixed point of (5.7.1). Our aim is to study the behavior of the non-linear system in the neighborhood of its equilibrium point x^*. To better illustrate the idea we consider $k = 2$, and look at the two-dimensional system

$$x_1(n+1) = f_1(x_1(n), x_2(n))$$
$$x_2(n+1) = f_2(x_1(n), x_2(n)). \tag{5.7.2}$$

Let x^* be the equilibrium solution so that

$$x_1^* = f_1(x_1^*, x_2^*), \quad x_2^* = f_1(x_1^*, x_2^*).$$

Then, our goal is to study the stability of these equilibrium solutions by considering and analyzing small deviations by perturbing x_1^* and x_2^* by an amount of $\eta_1(n)$ and $\eta_2(n)$. To do this, we must use Taylor series expansions of the functions of two variables f_1 and f_2. That is for

$$x_1(n) = x_1^* + \eta_1(n), \quad x_2(n) = x_2^* + \eta_2(n)$$

we have

$$x_1(n+1) = \eta_1(n+1) + x_1^* = f_1(x_1^*, x_2^*) + \frac{\partial f_1}{\partial x_1}(x_1^*, x_2^*) + \frac{\partial f_1}{\partial x_2}(x_1^*, x_2^*) + \cdots$$

$$x_2(n+1) = \eta_2(n+1) + x_2^* = f_2(x_1^*, x_2^*) + \frac{\partial f_2}{\partial x_1}(x_1^*, x_2^*) + \frac{\partial f_2}{\partial x_2}(x_1^*, x_2^*) + \cdots.$$

By making use of

$$f_1(x_1^*, x_2^*) = x_1^*, \quad f_2(x_1^*, x_2^*) = x_2^*,$$

we arrive at

$$\eta_1(n+1) = \frac{\partial f_1}{\partial x_1}(x_1^*, x_2^*) + \frac{\partial f_1}{\partial x_2}(x_1^*, x_2^*) + \cdots$$

$$\eta_2(n+1) = \frac{\partial f_2}{\partial x_1}(x_1^*, x_2^*) + \frac{\partial f_2}{\partial x_2}(x_1^*, x_2^*) + \cdots.$$

Let

$$a = \frac{\partial f_1}{\partial x_1}(x_1^*, x_2^*), \quad b = \frac{\partial f_1}{\partial x_2}(x_1^*, x_2^*),$$

$$c = \frac{\partial f_2}{\partial x_1}(x_1^*, x_2^*), \quad d = \frac{\partial f_2}{\partial x_2}(x_1^*, x_2^*).$$

In matrix notation the above system takes the form

$$\eta(n+1) = A\eta(n) \tag{5.7.3}$$

where

$$\eta(n) = \begin{pmatrix} \eta_1(n) \\ \eta_2(n) \end{pmatrix}, \quad A = \begin{pmatrix} a & b \\ c & d \end{pmatrix}$$

is called the *Jacobian matrix*. The characteristic equation of the matrix A is

$$\lambda^2 - (a+d)\lambda + ad - bc = 0.$$

By introducing

$$p = a + d,$$
$$q = ad - bc,$$

the characteristic equation becomes

$$\lambda^2 - p\lambda + q = 0.$$

If we let λ_1 and λ_2 be the roots of this equation, then we have

$$\lambda_{1,2} = \frac{1}{2}\left[p \pm \sqrt{p^2 - 4q}\right] = \frac{p}{2} \pm \sqrt{(\frac{p}{2})^2 - q}. \tag{5.7.4}$$

- If the roots are complex, then we want

$$1 > \left|\frac{p}{2} \pm i\sqrt{q - (\frac{p}{2})^2}\right| = \sqrt{(\frac{p}{2})^2 + q - (\frac{p}{2})^2}.$$

This implies that $1 > \sqrt{q}$, or $1^2 > (\sqrt{q})^2$. Consequently, one has

$$1 > |q|.$$

- The roots are real and distinct when $(\frac{p}{2})^2 - q > 0$. Moreover, for the roots to have magnitudes less than one, one must require that $\left|\frac{p}{2}\right| < 1$. With this in mind, the real roots have magnitudes less than one, translate into

$$\left|\frac{p}{2} \pm \sqrt{(\frac{p}{2})^2 - q}\right| < 1.$$

The above inequality is valid for

$$-1 < \frac{p}{2} \pm \sqrt{(\frac{p}{2})^2 - q} < 1.$$

With the greater real root and the upper boundary on the real line we have

$$1 > \frac{p}{2} + \sqrt{(\frac{p}{2})^2 - q}.$$

By rearranging terms and squaring both sides, we arrive at $-p > -q - 1$, or

$$q > p - 1. \tag{5.7.5}$$

Similarly, for the lesser real root and the lower boundary on the real line we have

$$-1 < \frac{p}{2} - \sqrt{(\frac{p}{2})^2 - q},$$

or

$$1 + \frac{p}{2} > \sqrt{(\frac{p}{2})^2 - q}.$$

By squaring both sides, we arrive at

$$q > -p - 1. \tag{5.7.6}$$

Thus, for stability of a given equilibrium solution for system (5.7.2), in addition to the conditions (5.7.5) and (5.7.6), we require

$$|q| < 1. \tag{5.7.7}$$

We note that conditions (5.7.5) and (5.7.6) give $-q - 1 < p < q + 1$, which is equivalent to

$$|p| < q + 1.$$

By making use of (5.7.7) we arrive at the concise necessary condition for stability

$$|p| < q + 1 < 2. \tag{5.7.8}$$

Thus, we have the following lemma.

Lemma 5.1. *Let (u^*, v^*) be an equilibrium solution of (5.7.2). Let a, b, c, and d be the matrix coefficient of the Jacobian matrix A given by (5.7.3). Then (u^*, v^*) is asymptotically stable if and only if (5.7.8) holds, where*

$$p = a + d, \ q = ad - bc.$$

Proof. Since all steps in the derivations of (5.7.5), (5.7.6) and (5.7.7) are reversible, the proof follows. $\qquad\square$

We shift back to the system (5.7.1) with each component f_i, $i = 1, 2, \cdots, k$ is continuously differentiable. We assume the vector $x^* = (x_1^*, x_2^*, \ldots, x_n^*)$ is a fixed point of (5.7.1). Define the linear part of f at x^* by $J(x^*)$, where

$$J(x^*) := \begin{pmatrix} \frac{\partial f_1}{\partial x_1}(x^*) & \frac{\partial f_1}{\partial x_2}(x^*) & \cdots & \frac{\partial f_1}{\partial x_k}(x^*) \\ \frac{\partial f_2}{\partial x_1}(x^*) & \frac{\partial f_2}{\partial x_2}(x^*) & \cdots & \frac{\partial f_2}{\partial x_k}(x^*) \\ \vdots & \vdots & \ddots & \vdots \\ \frac{\partial f_k}{\partial x_1}(x^*) & \frac{\partial f_k}{\partial x_2}(x^*) & \cdots & \frac{\partial f_k}{\partial x_k}(x^*) \end{pmatrix}. \tag{5.7.9}$$

$J(x^*)$ given by (5.7.9) is called the *Jacobian* of f at x^*. Since $f \in C^1(\mathbb{R}^k, \mathbb{R}^k)$, Taylor's theorem for functions of several variables says that

$$f(x) = J(x^*)(x - x^*) + g(x), \ \text{(we have used } f(x^*) = x^*) \tag{5.7.10}$$

where g is a function that is small in the neighborhood of x^* in the sense that

$$\lim_{x \to x^*} \frac{|g(x)|}{|x - x^*|} = 0. \tag{5.7.11}$$

It is noted that since $f(x^*) = x^*$, (5.7.10) implies that $g(x^*) = x^*$. Our hope is to prove that if $\mathcal{M}(\rho(J(x^*))) < 1$, or the origin for

$$y(n+1) = J(x^*)y(n)$$

is asymptotically stable, then the equilibrium point x^* of (5.7.1) is asymptotically stable. Note that by introducing the transformation $y = x - x^*$, we may assume without loss of generality that the fixed point is at the origin. Thus we have the following general theorem.

Theorem 5.7.1. *Assume* $\mathcal{M}(\rho(A)) < 1$ *where* A *is an* $k \times k$ *constant matrix (or the zero solution of the linear part is (AS)). Assume the continuous function* $g(x)$ *satisfies*

$$\lim_{x \to 0} \frac{|g(x)|}{|x|} = 0, \tag{5.7.12}$$

with $g(0) = 0$. *Then the origin for the perturbed non-linear system*

$$x(n+1) = Ax(n) + g(x(n)), \quad x(0) = x_0, \ n \geq 0 \tag{5.7.13}$$

is (AS).

Proof. By Theorem 5.4.3, there are positive constants G^* and $\xi \in (0, 1)$ such that $||A||^n \leq G^*\xi^n$. The proof is similar to the proof of Theorem 5.5.2 but with small changes. Thus,

$$|x(n)| \leq ||A||^n|x_0| + \sum_{s=0}^{n-1} ||A||^{n-s-1} |g(x(s))|.$$

Due to condition (5.7.12), for any given $\eta > 0$ there is a $\delta > 0$ so that

$$|g(x)| \leq \eta|x|, \quad \text{whenever} \ |x(n)| < \delta.$$

Hence,

$$|x(n)| \leq G^*\xi^n|x_0| + G^* \sum_{s=0}^{n-1} \xi^{n-s-1} \eta|x(s)|. \tag{5.7.14}$$

Factor G^* and multiply both sides with ξ^{-n} and get

$$\xi^{-n}|x(n)| \leq G^* \left[|x_0| + \sum_{s=0}^{n-1} \xi^{-1}\xi^{-s}\eta|x(s)| \right].$$

Set $y(n) = \xi^{-n}|x(n)|$. Then the above expression reduces to

$$y(n) \leq G^* \left[|x_0| + \sum_{s=0}^{n-1} \xi^{-1}\eta \, y(s) \right]$$

$$\leq |x_0| \prod_{s=0}^{n-1} \left(1 + G^* \xi^{-1} \eta\right)$$

$$= |x_0| \left(1 + G^* \xi^{-1} \eta\right)^n,$$

which implies that

$$||x(n)|| \leq ||x_0|| \left(\xi + \eta G^*\right)^n.$$

As a final step, we choose $\eta < \frac{1-\xi}{G^*}$, so that

$$\xi + \eta G^* < 1.$$

Thus, $||x(n)|| \leq ||x_0||$, $n \geq 0$. Consequently, (5.7.14) holds. From the inequality

$$||x(n)|| \leq ||x_0|| \left(\xi + \eta G^*\right)^n,$$

we obtain stability and $||x_0|| (\xi + \eta G^*)^n \to 0$, as $n \to \infty$, since $\xi + \eta G^* < 1$. This completes the proof. $\square$

Remark 5.3. Consider the autonomous system

$$x(n+1) = f(x(n)), \tag{5.7.15}$$

where $f : \mathbb{R}^k \to \mathbb{R}^k$ with each component being continuously differentiable with $f(0) = 0$. Let $A = J(0)$ be the Jacobian matrix of f at $x = 0$. Rewrite (5.7.15) as

$$x(n+1) = Ax(n) + g(x(n)) \tag{5.7.16}$$

where $g(x(n)) = f(x(n)) - Ax(n)$. Differentiability of f implies $g(x) = o(|x|)$, as $x \to \infty$. Therefore, $g(x)$ satisfies the condition (5.7.12).

The proof of the next theorem is left as an exercise. See Exercise 5.31.

Theorem 5.7.2. *Assume* $\mathcal{M}(\rho(A)) < 1$ *where* A *is an* $k \times k$ *constant matrix (or the zero solution of the linear part is (UAS)). Assume the function* $g(n, x)$ *satisfies*

$$\lim_{x \to 0} \frac{|g(n, x(n))|}{|x(n)|} = 0, \tag{5.7.17}$$

uniformly in $n \geq 0$ *with* $g(0) = 0$. *Then the origin for the perturbed non-linear system*

$$x(n+1) = Ax(n) + g(n, x(n)), \quad x(0) = x_0, \; n \geq 0 \tag{5.7.18}$$

is (ES).

Remark 5.4. (1). If the origin is stable for the linear system $x(n+1) = Ax(n)$ but not asymptotically stable, then it is not necessary that the origin is stable for the non-linear system (5.7.13).

(2). If the origin is unstable for the linear system $x(n+1) = Ax(n)$, then it is unstable for the non-linear system (5.7.13).

We reinforce this by stating the next theorem.

Theorem 5.7.3. *Assume that the zero solution of the homogeneous system $x(n+1) = Ax(n)$ is unstable and (5.7.17) hold. Then the origin is unstable for (5.7.18).*

Example 5.16. The zero solution of the non-linear non-autonomous system

$$x(n+1) = ax(n) + \frac{n}{n+1}x^2(n)$$

is (ES) for $|a| < 1$. First of all $x^* = 0$ is an equilibrium solution. It is obvious the zero solution of the linear part $x(n+1) = ax(n)$ is (AS). Let $g(n,x) = \frac{n}{n+1}x^2(n)$. Then g is continuously differentiable in x. Moreover,

$$\lim_{x \to 0} \frac{|g(n,x)|}{|x|} \le \lim_{x \to 0} \frac{x^2}{|x|} = 0, \text{ uniformly for } n \in [0, \infty).$$

Hence condition (5.7.17) is satisfied. Thus, by Theorem 5.7.1, the zero solution of the non-linear system is (ES). ☐

Linearization says nothing about the stability of the zero solution, as the next example shows.

Example 5.17. Consider the planar system

$$x(n+1) = y(n) - x^2(n)y(n)$$
$$y(n+1) = x(n) - x^3(n).$$

Clearly $(0, 0)$ is an equilibrium solution of the system. Theorem 5.7.1 is inconclusive since $(0, 0)$ is (S) but not (AS) for the linear part

$$x(n+1) = y(n)$$
$$y(n+1) = x(n).$$

To see this, we rewrite the system in the form

$$\begin{pmatrix} x(n+1) \\ y(n+1) \end{pmatrix} = \begin{pmatrix} 0 & 1 \\ 1 & 0 \end{pmatrix} \begin{pmatrix} x(n) \\ y(n) \end{pmatrix} + \begin{pmatrix} -x^2(n)y(n) \\ -x^3(n) \end{pmatrix}.$$

The eigenvalues of

$$A = J(0,0) = \begin{pmatrix} 0 & 1 \\ 1 & 0 \end{pmatrix}$$

are ± 1, which mean the origin is (S) for the linear system. Moreover, by Remark 5.3 the non-linear function g satisfies condition (5.7.12) and we conclude that Theorem 5.7.1 is inconclusive regarding the stability of the zero solution of the non-linear system. In Chapter 6, we will reconsider this example and show its zero solution is stable using Lyapunov function. $\qquad\qquad\square$

Next we provide an example in which we consider the predator-prey model.

Example 5.18. Consider the exponential discrete predator-prey model

$$\begin{cases} x(n+1) = x(n)e^{r_1 - a_{11}x(n) - a_{12}y(n)} \\ y(n+1) = y(n)e^{r_2 - a_{21}x(n) - a_{22}y(n)} \end{cases} \tag{5.7.19}$$

where $x(n) \geq 0$ and $y(n) \geq 0$ stands for population densities of a prey and a predator, respectively, at time n. The positive constants r_1, r_2 denote the intrinsic growth rates of the respective species and $a_{ij} > 0$ represents the *contact parameters* of species $j = 1, 2$ on species $i = 1, 2$. In the model (5.7.19), species quantity of the next generation is found by taking the current species quantity and multiplying it by a growth function, which is exponential and depends on both species. For the sake of reducing the number of parameters in the model, we write (5.7.19) in dimensionless form by letting

$$u(n) = a_{11}x(n), \quad v(n) = a_{22}y(n), \quad c_{12} = \frac{a_{12}}{a_{22}}, \quad c_{21} = \frac{a_{21}}{a_1}.$$

Then the first Eq. (5.7.19) is transformed into

$$\frac{u(n+1)}{a_{11}} = \frac{u(n)}{a_{11}}e^{r_1 - a_{11}\frac{u(n)}{a_{11}} - a_{12}\frac{v(n)}{a_{22}}}.$$

In a similar fashion, the second equation in the model is transformed into

$$\frac{v(n+1)}{a_{22}} = \frac{v(n)}{a_{22}}e^{r_2 - a_{21}\frac{u(n)}{a_{11}} - a_{22}\frac{v(n)}{a_{22}}}.$$

After simplifying terms we arrive at the nondimensionalized system

$$\begin{cases} u(n+1) = u(n)e^{r_1 - u(n) - \frac{a_{12}}{a_{22}}v(n)}, \\ v(n+1) = v(n)e^{r_2 - \frac{a_{21}}{a_{11}}u(n) - v(n)}. \end{cases}$$

Since this transformed system is symmetric, the system can be written as

$$\begin{cases} u(n+1) = u(n)e^{r-u(n)-c_1v(n)}, \\ v(n+1) = v(n)e^{r-c_1u(n)-v(n)}, \end{cases} \tag{5.7.20}$$

where $r_1 = r_2 = r$, and $c_1 = \frac{a_{12}}{a_{22}} = \frac{a_{21}}{a_{11}}$.

Next, we search for the equilibrium solutions (u^*, v^*) of (5.7.20). Set the first and the second equation in (5.7.20) to u^*, and v^*, respectively and obtain

$$u^*\left[1 - e^{r-u^*-c_1v^*}\right] = 0,$$
$$v^*\left[1 - e^{r-c_1u^*-v^*}\right] = 0.$$

From the first equation we get,

$$u^* = 0 \text{ or } r - u^* - c_1v^* = 0.$$

Similarly, from the second equation one has

$$v^* = 0 \text{ or } r - c_1u^* - v^* = 0.$$

If $u^* = 0$, then from the first term in the second equation we obtain $v^* = 0$ and we have the equilibrium solution $(0, 0)$. On the other hand, if we substitute $u^* = 0$ into the second term of the second equation we arrive at $v^* = r$. Consequently, we have the second equilibrium solution $(0, r)$. Similarly, if $v^* = 0$, then from the first term in the first equation we obtain $u^* = 0$, and we get the equilibrium solution $(0, 0)$ that we already have. If we substitute $v^* = 0$ into the second term of the first equation we arrive at $u^* = r$, and hence we have the third equilibrium solution $(r, 0)$. Next we find u^* and v^* as solutions of $r - u^* - c_1v^* = 0$ and $r - c_1u^* - v^* = 0$. Solving for u^* in $r - u^* - c_1v^* = 0$ and then substituting its value in $r - c_1u^* - v^* = 0$, we obtain $v^* = \frac{r(c_1-1)}{c_1^2-1} = \frac{r}{c_1+1}$. Making use of $u^* = r - c_1v^*$ we arrive at $u^* = \frac{r}{c_1+1}$. Thus, the equilibrium solution $(\frac{r}{c_1+1}, \frac{r}{c_1+1})$ is the only one that is of biological interest since both preys and predators coexist. Let

$$f_1(u, v) = ue^{r-u-c_1v}, \quad f_2(u, v) = ve^{r-c_1u-v}.$$

Then

$$\frac{\partial f_1}{\partial u} = (1-u)e^{r-u-c_1v}, \quad \frac{\partial f_1}{\partial v} = -c_1ue^{r-u-c_1v},$$
$$\frac{\partial f_2}{\partial u} = -c_1ve^{r-c_1u-v}, \quad \frac{\partial f_2}{\partial v} = (1-v)e^{r-c_1u-v}.$$

Let $(u^*, v^*) = (\frac{r}{c_1+1}, \frac{r}{c_1+1})$. Then

$$a = \frac{\partial f_1}{\partial u}(u^*, v^*) = 1 - \frac{r}{c_1+1}, \quad b = \frac{\partial f_1}{\partial v}(u^*, v^*) = -c_1\frac{r}{c_1+1},$$

$$c = \frac{\partial f_2}{\partial u}(u^*, v^*) = -c_1 \frac{r}{c_1 + 1}, \quad d = \frac{\partial f_2}{\partial v}(u^*, v^*) = 1 - \frac{r}{c_1 + 1}.$$

Moreover,

$$p = a + d = 2 - 2\frac{r}{c_1 + 1},$$

and

$$q = ad - bc = 1 - 2\frac{r}{c_1 + 1} + \frac{r^2}{(c_1 + 1)^2}(1 - c_1^2).$$

Next we apply condition (5.7.5), $q > p - 1$ and get

$$1 - 2\frac{r}{c_1 + 1} + \frac{r^2}{(c_1 + 1)^2}(1 - c_1^2) > 2 - 2\frac{r}{c_1 + 1} - 1,$$

which reduces to

$$(1 - c_1^2)\frac{r^2}{(c_1 + 1)^2} > 0.$$

This implies that

$$0 < c_1 < 1.$$

Applying condition (5.7.8), $|p| < 2$, one has

$$2\left|1 - \frac{r}{c_1 + 1}\right| < 2.$$

This simplifies to

$$0 < \frac{r}{c_1 + 1} < 2. \tag{5.7.21}$$

Note that since $\inf\{c_1 : 0 < c_1 < 1\} = 0$, then for (5.7.21) to hold, we must have

$$\frac{r}{c_1 + 1} < \frac{r}{\inf\{c_1 + 1\}} = \frac{r}{0 + 1} < 2.$$

We conclude that

$$0 < r < 2.$$

Thus, by Lemma 5.1, the equilibrium solution $(\frac{r}{c_1+1}, \frac{r}{c_1+1})$ is asymptotically stable if and only if $0 < c_1 < 1$ and $0 < r < 2$. $\square$

We have seen that linearization failed to yield any results regarding stability, which forces one to seek other alternatives. Polar transformation is another alternative that, in some cases, can be used to reduce a complicated problem to a simpler one in terms of polar coordinates. The method is illustrated in the next example.

Example 5.19. For $n \in \mathbb{Z}^+$, consider the non-linear system

$$
\begin{cases}
u(n+1) = 2u(n)v(n), \\
v(n+1) = v^2(n) - u^2(n).
\end{cases}
\tag{5.7.22}
$$

Clearly $(0,0)$ is the only equilibrium solution of the system (5.7.22) and the method of linearization does not apply here. We resort to transforming the problem using polar coordinates

$$
u(n) = r(n)\cos(\theta(n)), \quad v(n) = r(n)\sin(\theta(n)),
$$

where

$$
r(n) = \sqrt{u^2(n) + v^2(n)}, \quad \theta(n) = \arctan\left(\frac{v(n)}{u(n)}\right).
$$

Next we write $u(n+1)$ and $v(n+1)$ in terms of r and θ.

$$
u(n+1) = 2r^2(n)\cos(\theta(n))\sin(\theta(n)) = r^2(n)\sin(2\theta(n)),
$$

and

$$
v(n+1) = r^2(n)\sin^2(\theta(n)) - r^2(n)\cos^2(\theta(n)).
$$

Using the identity $\cos(2z) = \cos^2(z) - \sin^2(z)$, we obtain

$$
v(n+1) = -r^2(n)\cos(2\theta(n)).
$$

Consequently,

$$
\begin{aligned}
r(n+1) &= \sqrt{u^2(n+1) + v^2(n+1)} \\
&= \sqrt{\left(r^2(n)\sin(2\theta(n))\right)^2 + \left(-r^2(n)\cos(2\theta(n))\right)^2} \\
&= r^2(n).
\end{aligned}
$$

In addition,

$$
\theta(n+1) = \arctan\left(\frac{v(n+1)}{u(n+1)}\right) = \arctan(-\cot(2\theta(n))).
$$

Let $r(0) = r_0$, and our aim now is to solve

$$
r(n+1) = r^2(n), \quad r(0) = r_0, \quad n \in \mathbb{Z}^+.
$$

We iterate on $n = 0, 1, \ldots$ and obtain,

$$r(1) = r_0^2$$
$$r(2) = (r(1))^2 = r_0^{2^2}$$
$$\vdots$$
$$r(k) = r_0^{2^k}.$$

Thus, we can see that the general solution is given by

$$r(n) = r_0^{2^n}, \quad n \in \mathbb{Z}^+.$$

- If $0 < r_0 < 1$, then

$$\lim_{n \to \infty} r(n) = 0,$$

which in turn means that $\sqrt{u^2(n) + v^2(n)} \to 0$, as $n \to \infty$. Consequently, solutions starting inside the unit circle will spiral toward the equilibrium solution $(0, 0)$ and hence the origin is asymptotically stable.

- If $r_0 > 1$, then

$$\lim_{n \to \infty} r(n) = \infty,$$

which implies that $\sqrt{u^2(n) + v^2(n)} \to \infty$, as $n \to \infty$. Consequently, solutions starting outside the unit circle spiral away from $(0, 0)$ and hence the origin is unstable.

- If $r_0 = 1$, then

$$\lim_{n \to \infty} r(n) = 1,$$

which implies that $\sqrt{u^2(n) + v^2(n)} \to 1$, as $n \to \infty$. Consequently, solutions starting on the unit circle will stay on it and in this case $(0, 0)$ is stable. The expression of the angle θ is complex. However, the angle $\theta(n)$ influences the behavior of both $u(n + 1)$ and $v(n + 1)$ by determining the balance between cosine and sine terms in the equations. The specific role of $\theta(n)$ depends on its value at each step in the sequence and how it affects the magnitudes and directions of $u(n + 1)$ and $v(n + 1)$. The evolution of $\theta(n)$ will determine the overall behavior of the system as it progresses through the sequence. □

We end this section with the remark that another type of method must be used when the equilibrium solution of an associated linear difference equation is a center, and therefore nothing can be concluded from the method of linearization regarding the stability of the associated non-linear system. The method that we will look at in Chapter 6 is called the method of Lyapunov functions. The theory of such a method will be developed and studied in broad detail to suit autonomous and non-autonomous systems.

5.7.1 Exercises

Exercise 5.30. Construct a 2×2 matrix A so that condition (5.7.5) is satisfied.

Exercise 5.31. Assume $\mathcal{M}(\rho(A)) < 1$ where A is an $k \times k$ constant matrix (or the zero solution of the linear part is (UAS)). Assume the $g(n, x)$ satisfies

$$\lim_{x \to 0} \frac{|g(n, x(n))|}{|x(n)|} = 0$$

uniformly in $n \geq 0$ with $g(0) = 0$. Then the origin for the perturbed non-linear system

$$x(n+1) = Ax(n) + g(n, x(n)), \quad x(0) = x_0, \; n \geq 0$$

is (ES).

Exercise 5.32. For constants a and b, show the origin is (AS) for the non-linear system

$$x_1(n+1) = \frac{1}{3} x_1(n) + a x_1(n) x_2(n),$$

$$x_2(n+1) = \frac{1}{2} x_2(n) + b x_1(n) x_2(n).$$

Exercise 5.33. For nonzero constant α, show the zero solution is unstable for the non-linear system

$$x(n+1) = \alpha \sin(x(n)) + y(n) + x(n) y^3(n),$$
$$y(n+1) = -x(n) \cos(y(n)) + \alpha(e^{y(n)} - 1).$$

Exercise 5.34. Let α be a constant and consider the system

$$x_1(n+1) = (1 - x_1(n)) x_1(n) + \alpha x_2(n)$$
$$x_2(n+1) = \frac{1}{2} x_2(n) + x_1(n).$$

(a) Show the two equilibrium solutions of the system are $(0, 0)$ and $(2\alpha, 4\alpha)$.
(b) Determine the matrix $A = J(0, 0)$ and find its eigenvalues.
(c) Show that if $\alpha < -\frac{1}{2}$, then the origin is unstable.
(d) Show that if $-\frac{1}{2} < \alpha < 0$, then the origin is (AS).
(e) Show that if $0 < \alpha < 3$, then the origin is unstable.
(f) Show that if $\alpha > 3$, then the origin is unstable.
(g) Show that Theorem 5.7.1 is inconclusive regarding the stability of the zero solution for $\alpha = -\frac{1}{2}$, $\alpha = 0$, or $\alpha = 3$.

Exercise 5.35. Consider the system in Exercise 5.34.

(a) Compute the eigenvalues λ_1, λ_2 of the matrix

$$A = J(2\alpha, 4\alpha).$$

(b) Use condition (5.7.5) to put condition(s) on α so that $|\lambda_1| < 1$ and $|\lambda_2| < 1$.

Exercise 5.36. For $n \in \mathbb{Z}^+$, consider the non-linear system

$$\begin{cases} u(n+1) = u^2(n) - v^2(n), \\ v(n+1) = 2u(n)v(n). \end{cases}$$

Use polar coordinates to study the stability of the origin. In addition, find an equation that involves $\theta(n+1)$ and $\theta(n)$ and find a close form for $\theta(n)$.

Exercise 5.37. Let all parameters in this example be defined as in Example 5.18 and consider the prey-predator model

$$\begin{cases} u(n+1) = ru(n)[1 - u(n) - c_1v(n)], \\ v(n+1) = rv(n)[1 - c_1u(n) - v(n)]. \end{cases}$$

Find all four equilibrium solutions and provide an analysis of the stability of each one of them by placing conditions on some of the parameters involved.

Exercise 5.38 (Host-parasitoid model). This exercise is about host-parasitoid model, which describes how parasites interact with their host organisms, how they affect one another, and how these interactions can influence various ecological and evolutionary processes. The host is the organism that provides a habitat and resources for the parasite. It can be any living organism, such as animals, plants, or even microorganisms. The host is usually the organism that suffers the detrimental effects of the parasite's presence and activities. The parasite is an organism that lives in or on another organism (the host) and relies on the host for resources and/or a place to live. Parasites come in various forms, including viruses, bacteria, fungi, protozoa, helminths (worms), and more. Additionally, the model assumes that parasitoids are randomly distributed among the available hosts and that the number of interactions between host larvae and parasitoids is proportional to both host population size and parasitoids population size. Then, the host-parasitoid model is

$$\begin{cases} u(n+1) = ru(n)e^{-dv(n)}, \\ v(n+1) = ru(n)(1 - e^{-dv(n)}), \end{cases}$$

where $u(n)$ denotes the host population size and $v(n)$ denotes the parasitoid population size. The model assumes that the host lays enough eggs to produce $r > 0$ larvae. The term $e^{-dv(n)}$ is the fraction of host larvae surviving to adulthood for a positive constant b.

(a) Show the two equilibrium solutions are $(0, 0)$ and $(\frac{1}{d(r-1)} \log(r), \frac{1}{d} \log(r))$. Make use of condition (5.7.8) to show that:

(b) the extinction solution $(0, 0)$ is asymptotically stable for $r < 1$,

(c) the equilibrium solution $(\frac{1}{d(r-1)} \log(r), \frac{1}{d} \log(r))$ is unstable for $r > 1$, and hence both populations will oscillate.

Exercise 5.39 (Revisited Host-parasitoid model). We just saw that the model in Exercise 5.38 does not allow for coexistence. To have coexistence we make the additional assumption that adult hosts may live longer than one generation. With this in mind, if we let $a > 0$ be the portion of hosts that survive to the next generation, then the host-parasitoid model becomes

$$\begin{cases} u(n+1) = ru(n)e^{-dv(n)} + au(n), \\ v(n+1) = ru(n)(1 - e^{-dv(n)}). \end{cases}$$

(a) Show the two equilibrium solutions are the extinction one $(0, 0)$ and co-existence one $\left(\frac{1}{d(a+r-1)} \log(\frac{r}{1-a}), \frac{1}{d} \log(\frac{r}{1-a})\right)$, which exist if and only if $r + a > 1$. Conclude that $0 < a < 1$. Make use of condition (5.7.8) to show that,

(b) the extinction solution $(0, 0)$ is asymptotically stable if and only if $a + r < 1$,

(c) the equilibrium solution $\left(\frac{1}{d(a+r-1)} \log(\frac{r}{1-a}), \frac{1}{d} \log(\frac{r}{1-a})\right)$, is asymptotically stable if and only if

$$\frac{(a+r)(1-a)}{a+r-1} \log(\frac{r}{1-a}) < 1.$$

5.8 Floquet theory

Floquet theory is used to analyze the behavior of periodic linear and non-linear dynamical systems. It is particularly useful in understanding the stability and behavior of discrete systems that exhibit periodicity. Floquet theory is named after Gaston Floquet, a French mathematician, and it deals with the study of periodic solutions in differential equations.

In this section we take up the study of periodic linear systems with time varying matrices. Thus, we consider the homogeneous system

$$x(n+1) = A(n)x(n), \tag{5.8.1}$$

where the matrix $A(n)$ is periodic of period T (constant), that is,

$$A(n+T) = A(n), \text{ for all } n \in \mathbb{N} \text{ and } T > 0,$$

where T is the least number. For the rest of this section we assume $A(n)$ is periodic matrix with period T. When A is constant, we see that its fundamental matrix solution is formed from the eigenvalues of A and their corresponding eigenvectors. This is not the case when the matrix A depends on n. However, the eigenvalues will play an important role in the stability when the matrix $A(n)$ is periodic. Consequently, our ultimate goal here is to find a close expression for the fundamental matrix. To motivate the subject, we consider the scalar difference equation

$$x(n+1) = a(n)x(n) \tag{5.8.2}$$

where $a : \mathbb{N} \to \mathbb{N}$, is periodic with minimal period T. Then every solution $\phi(n)$ of (5.8.2) has the form

$$\phi(n) = c \prod_{i=0}^{n-1} a(i). \tag{5.8.3}$$

Set the constant $c = 1$ and for a constant λ, we observe that

$$\prod_{i=n}^{n+T-1} a(i) = \lambda^T, \quad \text{a constant}$$

for all $n \in \mathbb{N}$, since $a(n+T) = a(n)$. Using (5.8.3) with $c = 1$ we have

$$\phi(n+T) = \prod_{i=0}^{n-1} a(i) \prod_{i=n}^{n+T-1} a(i) = \phi(n)\lambda^T. \tag{5.8.4}$$

The number λ in (5.8.4) has a profound effect on the behavior of the solution ϕ. If ϕ is periodic of period T, that is $\phi(n+T) = \phi(n)$ then (5.8.4) implies that

$$\lambda^T = 1.$$

We will say much more later on the number λ. The above statement can be better illustrated with the following example of a scalar difference first-order equation.

Example 5.20. Consider the 2-periodic ($T = 2$) scaler difference equation

$$x(n+1) = \cos(n\pi)x(n), \quad n \in \mathbb{N}.$$

Then, for a constant c its solution is given by

$$\phi(n) = c(-1)^{\frac{n(n-1)}{2}} = c(-1)^{\frac{n^2-n}{2}}.$$

On the other hand,

$$\phi(n+2) = c(-1)^{\frac{(n+2)^2-n-2}{2}}$$
$$= c(-1)^{\frac{(n+2)^2-n-2}{2}}$$

$$= \phi(n)(-1)^{2n}(-1)$$
$$= \phi(n)(i)^2.$$

Hence, in this example, $\lambda = i$ and $T = 2$. □

A deeper understanding of the form of the solutions of (5.8.2) can be revealed by letting the new function

$$p(n) = \phi(n)\lambda^{-n}.$$

Then the function p is periodic with period T since

$$p(n+T) = \phi(n+T)\lambda^{-n-T} = \phi(n)\lambda^T\lambda^{-n-T} = \phi(n)\lambda^{-n} = p(n).$$

For now, this shows that any solution of (5.8.2) can be written in the form

$$\phi(n) = p(n)\lambda^n. \tag{5.8.5}$$

Eq. (5.8.5) is significant in the sense that it permits us to study the behavior of all solutions on $-\infty < n < +\infty$. Now we go back to the periodic system (5.8.1)

Theorem 5.8.1. *Let $A(n+T) = A(n)$ for all $n \in \mathbb{N}$. If $\Phi(n)$ is a fundamental matrix of (5.8.1) then so is $\Phi(n+T)$ and there exists a nonsingular constant matrix B such that*

(a) $\Phi(n+T) = \Phi(n)B$ *for all $n \in \mathbb{N}$*
(b) $B = A(T-1)A(T-2)\dots A(0)$.

Proof. (a) Set $Z(n) = \Phi(n+T)$. Then

$$Z(n+1) = \Phi(n+T+1) = A(n+T)\Phi(n+T) = A(n)Z(n).$$

In addition, $\det(Z(n)) = \det(\Phi(n+T)) \neq 0$, since $\Phi(n)$ is a fundamental matrix. This proves that $\Phi(n+T)$ is also a fundamental matrix. The existence of the matrix B follows from Theorem 4.4.2. This completes the proof of (a). For the proof of (b) we iterate over $n = 0, 1, \dots, T-1$, the equation

$$\Phi(n+1) = A(n)\Phi(n).$$

We begin with $n = T - 1$, and go down the list until we reach $n = 0$.

$$\Phi(T) = A(T-1)\Phi(T-1)$$
$$\Phi(T-1) = A(T-2)\Phi(T-2)$$
$$\Phi(T-2) = A(T-3)\Phi(T-3)$$
$$\vdots$$
$$\Phi(2) = A(1)\Phi(1)$$
$$\Phi(1) = A(0)\Phi(0).$$

By repeatedly substituting $\Phi(T-1)$ and then $\Phi(T-2)$ and so on until $\Phi(1)$ into $\Phi(T) = A(T-1)\Phi(T-1)$, we obtain

$$\Phi(T) = A(T-1)A(T-2)\ldots A(2)A(1)A(0)\Phi(0).$$

Now, if we let $n = 0$ in $\Phi(n+T) = \Phi(n)B$, we obtain

$$\Phi(T) = \Phi(0)B.$$

Equating this equation with $\Phi(T)$, that was obtained from the prior iteration, we have

$$\Phi(0)B = A(T-1)A(T-2)\ldots A(2)A(1)A(0)\Phi(0).$$

Since $\Phi(0)$ is invertible we arrive at

$$B = A(T-1)A(T-2)\ldots A(2)A(1)A(0).$$

This completes the proof. $\qquad\qquad\qquad\qquad\qquad\qquad\qquad\qquad\square$

The next definition is about Floquet multipliers.

Definition 5.8.1. Let $\Phi(n)$ be the fundamental matrix for the Floquet system (5.8.1). Then the eigenvalues $\mu_1, \mu_2, \ldots, \mu_k$ of

$$B = A(T-1)A(T-2)\ldots A(0)$$

are called the *Floquet multipliers* of the Floquet system (5.8.1).

Corollary 5.4. *The Floquet multipliers μ_j are an intrinsic property of the system (5.8.1) and do not depend on the choice of the fundamental matrix $\Phi(n)$.*

Proof. Suppose $\Psi(n)$ is another fundamental matrix. Then by Theorem 5.8.1 there is a nonsingular matrix B such that

$$\Psi(n+T) = \Psi(n)B.$$

Now by Theorem 4.4.2

$$\Psi(n) = \Phi(n)C,$$

where C is a nonsingular matrix. Thus,

$$\Psi(n+T) = \Phi(n+T)C,$$

or

$$\Psi(n)B = (\Phi(n)B)C.$$

This gives

$$\Phi(n)CB = \Phi(n)BC, \quad \text{or } CB = BC.$$

As a consequence,

$$CBC^{-1} = B,$$

so the eigenvalues of B and C are the same. This completes the proof. $\square$

We provide the following example.

Example 5.21. Consider the system

$$x(n+1) = \begin{pmatrix} 0 & \frac{2+(-1)^n}{2} \\ \frac{2+(-1)^n}{2} & 0 \end{pmatrix} x(n).$$

The matrix A is periodic with period $T = 2$, and

$$B = A(1)A(0) = \begin{pmatrix} 0 & \frac{1}{2} \\ \frac{1}{2} & 0 \end{pmatrix} \begin{pmatrix} 0 & \frac{3}{2} \\ \frac{3}{2} & 0 \end{pmatrix} = \begin{pmatrix} \frac{3}{4} & 0 \\ 0 & \frac{3}{4} \end{pmatrix}.$$

Hence, the Floquet multipliers are $\mu_1 = \frac{3}{4}$, $\mu_2 = \frac{3}{4}$. $\square$

The next result is essential to proving the Floquet Theorem.

Lemma 5.2. *If B is a $k \times k$ matrix and nonsingular, then there is a matrix C so that $C^m = B$.*

Proof. We will only do the proof when the matrix A is diagonalizable. That is, there exists an invertible matrix P and a diagonal matrix D such that

$$B = PDP^{-1}$$

where the matrix D has the eigenvalues of B along its diagonal, and P is a matrix whose columns are the corresponding eigenvectors of B.

Now, we want to find a matrix C such that $C^m = B$. We can achieve this by taking the mth root of both sides of the diagonalized form of B. That is

$$C^m = (PDP^{-1})^m.$$

Or,

$$C^m = (PDP^{-1})(PDP^{-1})\dots(PDP^{-1}) = PD^m P^{-1}$$

where

$$D^m = \begin{pmatrix} \lambda_1^m & 0 & \cdots & 0 \\ 0 & \lambda_2^m & \cdots & 0 \\ \vdots & \vdots & \ddots & \vdots \\ 0 & 0 & \cdots & \lambda_k^m \end{pmatrix}$$

with λ_i, $i = 1, 2, \ldots, k$ are the eigenvalues of B. So, by setting $C = PD^{1/m}P^{-1}$, we have a matrix C such that $C^m = B$, where

$$
D^{1/m} = \begin{pmatrix}
\lambda_1^{1/m} & 0 & \cdots & 0 \\
0 & \lambda_2^{1/m} & \cdots & 0 \\
\vdots & \vdots & \ddots & \vdots \\
0 & 0 & \cdots & \lambda_k^{1/m}
\end{pmatrix}.
$$

This completes the proof. $\qquad\qquad\square$

For a proof of the case when the matrix A is not diagonalizable we refer to any graduate textbook on linear algebra. Such a scenario happens when A has repeated eigenvalues, and one is not able to generate k linearly independent eigenvectors.

Now we are ready to state and prove Floquet's Theorem.

Theorem 5.8.2 (Floquet's Theorem). *Every fundamental matrix of* (5.8.1) *has the form*

$$
\Phi(n) = P(n)C^n,
$$

where $P(n)$ is T-periodic matrix, and C is a constant $k \times k$ matrix.

Proof. Let $\Phi(n)$ be a fundamental matrix of (5.8.1). Then

$$
\begin{aligned}
\triangle(\Phi^{-1}(n)\Phi(n+T)) &= \Phi^{-1}(n+1)\Phi(n+1+T) - \Phi^{-1}(n)\Phi(n+T) \\
&= \Phi^{-1}(n)A^{-1}(n)A(n+T)\Phi(n+T) - \Phi^{-1}(n)\Phi(n+T) \\
&= \Phi^{-1}(n)\Phi(n+T) - \Phi^{-1}(n)\Phi(n+T) \\
&= 0.
\end{aligned}
$$

Thus,

$$
\Phi^{-1}(n)\Phi(n+T) = B \quad \text{(Constant matrix)},
$$

or

$$
\Phi(n+T) = \Phi(n)B.
$$

Since $\Phi^{-1}(n)$ and $\Phi(n+T)$ are nonsingular, so B is nonsingular. Thus, by Lemma 5.2 $B = C^T$ for some constant matrix C. Left to show $P(n)$ is periodic of period T. We define $P(n) = \Phi(n)C^{-n}$. Then

$$
\begin{aligned}
P(n+T) &= \Phi(n+T)C^{-n-T} \\
&= \Phi(n)BC^{-T}C^{-n} \\
&= \Phi(n)C^{-n}
\end{aligned}
$$

$$= P(n).$$

This completes the proof. □

Theorem 5.8.3. *Let μ be a Floquet multiplier of (5.8.1). Then there exists a nontrivial solution $x(n)$ of (5.8.1) such that $x(n + T) = \mu x(n)$.*

Proof. Let μ be an eigenvalue of the matrix B with corresponding eigenvector b. Assume $\Phi(n)$ is a fundamental matrix of the Floquet system given by (5.8.1) such that $x(n) = \Phi(n)b$. Then

$$x(n + T) = \Phi(n + T)b$$
$$= \Phi(n)Bb$$
$$= \mu\Phi(n)b$$
$$= \mu x(n).$$

This completes the proof. □

Example 5.22. In this example we make use of Theorem 5.8.3 to find the Floquet multiplier of the scalar difference equation

$$x(n + 1) = \cos(n\pi)x(n), \quad n \in \mathbb{N}.$$

First of all, the equation is Floquet with $T = 2$. In Example 5.20 we found the solution to be

$$x(n) = c(-1)^{\frac{n(n-1)}{2}} = c(-1)^{\frac{n^2-n}{2}}.$$

Thus, for nonzero constant c, we have from Theorem 5.8.3 that

$$\mu = x^{-1}(0)x(2) = \frac{1}{c}c(-1)^{2/2} = -1. \quad □$$

The next lemma is necessary for the next results regarding stability.

Lemma 5.3. *Let $\Phi(n) = P(n)C^n$ as defined in Theorem 5.8.2. Then $y(n)$ is a solution of (5.8.1) if and only if $z(n) = P^{-1}(n)y(n)$ is a solution of $z(n + 1) = Cz(n)$.*

Proof. Assume $y(n)$ is a solution of (5.8.1) and there is a column vector x_0 so that

$$y(n) = \Phi(n)x_0 = P(n)C^n x_0.$$

Define $z(n) = P^{-1}(n)y(n) = C^n x_0$. Then

$$z(n + 1) = C^{n+1}x_0$$
$$= C^n C x_0$$

$$= Cz(n).$$

Hence, $z(n)$ is a solution of $z(n+1) = Cz(n)$. Since every step can be reversed, the converse holds. This completes the proof. $\qquad\square$

We end this section by taking advantage of Floquet theory that we developed to provide results concerning the stability and behavior of discrete systems that exhibit periodicity.

Theorem 5.8.4. *The zero solution of* (5.8.1) *is*

    (i) *stable if and only if the Floquet multipliers μ satisfy $|\mu| \leq 1$ and there are a complete set of eigenvectors for any multiplier of modulus 1,*

    (ii) *asymptotically stable if and only if $|\mu| < 1$ for every μ,*

    (iii) *unstable if there is at least one Folquet multiplier, μ such that $|\mu| > 1$.*

Proof. Set $\Phi(n) = P(n)C^n$ as was defined in Theorem 5.8.2. It follows from Lemma 5.3 that the Floquet system is equivalent to the autonomous system $z(n+1) = Cz(n)$. Consequently, the stability of the zero solution follows from the results of Section 5.4, since the eigenvalues of C are the mth root of the matrix B, from Lemma 5.2. This completes the proof. $\qquad\square$

5.8.1 Non-homogeneous systems

In this subsection, we extend the Floquet theory of Section 5.8 to non-homogeneous periodic linear systems of the form

$$x(n+1) = A(n)x(n) + g(n) \qquad (5.8.6)$$

where $A(n)$ is an $k \times k$ matrix and $g(n)$ is an $k \times 1$ vector that are defined on $\mathbb{N}$. We assume $A(n)$ and $g(n)$ to be T-periodic. That is, there is a least positive integer T such that

$$A(n+T) = A(n), \quad g(n+T) = g(n), \quad n \in \mathbb{N}. \qquad (5.8.7)$$

We begin with the following theorem.

Theorem 5.8.5. *Let $A(n)$ and $g(n)$ be as defined in* (5.8.7). *Then* (5.8.6) *has a unique solution $x(n)$ with $x(n_0) = \eta$ for all $n \in \mathbb{N}_{n_0}$.*

Proof. Let $\| \cdot \|$ be a suitable matrix norm. For existence, it suffices to show that solutions remain bounded for all $n \in [0, T-1]$. It is easy to see that if x is a solution of (5.8.6), then it is given by

$$x(n) = \eta + \sum_{s=n_0}^{n-1} [A(s)x(s) + g(s)]$$

with $x(n_0) = \eta$. Let $n_0 = 0$ and $n \in [0, T - 1]$ and set

$$K_1 = |\eta|, \quad K_2 = \max_{n \in [0, T-1]} |g(n)| T, \text{ and } K_3 = \max_{n \in [0, T-1]} ||A(n)||.$$

Then, for $n \in [0, T - 1]$ we have

$$|x(n)| \leq |\eta| + \left| \sum_{s=n_0}^{n-1} [A(s)x(s) + g(s)] \right|$$

$$\leq K_1 + \sum_{s=0}^{T-1} ||A(s)|| |x(s)| + \sum_{s=0}^{T-1} |g(s)|$$

$$\leq K_1 + \max_{n \in [0, T-1]} ||A(s)|| \sum_{s=0}^{T-1} |x(s)| + \max_{n \in [0, T-1]} |g(n)| \sum_{s=0}^{T-1} 1$$

$$\leq K_1 + K_3 \sum_{s=0}^{T-1} |x(s)| + T K_2$$

$$= \left(K_1 + T K_2 \right) + K_3 \sum_{s=0}^{T-1} |x(s)|. \tag{5.8.8}$$

Applying Theorem 5.5.1 (Gronwall's inequality) we arrive at the inequality

$$|x(n)| \leq \left(K_1 + K_2 \right) e^{\sum_{s=0}^{T-1} K_3}$$

$$= \left(K_1 + K_2 \right) e^{T K_3}. \tag{5.8.9}$$

From (5.8.9) we see that solutions are bounded for all $n \in [0, T - 1]$ and hence exist. This completes the proof of existence. For the uniqueness we assume the existence of two solutions ϕ_1 and ϕ_2 of (5.8.6) passing through $(0, \eta)$. Then imitating the proof the existence we arrive at

$$|\phi_1(n) - \phi_2(n)| \leq \left| \sum_{s=0}^{T-1} A(s)[\phi_1(s) - \phi_2(s)] \right|$$

$$\leq K_3 \sum_{s=0}^{T-1} |\phi_1(s) - \phi_2(s)|.$$

Applying Gronwall's inequality it follows that

$$|\phi_1(n) - \phi_2(n)| = 0 \text{ for all } n \in [0, T - 1]$$

and hence $\phi_1(n) = \phi_2(n)$ for all $n \in [0, T - 1]$ and the proof is complete. $\quad \square$

Theorem 5.8.6. *A solution x of the periodic system (5.8.6) is periodic, if and only if*

$$x(0) = x(T), \quad \text{for all } T \geq 1.$$

Proof. Since the solution x is periodic, it follows that $x(0) = x(T)$. For the converse, we assume x is a solution of (5.8.6) with $x(n) = x(0)$ for $n \in \mathbb{N}$. Define another solution y of (5.8.6) by $y(n) = x(n + T)$. Then it follows from $x(n) = x(0)$ that

$$y(0) = x(T) = x(0).$$

Hence, the two solutions have the same initial values, and by uniqueness it follows from Theorem 5.8.5, that $x(n) = y(n) = x(n + T)$, for all $n \in \mathbb{N}$. This completes the proof. $\square$

To better understand the nature of solutions of the non-homogeneous system (5.8.6), we make the following definition regarding its corresponding homogeneous system.

Definition 5.8.2. The linear homogeneous system

$$x(n + 1) = A(n)x(n) \tag{5.8.10}$$

with $A(n + T) = A(n))$, where A is an $k \times k$ matrix, is said to be *noncritical* with respect to T, if it has *no periodic* solution of period T except the trivial solution $x = 0$. Otherwise, system (5.8.10) is said to be *critical*.

Example 5.23. Consider the system

$$x(n + 1) = \begin{pmatrix} 0 & 2 + (-1)^n \\ 2 - (-1)^n & 0 \end{pmatrix} x(n).$$

Then $A(n + 2) = A(n)$ and the eigenvalues of A are $\pm\sqrt{3}$ for all $n \in \mathbb{N}$. Hence the system has no 2-periodic solution except $x(n) = 0$, and therefore the system is noncritical. $\square$

One more definition that distinguishes between periodic solution and oscillatory solution.

Definition 5.8.3. A solution of (5.8.10) is said to be non-oscillatory if there is at most one change of sign in the solution. If there is an infinite change in sign, then the solution is said to be oscillatory.

The next theorem provides a useful criterion for the periodic solutions of the non-homogeneous system (5.8.6).

Theorem 5.8.7. *Assume (5.8.7). Then the non-homogeneous system (5.8.6) has a T-periodic solution if and only if its corresponding homogeneous system (5.8.10) is noncritical.*

Proof. Let Φ be the fundamental matrix of the homogeneous system (5.8.10) with $\Phi^{-1}(0) = I$. Then by Theorem 4.5.1 the solution of (5.8.6) is given by

$$x(n) = \Phi(n)x_0 + \sum_{s=0}^{n-1} \Phi(n)\Phi^{-1}(s+1)g(s).$$

By Theorem 5.8.6, the solution x will be periodic if and only if

$$x(0) = x_0 = x(T).$$

Thus,

$$x(0) = x_0 = \Phi(T)x_0 + \sum_{s=0}^{T-1} \Phi(T)\Phi^{-1}(s+1)g(s).$$

Factoring x_0 gives

$$(I - \Phi(T))x_0 = \Phi(T) \sum_{s=0}^{T-1} \Phi^{-1}(s+1)g(s).$$

The above equation is an algebraic non-homogeneous system that needs to be solved components wise for every nonzero periodic vector g. This is possible if and only if

$$\det(I - \Phi(T)) \neq 0.$$

This is equivalent to the equation $\Phi(T)x_0 = x_0$, which has only the trivial solution x_0. Let

$$y(n) = \Phi(n)x_0.$$

Then

$$y(T) = \Phi(T)x_0 = y(0) = x_0,$$

can only be satisfied by the trivial solution $y(n) = 0$. Now by Theorem 5.8.2 this is equivalent to the assertion that (5.8.10) has only the trivial solution as a periodic solution of period T. This completes the proof. $\square$

5.8.2 Exercise

Exercise 5.40. Let B be a 2×2 matrix and nonsingular. Show that there is a matrix C such that

$$C^m = B = P^{-1}JP,$$

for the cases:

(a) $J = \begin{pmatrix} \lambda & 1 \\ 0 & \lambda \end{pmatrix}$. (The matrix B has one eigenvalue λ with a single indepen-

dent eigenvector).

(b) $J = \begin{pmatrix} \alpha & \beta \\ -\beta & \alpha \end{pmatrix}$. (The matrix B has complex conjugate eigenvalues of the

form $\alpha \pm i\beta$).

Exercise 5.41. Find the Floquet multiplier for

$$x(n+1) = \cos(n\pi)x(n), \quad n \in \mathbb{N}.$$

Exercise 5.42. Consider the Floquet system

$$x(n+1) = \begin{pmatrix} 0 & \frac{2+(-1)^n}{2} \\ \frac{2-(-1)^n}{2} & 0 \end{pmatrix} x(n).$$

(a) Compute the eigenvalues of the matrix A. Can anything be concluded about the stability of the zero solution from the eigenvalues?
(b) Find the Floquet multipliers and discuss the stability of the zero solution.

Exercise 5.43. Find the Floquet multipliers and discuss the stability of the zero solution of

$$x(n+1) = \begin{pmatrix} \frac{\cos(n\pi)+\sin(n\pi)}{2-\cos(n\pi)} & 0 \\ 1 & 1 \end{pmatrix} \begin{pmatrix} x_1(n) \\ x_2(n) \end{pmatrix}.$$

Exercise 5.44. Find the Floquet multipliers and discuss the stability of the zero solution of

$$x(n+1) = \begin{pmatrix} 1 + \frac{\cos(n\pi)}{2+\sin(n\pi)} & 0 \\ 1 & -1 \end{pmatrix} \begin{pmatrix} x_1(n) \\ x_2(n) \end{pmatrix}.$$

Exercise 5.45. Find the Floquet multipliers and discuss the stability of the zero solution of each of the following equations:

(a)

$$x(n+1) = \frac{8}{9}\cos(2n\pi)x(n)$$

(b)

$$x(n+1) = \Big(\sin(n\pi) + \cos(n\pi)\Big)x(n)$$

(c)

$$x(n+1) = \Big(-1 + \cos(2n\pi)\Big)x(n).$$

Exercise 5.46. Show the system

$$x(n+1) = \begin{pmatrix} 1 & 1 + \cos(n\pi) \\ 0 & -1 \end{pmatrix} x(n)$$

is noncritical.

Exercise 5.47. Let $a(n+2) = a(n)$ and $b(n+2) = b(n)$ for all $n \in \mathbb{N}$ and consider the periodic system

$$x(n+1) = \begin{pmatrix} 0 & 1 \\ -b(n) & -a(n) \end{pmatrix} x(n).$$

Show that the Floquet multipliers μ_1, μ_2 are solutions of

$$\mu^2 + \Big(b(0) + b(1) - a(0)a(1)\Big)\mu + b(0)b(1) = 0.$$

Exercise 5.48. Consider the non-homogeneous system (5.8.6) and assume (5.8.7). Let Φ be the fundamental matrix of the homogeneous system (5.8.10) with $\Phi(0) = I$.

(a) Show that $\det \Phi(n) = \left(\sum_{i=0}^{n-1} \det A(i)\right) \det \Phi(0) \neq 0$. Note that part (a) holds without the periodicity on $A(n)$.

(b) Show that $\Phi(n+T) = \Phi(n)\Phi(T)$.

(c) Show that

$$x(n) = \Phi(n)\Big(\Phi^{-1}(T) - I\Big)^{-1} \sum_{j=n}^{n+T-1} \Phi^{-1}(j+1)g(j),$$

is a solution of (5.8.6).

(d) Show that $x(n+T) = x(n)$.

Exercise 5.49. Suppose $A(n)$ and $g(n)$ are periodic of period T. Show that if the non-homogeneous system (5.8.6) does not have periodic solutions with period T, it cannot have bounded solutions.

Exercise 5.50. Consider the scalar equation

$$x(n+1) = a(n)x(n), \quad n \in \mathbb{N}, \quad a(n) \neq 0$$

with $a(n+T) = a(n)$. Let

$$c = \left(\prod_{u=0}^{T-1} a(u)\right)^{\frac{1}{T}}.$$

Show that if $x(n)$ is a solution of the scalar equation then

$$y(n) = x(n)c^{-n}$$

satisfies $y(n + T) = y(n)$ for all $n \in \mathbb{N}$.

Exercise 5.51. Consider the discrete Hill's equation

$$x(n + 2) + (\alpha + \beta h(n)) x(n + 1) + x(n) = 0, \; n \in \mathbb{N},$$

where α and β are constants.

(a) Write the equation into a system in terms of x_1, x_2.

(b) For $\beta = 0$, show that Hill's equation is non-oscillatory for the values $\alpha \leq -2$, and oscillatory for $|\alpha| < 2$, and $\alpha \geq 2$.

(c) Suppose $h(n) = (-1)^n$. Show that the zero solution of the Hill's equation is stable for

$$|2 - \alpha^2 + \beta^2| \leq 2,$$

and unstable for

$$|2 - \alpha^2 + \beta^2| > 2.$$

Chapter 6

Lyapunov functions

Lyapunov function is named after Aleksandr Mikhailovich Lyapunov, a Russian mathematician who defended the thesis *The General Problem of Stability of Motion* at Kharkov University in 1892. Lyapunov was a pioneer in creating the global approach to the analysis of the stability of non-linear dynamical systems. His work, initially published in Russian and then translated into French, received little attention for many years. The current use of Lyapunov functions has proved that Lyapunov was ahead of his time in developing his theory that is being used in many different areas of science. In the theory, Lyapunov functions are scalar functions that may be used to prove the stability of equilibrium of a given dynamical system. Since the inception of Lyapunov functions, their successful usage has been extended to integral equations, integro-differential equations, functional differential equations and discrete dynamical systems. The literature on Lyapunov functions and functionals in differential equations is vast, and this is not the case for discrete dynamical systems. This chapter contains advanced material that we do not expect to be taught in classroom format. Sections 6.2, 6.3, 6.5, and 6.6 can be used as part of a reading course on special topics in discrete dynamical systems.

6.1 Introduction to Lyapunov functions

The existence of Lyapunov functions is a necessary and sufficient condition for stability in difference equations and differential equations. There are no concrete procedures on how to find Lyapunov functions, but in some cases, the construction of Lyapunov functions is known. In the particular case of homogeneous autonomous systems with constant coefficients, the Lyapunov function can be found as a quadratic form. As we shall see later, Lyapunov functions allow us to study stability of non-linear dynamical systems when the equilibrium solution of an associated linear difference equation is a center.

Let $G \subset \mathbb{R}^k$ be an open set and consider the system of autonomous difference equations

$$x(n+1) = f(x(n)) \tag{6.1.1}$$

where $f : G \to \mathbb{R}^k$ is continuous . We assume that x^* is an equilibrium solution of (6.1.1), that is $f(x^*) = x^*$.

Definition 6.1.1. Let the function $V : \mathbb{R}^k \to \mathbb{R}$ be continuous.

Difference Equations and Applications. https://doi.org/10.1016/B978-0-44-331492-6.00012-1

(i) The variation of V with respect to (6.1.1) is defined as

$$\Delta V(x(n)) = V(f(x(n))) - V(x(n)) = V(x(n+1)) - V(x(n)).$$

(ii) The function V is said to be a *Lyapunov function/functional* on a subset H of $\mathbb{R}^k$ if
  i) $V(x^*) = 0$, and $V(x) > 0$, for $x \neq x^*$ and
  ii) $\Delta V(x) \leq 0$, whenever x and $f(x)$ belong to the set H.

(iii) The function V is said to be a strict Lyapunov function/functional on a subset H of $\mathbb{R}^k$ if $\Delta V(x) < 0$.

(iv) Let $B(x, \gamma)$ denote the open ball in $\mathbb{R}^k$ of radius γ and center x defined by $B(x, \gamma) = \{y \in \mathbb{R}^k \mid ||y - x|| < \gamma\}$. We say V is positive definite at x^* if $V(x^*) = 0$, and $V(x^*) > 0$, for all $x \in B(x^*, \gamma), x \neq x^*$, for some $\gamma > 0$.

We provide this simple example.

Example 6.1. Consider the scalar difference equation

$$x(n+1) = ax(n), \quad x(0) = x_0.$$

Clearly, $x^* = 0$ is the only equilibrium solution. Consider the function

$$V(x) = |x(n)|.$$

Then, $V(x^*) = V(0) = 0$, and $V(x) > 0$, for $x \neq 0$. Moreover,

$$\Delta V(x) = |x(n+1)| - |x(n)| = |ax(n)| - |x(n)| = (|a| - 1)|x(n)|.$$

We already know, that the zero solution is stable if $a = 0$, unstable if $|a| > 1$, and exponentially stable if $|a| < 1$. We shall see later that the same results will be deduced from the inequality

$$\Delta V(x) = (|a| - 1)|x(n)|. \quad \square$$

The next examples show that there are more than one Lyapunov function to obtain the same results.

Example 6.2. Consider the scalar difference equation

$$x(n+1) = x^2(n), \quad x(0) = x_0.$$

Clearly, $x^* = 0, 1$ are the equilibrium solutions. Here we only address the equilibrium solution $x^* = 0$. Consider the function

$$V(x) = x^2(n).$$

Then, $V(x^*) = V(0) = 0$, and $V(x) > 0$, for $x \neq 0$. Moreover,

$$\Delta V(x) = x^2(n+1) - x^2(n) = x^4(n) - x^2(n) = x^2(n)(x^2(n) - 1).$$

Thus, $\triangle V(x) < 0$, on the set

$$H = \{x \in \mathbb{R} : (-1, 0) \cup (0, 1)\}.$$

This means that any solution emanating from the set H will converge to $x^* = 0$. Whereas, any solution with $|x(0)| = |x_0| > 1$, will diverge away from $x^* = 0$ to infinity. Now we try another function

$$V(x) = |x(n)|.$$

Then, $V(x^*) = V(0) = 0$, and $V(x) > 0$, for $x \neq 0$. Moreover,

$$\triangle V(x) = |x(n+1)| - |x(n)| = x^2(n) - |x(n)| = (|x(n)| - 1)\,|x(n)|.$$

This gives the same information as the other function did that $\triangle V(x) < 0$, for $x \in H$. Since we are here, let's display a Lyapunov function for the equilibrium solution $x^* = 1$. Set

$$V(x) = (x(n) - 1)^2.$$

Then, $V(x^*) = V(0) = 0$, and $V(x) > 0$, for $x \neq 1$. Moreover, for $x \neq 1$,

$$\begin{aligned}
\triangle V(x) &= (x(n+1) - 1)^2 - (x(n) - 1)^2 \\
&= (x^2(n) - 1)^2 - (x(n) - 1)^2 \\
&= (x(n) - 1)^2[(x(n) + 1)^2 - 1].
\end{aligned}$$

We may easily see that $\triangle v < 0$ on the set $H = \{x \in \mathbb{R} : -2 < x < 0\}$. Let's take a closer look on the implication of $\triangle v < 0$ on the set H. It implies that any solution starting in H will converge to the equilibrium solution $x^* = 1$, but this can not happen since solutions would have to cross the other equilibrium solution $x^* = 0$, to reach $x^* = 1$, which violates the uniqueness of the solution. Thus, $\triangle v > 0$ for all $x \notin H$, and $x \neq 0, 1$. Remember, according to Theorem 5.1.1, $x^* = 0$ is (AS) and $x^* = 1$ is unstable. The same results are obtained from the staircase method. □

6.1.1 Stability of autonomous systems

Let D be an open subset of $\mathbb{R}^k$ containing 0 and $f : D \rightarrow \mathbb{R}^k$ with $f(0) = 0$. Assume the existence of the unknown solution $x : [n_0, \infty) \rightarrow D$ on each $(n^*, x_0) \in [n_0, \infty) \times D$ and consider the system

$$x(n+1) = f(x(n)), \quad x(0) = x_0, \, n \geq 0. \tag{6.1.2}$$

We make the following definitions regarding an open ball so that we may state the stability notion in terms of such balls.

Definition 6.1.2. Let $x^* \in \mathbb{R}^n$ and $n > 0$. Then $B_r(x^*)$ denotes the open ball centered at x^* with radius r. That is

$$B_r(x^*) = \{x \in \mathbb{R}^n : ||x - x^*|| < r\}.$$

Definition 6.1.3. The equilibrium point x^* of (6.1.2);

(a) is *stable* (S) if for each ball $B_\epsilon(x^*)$, there is a ball $B_\delta(x^*)$ $(\delta \le \epsilon)$ so that if $x \in B_\delta(x^*)$, then $x(n, x_0)$ remains in the ball $B_\epsilon(x^*)$ for $n \ge 0$,

(b) is *asymptotically stable* (AS) if it is *stable* and there is a ball $B_c(x^*)$ so that for each $x \in B_c(x^*)$, $x(n, x_0) \to 0$, as $n \to \infty$.

We warm up with the following example.

Example 6.3. The origin $(0, 0)$ is the only equilibrium solution and is a node for the system

$$x(n + 1) = \frac{1}{2}x(n)$$

$$y(n + 1) = \frac{1}{2}y(n).$$

We apply Definition 6.1.3. The solution is easily computed,

$$\begin{pmatrix} x \\ y \end{pmatrix} = \begin{pmatrix} c_1(\frac{1}{2})^n \\ c_2(\frac{1}{2})^n \end{pmatrix}$$

for $x(0) = c_1$ and $y(0) = c_2$. We apply Definition 6.1.3. For all $\epsilon > 0$, let $\delta = \epsilon > 0$. Then for all $c \in B_\delta(0) = \{(x, y) \in \mathbb{R}^2 : \sqrt{x^2 + y^2} < \delta\}$ and $n > 0$ we have

$$||(x, y)|| = \sqrt{x^2(n) + y^2(n)} = \sqrt{(c_1^2 + c_2^2)(\frac{1}{2})^n} \le \delta = \epsilon.$$

It follows that

$$(x(n, 0, c_1), y(n, 0, c_2)) \in B_\delta(0),$$

and hence $(0, 0)$ is stable. Moreover,

$$(x(n, 0, c_1), y(n, 0, c_2)) = \left(c_1(\frac{1}{2})^n, c_2(\frac{1}{2})^n\right) \to (0, 0) \text{ as } n \to \infty$$

shows that $(0, 0)$ is (AS). □

Theorem 6.1.1. *Let $V(x)$ be a Lyapunov function in D. That is, $V : D \to [0, \infty)$*

$$V(0) = 0, \quad and \quad V(x) > 0, \quad and \quad \Delta V \le 0 \quad if x \ne 0.$$

Then the zero solution of (6.1.2) is stable.

Proof. Pick an $\epsilon > 0$ so that the sphere $B_\epsilon(0)$ is in D. Let S_ϵ be the boundary of $B_\epsilon(0)$. Due to the continuity of V at the origin and $V(0) = 0$, there must be a point on S_ϵ where V attains its minimum that we denote by $m = \min_{x \in S_\epsilon} V(x)$, where $m > 0$ since $V(x) > 0$. Let $B_\delta(0)$ be another sphere in D. Since $V(0) = 0$ and V is continuous we can choose $\delta > 0$ such that $V(x) < m$ for all $x \in B_\delta(0)$. Our goal is to show that if $x_0 \in B_\delta(0)$ then $x(n, 0, x_0) \in B_\epsilon(0)$ for all $n > 0$. Since $\triangle V \leq 0$, we have

$$V(x(n+1)) \leq V(x(n)) \leq \cdots V(x(0)) \leq m,$$

for all $n \geq 0$. In other words, V does not increase along the orbits with the initial condition x_0. This, in turn, implies that $x(n, 0, x_0)$ can not cross S_ϵ since $V(x) \geq m$ for all $x(n, 0, x_0) \in S_\epsilon$. This implies the origin is stable. This completes the proof. □

Theorem 6.1.2. *Let $V(x)$ be a strict Lyapunov function in D. That is, $V : D \to [0, \infty)$*

$$V(0) = 0, \quad V(x) > 0, \text{ and } \triangle V(x) < 0 \quad \text{for } x \neq 0.$$

Then the zero solution of (6.1.2) is asymptotically stable.

Proof. Since the origin is stable and $\triangle V(x) < 0$ along the solutions of (6.1.2), it follows $V(x)$ decreases to the value c as $n \to \infty$. The proof will be complete if we can show $c = 0$. Assume for the sake of contradiction that $c > 0$. Then there exists $\beta < \epsilon$ (ϵ from the stability theorem) such that $V(x) < c$ for all $x \in B_\beta(0)$. But then $x(n)$ cannot enter $B_\beta(0)$. Let

$$m = \min\{-\triangle V : \beta \leq |x| \leq \epsilon\}.$$

As $-\triangle V > 0$, we have that $\triangle V(x) \leq -m$ for all $n \geq 0$. A summation of $\triangle V \leq -m$ from 0 to $n - 1$ yields

$$V(x(n)) - V(x_0) \leq -mn + mn_0.$$

Thus $\lim_{n \to \infty} V(x(n)) = -\infty$, which contradicts the assumption that V is positive definite in D and equal c as $n \to \infty$. Hence c must vanish. We next claim that $x(n) \to 0$ as $n \to \infty$. If this is not the case, then there are constants $\alpha > 0$ and a sequence $n_k \to \infty$ such that $|x(n_k)| \geq \alpha$ for all k. Since the closure $\bar{B}_\beta(0)$ is compact and all the points $x(n_k) \in \bar{B}_\beta(0)$, there is a subsequence that converges to a point $z \in B_\beta(0)$. We may assume that $x(n_k) \to z$ and we must have $0 \neq |z| \geq \alpha$. By continuity of V, we have that $V(x(n_k)) \to V(z) > 0$. However, since $n_k \to \infty$ we must have $V(x(n_k)) \to c = 0$, which is a contradiction. Thus the origin is asymptotically stable. This completes the proof. □

Definition 6.1.4. A solution $x(n)$ of (6.1.2) is said to be bounded if for any $n_0 \in \mathbb{Z}^+$ and number r there exists a number $\alpha(n_0, r)$ depending on n_0 and r such that $||x(n, n_0, x_0)|| \leq \alpha(n_0, r)$ for all $n \geq n_0$ and x_0, $|x_0| < r$. It is uniformly bounded if α is independent of the initial time n_0.

We want the reader to be aware that there is no correlation between boundedness of all solutions and stability of a solution we shall demonstrate in the next example.

Example 6.4. Consider the linear difference equation

$$x(n + 1) = x(n) + 1, \quad x(n_0) = x_0. \tag{6.1.3}$$

It is easy to check that $x(n) = x_0 + (n - n_0)$ is the solution of (6.1.3). If $y(n)$ is another solution with $y(n_0) = y_0$, then we have $y(n) = y_0 + (n - n_0)$. For any $\epsilon > 0$, let $\delta = \epsilon$. Then

$$||x(n) - y(n)|| = ||x_0 + (n - n_0) - y_0 - (n - n_0)|| = ||x_0 - y_0)|| < \epsilon$$

whenever, $||x_0 - y_0|| < \delta$. Hence, the solution $x(n)$ is stable, but unbounded. This simple example shows that the properties of boundedness of all solutions and stability of a solution do not coincide. On the other hand, the difference equation

$$x(n + 1) = x^{1/3}(n),$$

has its solution $x(n)$ satisfies $|x(n)| \to 1$ for all initial values $x_0 \neq 0$, whereas the solution $x(n) = 0$ is unstable, since if we set $f(x) = x^{1/3}$, then $f'(x) = \frac{1}{3}x^{-\frac{2}{3}}$ becomes positively large as x approached zero from either side. □

Example 6.5. Consider the scalar non-linear difference equation

$$x(n + 1) = ax(n) + b\frac{x(n)}{1 + x^2(n)}, \quad x(n_0) = x_0, n \geq n_0 \geq 0. \tag{6.1.4}$$

• If

$$|a| + |b| \leq 1,$$

then all solutions of (6.1.4) are bounded and the zero solution is (S).
• If

$$|a| + |b| < 1,$$

then all solutions of (6.1.4) are bounded and the zero solution is (AS).

The system has the only equilibrium solution $x^* = 0$. Consider the Lyapunov function $V(x(n)) = |x(n)|$. Then along the solutions of (6.1.4) we have

$$\triangle V(x(n)) = |x(n + 1)| - |x(n)| \leq (-1 + |a|)|x(n)| + |b|\frac{|x(n)|}{1 + x^2(n)}$$

$$\leq (-1 + |a| + |b|)|x(n)|.$$

The results for stability follow from Theorems 6.1.1 and 6.1.2. With respect to boundedness, we know that $\Delta V(x(n)) \leq 0$, and hence

$$\sum_{s=n_0}^{n-1} \Delta V(x(s)) = V(x(n)) - V(x(n_0)) \leq 0.$$

That is

$$V(x(n)) \leq V(x(n_0)).$$

But, $V(x(n)) = |x(n)|$ and so we have

$$|x(n)| = V(x(n)) \leq V(x(n_0)) = |x(n_0)| = |x_0|.$$

Hence we have boundedness on all solutions. $\qquad\square$

We would like to mention that the stability definitions of 5.2.2 are the ones that we use throughout this chapter, unless, mentioned otherwise.

Definition 6.1.5. Assume the zero solution of (6.1.2) is stable. If $|x(n, n_0, x_0)| \to 0$ as $n \to \infty$ for every $x(n_0) = x_0$, then we say the zero solution is *globally asymptotically stable* (GAS).

Definition 6.1.6. A function $V : D \to \mathbb{R}^+$ is said to be *radially unbounded* if

$$V(x) \to \infty, \quad \text{as } ||x|| \to \infty.$$

The previous definition is very deceiving and careful care must be taken when it is applied.

Example 6.6. We examine in detail why

$$V(x_1, x_2) = \frac{x_1^2}{1 + x_1^2} + x_2^2$$

is not radially unbounded. The condition that $V(x)$ is radially unbounded means that all level sets are closed. Let $x_2 = 0$ and we try to determine the level set $V(x) = 1$. Or, $V(x_1, 0) = \frac{x_1^2}{1 + x_1^2} = 1$, which can not hold for finite x_1. Thus, V is not radially unbounded. Next we examine the definition of being radially unbounded in the context of

$$V(x) \to \infty, \quad \text{as } ||x|| \to \infty.$$

This means that any combinations of x_1, x_2 that make $||x|| \to \infty$ also has to make $V(x) \to \infty$. The Euclidean norm

$$||(x_1, 0)||_2 = \sqrt{x_1^2 + 0^2} = \sqrt{\infty^2 + 0^2} \to \infty.$$

However, in this case, we get for $V(x)$ that

$$\lim_{x_1 \to \infty} V(x_1, 0) = 1 \neq \infty.$$

It follows that V is not radially unbounded.   □

We state the following theorem regarding (GAS).

Theorem 6.1.3. *Assume the hypothesis of Theorem 6.1.2 holds with $D = \mathbb{R}^n$. In addition, if $V(x)$ is radially unbounded; that is*

$$V(x) \to \infty, \quad \text{as } ||x|| \to \infty, \tag{6.1.5}$$

then the zero solution of (6.1.2) is (GAS).

Proof. By Theorem 6.1.1, the zero solution is stable. Suppose there is a solution or trajectory that does not converge to zero. Since V is decreasing and nonnegative, it converges say to γ as $n \to \infty$. Since x does not converge to zero, we have $\gamma \leq V(x(n)) \leq V(x_0)$. Let

$$L = \{z : \gamma \leq V(z) \leq V(x_0)\}.$$

Then L is closed and bounded, hence compact. Therefore, $\triangle V$ attains its supremum on L. That is $\sup_{z \in L} \triangle V = -d < 0$. Or, $\triangle V(x(n)) \leq -d$ for all $n \geq n_0$. Summing $\triangle V \leq -d$ from 0 to $T - 1$, yields

$$V(x(T)) \leq V(x_0) - dT.$$

This implies that $V(x_0) < 0$, for $T > V(x_0)/d$, which is a contradiction and hence every trajectory converges to zero. As for the global stability, it suffices to show that all solutions are bounded. Suppose the contrary; that is there is a subsequence n_j with $n_j \to \infty$, so that $\{x(n_j)\} \to \infty$. But then, condition (6.1.5) would imply that $V(x(n_j)) \to \infty$ as $x(n_j) \to \infty$, which contradicts the fact that $V(x(n_j)) < V(x_0)$ for all j. This completes the proof.   □

The purpose of the radial unboundedness condition on V is to ensure that $x(n)$ remains at all times within the bounded region defined by $V(x) \leq V(x_0)$. If V were not radially unbounded, then not all level curves given by $V(x) \leq c$ for positive constant c would be closed curves, and it would be possible for $x(n)$ to drift away from the equilibrium even though $\triangle V < 0$.

Example 6.7. Let $|a| < 1$, and consider the scalar difference equation

$$x(n + 1) = ax(n) \sin(x(n)), \quad x(0) = x_0.$$

Clearly, $x^* = 0$ is an equilibrium solution. Consider the function

$$V(x) = x^2(n).$$

Then, $V(x^*) = V(0) = 0$, and $V(x) > 0$, for $x \neq 0$. Moreover,

$$\Delta V(x) = x^2(n+1) - x^2(n)$$
$$= a^2 x^2(n) \sin^2(x(n)) - x^2(n)$$
$$\leq a^2 x^2(n) - x^2(n) = x^2(n)(a^2 - 1) < 0,$$

for all $x \in \mathbb{R} - \{0\}$. Since

$$V(x) \to \infty \quad \text{as} \quad ||x|| \to \infty,$$

we have $x^* = 0$ is (GAS) by Theorem 6.1.3. $\qquad \square$

Example 6.8. In Example 5.17 we concluded that Theorem 5.7.1, or linearization gave no information regarding the stability of the zero solution of the non-linear system

$$x(n+1) = y(n) - x^2(n)y(n)$$
$$y(n+1) = x(n) - x^3(n).$$

In this example, we shall display a Lyapunov function and show that indeed the zero solution is stable. In the computations to follow, we will suppress the independent variable n. We consider the Lyapunov function

$$V(x, y) = x^2(n) + y^2(n).$$

Then, $V(x^*, y^*) = V(0, 0) = 0$, and $V(x, y) > 0$, for $(x, y) \neq 0$. Moreover,

$$\Delta V(x) = x^2(n+1) + y^2(n+1) - x^2(n) - y^2(n)$$
$$= (y - x^2 y)^2 + (x - x^3)^2 - x^2 - y^2$$
$$= y^2(1 - x^2) + x^2(1 - x^2)^2 - x^2 - y^2$$
$$= (1 - x^2)^2[y^2 + x^2] - (x^2 + y^2)$$
$$= (x^2 + y^2)[(1 - x^2)^2 - 1]$$
$$= x^2(x^2 + y^2)(x^2 - 2).$$

Thus, V is a strict Lyapunov function on the set

$$H = \{(x, y) \neq (0, 0) : -\sqrt{2} < x < \sqrt{2}\},$$

which implies the zero solution is (AS). We note that the zero solution is not (GAS) since $H \neq \mathbb{R}^k$. In other words, the initial condition $(x(n_0), y(n_0)) = (x_0, y_0)$ has to be in the set H in order for the solution $|x(n, n_0, x_0)| \to (0, 0)$ as $n \to \infty$. $\qquad \square$

Next we address the concept of domain of attraction. In engineering applications it is often important to get an estimate of the size of the domain of attraction.

Definition 6.1.7 (Domain of attraction). Assume $V(x)$ to be a strict Lyapunov function in D. Then the *domain of attraction* of an equilibrium consists of all points such that a solution starting at them tends to the equilibrium. Thus, if V is a strict Lyapunov function on a set $D \in \mathbb{R}^n$ then the origin is asymptotically stable and the *domain of attraction*

$$\bar{N} = \{x \in D : V(x) \leq C\}$$

for some positive constant C.

Example 6.9. Consider the non-linear system

$$x_1(n+1) = x_2(n) - x_2(n)\sqrt{x_1^2(n) + x_2^2(n) + 1}$$
$$x_2(n+1) = x_1(n) - x_1(n)\sqrt{x_1^2(n) + x_2^2(n) + 1}.$$

Clearly $(0,0)$ is an equilibrium point of the system. Notice that the Linearization theorem, Theorem 5.7.1 is inconclusive. Let

$$V(x_1, x_2) = x_1^2(n) + x_2^2(n).$$

Clearly $V(0,0) = 0$ and $V(x_1, x_2) > 0$ for $(x_1, x_2) \neq (0,0)$. After some computations, it can be shown that

$$\Delta V = (x_1^2(n) + x_2^2(n))\sqrt{x_1^2(n) + x_2^2(n) + 1}\left(\sqrt{x_1^2(n) + x_2^2(n) + 1} - 2\right).$$

To obtain (AS) we require $\Delta V(x_1, x_2) < 0$. Or,

$$\sqrt{x_1^2(n) + x_2^2(n) + 1} - 2 < 0,$$

which holds for $x_1^2(n) + x_2^2(n) < 3$. On the other hand, to keep the circles $V(x, y) = x_1^2 + x_2^2 = C$ inside $D = \{(x_1, x_2) : \sqrt{x_1^2(n) + x_2^2(n) + 1} - 2\}$, we must take $0 < C < 3$. Thus for any value of such C the origin is asymptotically stable and the domain of attraction is

$$\bar{N} = \{(x_1, x_2) \in D : V(x_1, x_2) \leq C < 3\}. \quad \square$$

The next theorem provides criteria of the instability of the zero solution.

Theorem 6.1.4. *Let x^* be an equilibrium solution of (6.1.2). Let U be a neighborhood of x^*. Suppose $V : U \to \mathbb{R}$ is continuous on U. If $V(x^*) = 0$, but there are points arbitrarily close to x^* where V is strictly positive, and that $\Delta V(x) > 0$ on $U \setminus \{x^*\}$, then x^* is unstable for (6.1.2).*

Proof. Without loss of generality, we take $x^* = 0$. Let $B_\beta(0) \subseteq U$, $\beta > 0$. We may choose a point $x_0 \in B_\beta(0)$ that is close to 0, such that $V(x_0) > 0$. The proof will be completed if we can show that $x(n)$ is outside the ball $\bar{B}_\beta(0)$. Assume the contrary. That is, solutions stay in $B_\beta(0)$ for all $n \geq 0$. Since $\bar{B}_\beta(0)$ is compact, V achieves its maximum; say at $V_0 \in \bar{B}_\beta(0)$. Thus,

$$V(x(n)) \leq V_0, \quad n \geq 0. \tag{6.1.6}$$

Since $\Delta V > 0$ and V is increasing along the solutions and $V(x) \geq V(x_0) > 0$. We can find some $\rho > 0$ so that $V(x) < V(x_0)$ for $x \in B_\rho(0)$. Then $x(n)$ is trapped in the compact annulus $\rho \leq |x| \leq \beta$. Let

$$m = \min\{\Delta V : \rho \leq |x| \leq \beta\}.$$

Summing $\Delta V \geq m$ from 0 to $n-1$, yields

$$V(x(n, x_0)) \geq V(x_0) + mn,$$

which contradicts (6.1.6), since the right side goes to infinity as n goes to infinity. Thus, the solution $x(n)$ must escape $B_\beta(0)$ and the origin is unstable. This completes the proof. $\qquad\square$

Example 6.10. Consider the non-linear system

$$x(n+1) = y - x^4 y$$
$$y(n+1) = -x - x^3 y^2.$$

Then $(0, 0)$ is an equilibrium solution. Set $V(x, y) = x^2 + y^2$. Then, $V(0, 0) = 0$, and $V(x, y) > 0$, for $(x, y) \neq 0$.

$$\begin{aligned}
\Delta V(x) &= x^2(n+1) + y^2(n+1) - x^2(n) - y^2(n) \\
&= (y - x^4 y)^2 + (x + x^3 y^2)^2 - x^2 - y^2 \\
&= x^8 y^2 + x^6 y^4 > 0,
\end{aligned}$$

for $(x, y) \neq (0, 0)$. Thus, by Theorem 6.1.4 the zero solution is unstable. It is worth noting that linearization, and in particular Theorem 5.7.1, is inconclusive here since the eigenvalues of the linearized matrix have magnitude one. $\qquad\square$

6.1.2 Exercises

Exercise 6.1. Show the zero solution of

$$x(n+1) = y(n) - y^3(n)$$
$$y(n+1) = x(n) - x(n)y^2(n)$$

is (AS) but not (GAS).

Exercise 6.2. Show the zero solution of

$$x(n+1) = ax(n)\sin(x(n)) + \frac{bx(n)}{1 + |x(n)|}$$

is (GAS) if

$$|a| + |b| < 1.$$

Exercise 6.3. Suppose f is continuous and $f(0, 0) = 0$, and consider the non-linear system

$$x(n+1) = y - xf(x, y)$$
$$y(n+1) = -x - yf(x, y).$$

Display a Lyapunov function to show the origin is unstable.

Exercise 6.4. Use Exercise 6.3 to show the origin is unstable for the system

$$x(n+1) = y - x^3$$
$$y(n+1) = -x - x^2 y.$$

Exercise 6.5. Consider the non-linear system

$$x_1(n+1) = x_2(n) - x_2(n)\left(x_1^2(n) + x_2^2(n)\right)$$
$$x_2(n+1) = x_1(n) - x_1(n)\left(x_1^2(n) + x_2^2(n)\right).$$

Examine the stability of the origin $(0, 0)$ using:
(a) Theorem 5.7.1.
(b) Lyapunov function.
(c) Determine the domain of attraction.

Exercise 6.6. Consider the non-linear system

$$x_1(n+1) = \frac{x_2(n)}{2} + x_2(n)x_1^2(n)$$
$$x_2(n+1) = 2x_1(n) - 2x_1(n)x_2^2(n).$$

(a) Find all equilibrium solutions.
    Examine the stability of the origin $(0, 0)$ using:
(b) Theorem 5.7.1.
(c) Lyapunov function. ($V = 4x_1^2(n) + x_2^2(n)$).
(d) Determine the domain of attraction.

Exercise 6.7. Show the zero solution is unstable for the system

$$x_1(n+1) = \frac{x_2(n)}{2} + x_2(n)x_1^2(n)$$
$$x_2(n+1) = 4x_1(n) - 2x_1(n)x_2^2(n).$$

Hint: Set $V = 16x_1^2(n) + x_2^2(n)$.

Exercise 6.8. Consider the non-linear system

$$x_1(n+1) = x_1(n)$$
$$x_2(n+1) = x_2(n) - x_1^2(n)x_2^3(n).$$

Examine the stability of the origin $(0, 0)$ using:

(a) Theorem 5.7.1.
(b) Lyapunov function.
(c) Determine the domain of attraction.

Exercise 6.9. For Exercise 6.6 to determine the stability for the other two equilibrium solutions

$$\left(\frac{1}{\sqrt{2}}, \frac{1}{\sqrt{2}}\right), \quad \left(-\frac{1}{\sqrt{2}}, -\frac{1}{\sqrt{2}}\right).$$

Exercise 6.10. Decide if the functions given below are radially unbounded or not.

a)
$$V(x_1, x_2) = (x_1 - x_2)^2.$$

b)
$$V(x_1, x_2) = \frac{x_1^2 + x_2^2}{1 + x_1^2 + x_2^2} + (x_1 - x_2)^2.$$

Exercise 6.11. Consider the planar system

$$x(n+1) = (x - y)(x^2 + y^2 - 1)$$
$$y(n+1) = (x + y)(x^2 + y^2 - 1).$$

Show that the origin is (AS) and find the domain of attraction.

Exercise 6.12. Suppose f and g are continuous with $f(0,0) = g(0,0) = 0$, and consider the non-linear system

$$x(n+1) = f(x, y)$$
$$y(n+1) = g(x, y).$$

(a) Show that if

$$f(x, y)g(x, y) > xy$$

for every point (x, y) in a neighborhood of the origin, then the origin is unstable.

(b) Give an example of such functions f and g.

Hint: Try $V = |x(n)||y(n)|$.

Exercise 6.13. Consider the non-linear scalar equation

$$x(n+1) = \frac{1}{2}x(n) + \alpha x^3(n),$$

where $\alpha > 0$.

(a) Find all equilibrium solutions.

(b) Use a Lyapunov function to show the zero solution is stable.

(c) Use Theorem 5.1.1 to study the stability of the other two equilibrium solutions. Feel free to construct a suitable Lyapunov function to do so for each of the equilibria.

6.2   LaSalle invariance principle

In this section we are concerned with the autonomous system given by (6.1.1). Two crucial ideas in dynamical systems are the *invariant set* and the *limit set*. The limit set is the asymptotic behavior of a system of difference equations, or the limit behavior of $f^n(x_0)$, as $n \to \infty$. The existence of limit sets is implied by a compact invariant set.

Definition 6.2.1 (Limit point and limit set). Given a motion $f^n x$, $\omega \in \mathbb{R}^m$ is a limit point of the motion $f^n x$ if there exists a subsequence $\{n_k\}_{k \in \mathbb{N}}$ such that $n_k \to \infty$ and $f^{n_k}(x) \to \omega$ as $k \to \infty$. When there is no ambiguity about the map f, we also refer to ω as a limit point of x. The set of all the limit points of x is referred to as the limit set of x, denoted by $\Omega(x)$. That is,

$$\Omega(x) = \left\{ y \in \mathbb{R}^m : \exists\{n_k\} \subset \mathbb{N} \text{ such that } n_k \to \infty \text{ and } f^{n_k}(x) \to \omega \text{ as } k \to \infty \right\}.$$

Given a set $H \subset \mathbb{R}^m$, the limit set of the set H is denoted by $\Omega(H)$ and defined as

$$\Omega(H) = \Big\{ y \in \mathbb{R}^m : \exists\{n_k\} \subset \mathbb{N} \text{ and } \{\omega_k\} \subset H \text{ such that } n_k \to \infty \text{ and }$$
$$f^{n_k}(\omega_k) \to \omega \text{ as } k \to \infty \Big\}.$$

Alternately, the *limit set* of a trajectory $x(\cdot, n_0, x_0)$ is the set of all points $\omega \in \mathbb{R}^n$ such that

$$\lim_{n \to \infty} x(n_k, n_0, x_0) = \omega,$$

where the sequence n_k is strictly increasing.

Definition 6.2.2. (Invariant set) The set $M \in \mathbb{R}^k$ is said to be *invariant set* if for all $y \in M$ and $n_0 \geq 0$, we have

$$x(n, n_0, y) \in M, \quad \text{for all } n \geq n_0.$$

The following lemma presents some important properties of the limit set of any set H, i.e., $\Omega(H)$.

Lemma 6.1 (Invariance and asymptotic properties of limit set $\Omega(H)$). *For any continuous map $T : \mathbb{R}^m \to \mathbb{R}^m$ and any set $H \subset \mathbb{R}^m$, the following statements for the limit set $\Omega(H)$ hold:*

1. *For any $x \in H$, $\Omega(x) \subset \Omega(H)$;*
2. *$\Omega(H)$ is closed and positively invariant;*
3. *If $\cup_{n=0}^{\infty} T^n(H)$ is bounded, then*
   a. *$\Omega(H)$ is non-empty, compact, and invariant;*
   b. *for any $x \in H$, $T^n(x)$ approaches $\Omega(H)$;*
   c. *$\Omega(H)$ is the smallest set that $T^n(H)$ approaches as $n \to \infty$.*

Theorem 6.2.1. *Assume V is a Lyapunov function of (6.1.1) on a bounded open set $U \in \mathbb{R}^k$ with $x^* \in U$. Let c be a positive constant and set*

$$S = \{x \in U : V(x) \leq c\},$$

which is closed in $\mathbb{R}^k$. Then S is invariant for $n \geq n_0 \geq 0$.

Proof. Let $x \in S$. Since V is a Lyapunov function on U, we have that $V(x(n, n_0, x_0)) \leq V(x_0) \leq c$ for $n \geq n_0 \geq 0$, as long as the solutions exist. Now since S is closed and bounded, the existence of solutions holds for all $n \geq n_0 \geq 0$. As a result, $x(n, n_0, x_0)$ remains in S for all $n \geq n_0 \geq 0$. This shows the set S is invariant. This completes the proof. $\qquad\square$

In Example 6.9 we considered $V(x, y) = x_1^2 + x_2^2$ and obtained $\triangle V = 0$. As a result of Theorem 6.2.1, we have the set

$$S = \{(x, y) \in \mathbb{R}^2 : x_1^2 + x_2^2 \leq c\}, \quad \text{for } 0 < c < 3$$

is invariant.

Unlike previous theorems on (AS), LaSalle invariance principle doers not assume the Lyapunov function to be positive definite.

Theorem 6.2.2 (LaSalle invariance principle). *Suppose there exists $V : \mathbb{R}^k \to \mathbb{R}$ satisfying*

i) *$V(x)$ is continuous at any $x \in \overline{G}$;*
ii) *$\triangle V(x) \leq 0$ for any $x \in G$.*

For any $x_0 \in G$, if the solution $x(n)$ to (6.1.1) satisfies that $\cup_{n=0}^{\infty}\{x(n)\}$ is bounded and $x(n) \in G$ for any $n \in \mathbb{N}$, then there exists $c \in \mathbb{R}$ such that $x(n) \to M \cap V^{-1}(c)$ as $n \to \infty$, where $V^{-1}(c) = \{x \in \mathbb{R}^m : V(x) = c\}$ and M is the largest invariant set in $E = \{x \in \overline{G} : \triangle V(x) = 0\}$.

Proof. For $B = \cup_{n=0}^{\infty}\{x(n)\}$, we have that $\overline{B} \subset \overline{G}$ is compact. Since $V(x)$ is continuous on $\overline{G}$, $V(x)$ achieves its lower bound on $\overline{B}$. Moreover, since $V(x(n))$ is non-increasing with respect to n for any $n \in \mathbb{N}$, there exists $c \in \mathbb{R}$ such that $V(x(n)) \to c$ as $n \to \infty$. Since there exists $\{n_k\} \to \infty$ such that $x(n_k) = f^{n_k}(x_0) \to \omega$ as $k \to \infty$ and due to the continuity of V, we have $V(x_{n_k}) \to V(\omega)$, for any $\omega \in \Omega(x_0)$. Therefore,

$$V(\omega) = \lim_{k \to \infty} V(x(n_k)) = \lim_{n \to \infty} V(x(n)) = c, \quad \text{for any } \omega \in \Omega(x_0),$$

and consequently, we obtain $\Omega(x_0) \in V^{-1}(c)$.

Moreover, for any $\omega \in \Omega(x_0)$, since B is bounded, according to Theorem 6.2.1, $\Omega(x_0)$ is invariant. Therefore, $f(\omega) \in \Omega(x_0)$, which implies that $V(f(\omega)) = c$ and thus $V(f(\omega)) - V(\omega) = 0$ for any $\omega \in \Omega(x_0)$. Now we obtain $\Omega(x_0) \subset E$. Since $\Omega(x_0) \subset E$ and M is the largest invariant set of E, we have $\Omega(x_0) \subset M \cap V^{-1}(c)$. Finally, since $f^n(x)$ approaches $\Omega(x_0)$ as $n \to \infty$, $x(n) \to M \cap V^{-1}(c)$ as $n \to \infty$. This completes the proof. $\qquad\square$

We have the next theorem as an immediate consequence of Theorem 6.2.2.

Theorem 6.2.3. *Let $D \subset \mathbb{R}^k$ that contains the origin. Let $V(x) : D \to \mathbb{R}$ such that $\triangle V \leq 0$ in D. Let*

$$Z = \{x \in D : \triangle V(x) = 0\}.$$

Suppose that the only bounded solution of (6.1.1) which remains inside Z for all time is the equilibrium solution zero. Then the origin of (6.1.1) is asymptotically stable. In addition, if V is radially unbounded and positive definite, then the zero solution is (GAS).

We provide the following example.

Example 6.11. Consider the planar system

$$x(n + 1) = y(n)$$
$$y(n + 1) = -\frac{x(n)}{2} - y(n).$$

Clearly $(0, 0)$ is an equilibrium point of the system. Set

$$V(x, y) = |x(n)| + |y(n)|.$$

Clearly $V(0, 0) = 0$ and $V(x, y) > 0$ for $(x, y) \neq (0, 0)$. Along the solutions we have

$$\triangle V(x, y) = \left(|y(n)| - \frac{|x(n)|}{2}\right).$$

It is clear that $\triangle V(x, y) = 0$, when $y(n) = \pm \dfrac{x(n)}{2}$. Next we apply Theorem 6.2.2. We have $\triangle V = 0$, for $y = \frac{x}{2}$. We may define

$$E = \left\{(x, y) \in \overline{G} = \mathbb{R}^2 : y = \frac{x}{2}\right\}.$$

Substitute $y = \frac{x}{2}$ in the original system and get $x(n+1) = \frac{x(n)}{2}$ and $y(n+1) = -2y(n)$. By straight forward calculations we obtain the solutions,

$$x(n) = x(0) \left(\frac{1}{2}\right)^n \quad \text{and} \quad y(n) = y(0)(-2)^n,$$

respectively, for $n \in \mathbb{N}$. The relation $y(n) = \frac{x(n)}{2}$, must hold for all $n \in \mathbb{N}$. That is

$$y(0)(-2)^n = \frac{1}{2}x(0)\left(\frac{1}{2}\right)^n,$$

which can only hold for $x(0) = y(0) = 0$. Thus, the only bounded solution in the set E is the origin and

$$M = \{(x, y) \in E : x = 0, y = 0\}$$

is the largest invariant subset of E. Thus, the origin is asymptotically stable by Theorem 6.2.2. Since, V is radially unbounded, the origin is (GAS). In reference to Theorem 6.2.2, this implies that $c = \infty$ in $V^{-1}(c) = \{x \in \mathbb{R}^m : V(x) = c\}$.

Now we deal with the case $\triangle V(x, y) = 0$, when $y(n) = -\dfrac{x(n)}{2}$. Let

$$E = \left\{(x, y) \in \overline{G} = \mathbb{R}^2 : y = -\frac{x}{2}\right\}.$$

Substituting $y = -\frac{x}{2}$ in the original system gives

$$x(n+1) = -\frac{x(n)}{2}, \quad y(n+1) = 0.$$

The corresponding solutions are

$$x(n) = x(0)\left(-\frac{1}{2}\right)^n \quad \text{and} \quad y(n) = 0,$$

for $n \in \mathbb{N}$. The relation $y(n) = \frac{x(n)}{2}$, must hold for all $n \in \mathbb{N}$. That is

$$0 = -\frac{1}{2}x(0)\left(-\frac{1}{2}\right)^n,$$

which can only hold for $x(0) = 0$. Thus, there is no other bounded solution in the set E but the origin and

$$M = \{(x, y) \in E : x = 0, y = 0\}$$

is the largest invariant subset of E. Thus, the origin is asymptotically stable by Theorem 6.2.2. Since, V is radially unbounded, the origin is (GAS). In reference to Theorem 6.2.2, this implies that $c = \infty$ in $V^{-1}(c) = \{x \in \mathbb{R}^m : V(x) = c\}$.

As a matter of fact, linearization gives similar results since the eigenvalues of the system have magnitude less than one. Note that here, we were able to show (GAS). $\qquad\square$

6.2.1 Exercises

Exercise 6.14. Use Theorem 6.2.2 and discuss the (AS) and (GAS) of the zero solution of the system

$$x(n + 1) = y(n)$$
$$y(n + 1) = -x(n) - y(n).$$

Exercise 6.15. Use Theorem 6.2.2 and discuss the (AS) and (GAS) of the zero solution of the system

$$x(n + 1) = y(n) - y^3(n)$$
$$y(n + 1) = x(n) - y^2(n)x(n).$$

Exercise 6.16. Discuss the (S), (AS) and (GAS) of the zero solution of the system

$$x(n + 1) = y(n) + x(n)(\mu^2 - x^2(n) - y^2(n))$$
$$y(n + 1) = -x(n) + y(n)(\mu^2 - x^2(n) - y^2(n))$$

for $\mu \neq 0$.

Exercise 6.17. Discuss the (S), (AS) and (GAS) of the zero solution of the system

$$x(n + 1) = \frac{y(n)}{2} + x(n)(\mu^2 - x^2(n) - y^2(n))$$
$$y(n + 1) = -\frac{x(n)}{2} + y(n)(\mu^2 - x^2(n) - y^2(n))$$

for $\mu \neq 0$.

Exercise 6.18. Use Theorem 6.2.2 and discuss the (AS) and (GAS) of the zero solution of the system

$$x(n+1) = \frac{1}{2}(x(n) + y(n))$$

$$y(n+1) = \frac{x(n)}{2} + \frac{y(n)}{3}.$$

Hint: Try $V(x, y) = |x(n)| + |y(n)|$.

Exercise 6.19. Consider

$$x(n+2) = \frac{ax(n)}{1 + bx^2(n+1)}, \quad b > 0.$$

(a) Write the second order difference equation into a system and find all its equilibrium solutions.

(b) Show that for $a \le 1$, the origin $(0, 0)$ is the only equilibrium solution and then discuss its stability.

6.3 Non-autonomous systems

Now we turn our attention to systems that explicitly include the independent variable n. We consider the non-autonomous system

$$x(n+1) = f(n, x) \tag{6.3.1}$$

where $f \in C([0, \infty) \times D, \mathbb{R}^k)$, $D \subset \mathbb{R}^k$ that is open with $0 \in D$. We say a vector $x^* \in \mathbb{R}^k$ is an *equilibrium solution*, or *constant solution*, or *equilibrium point* of (6.3.1) if

$$f(n, x^*) = x^*.$$

This is the perfect place to make clear that the existence of equilibrium points for non-autonomous systems may depend on the initial time. To see this we consider

$$x(n+1) = \begin{pmatrix} \frac{1}{n} & n \\ 1 & n \end{pmatrix} \begin{pmatrix} x_1(n) \\ x_2(n) \end{pmatrix}, \quad n > 0, \ x(0) = x_0.$$

The system has a unique equilibrium solution $(0, 0)$ provided that the matrix is non-singular, which is the case for $n \ne 1$. Thus, in this case, the system attains an equilibrium solution if and only if the initial time $n_0 > 1$.

Definition 5.2.2 of stability carries over to (6.3.1). In addition, uniform asymptotic stability for (6.3.1) is equivalent to asymptotic stability for autonomous system.

Definition 6.3.1. A continuous function $W : [0, \infty) \to [0, \infty)$ with $W(0) = 0$, $W(s) > 0$ if $s > 0$, and W is strictly increasing is called a wedge. (In this book wedges are always denoted by W or W_i, where i is a positive integer.

Definition 6.3.2. Let D be an open set in $\mathbb{R}^k$ containing 0. A continuous autonomous function $V : [0, \infty) \times D \to [0, \infty)$ is *positive definite* if

$$V(n, 0) = 0 \quad \text{and} \quad V(n, x) > 0 \quad \text{if } x \neq 0, \text{ and for all } n \in [0, \infty).$$

Thus, $V(n, x)$ is positive definite if and only if there is a wedge independent of time n such that $V(n, x) \geq W(|x|)$ for all $x \in D$.

Definition 6.3.3. Let D be an open set in $\mathbb{R}^k$ containing 0. Let

$$V : [0, \infty) \times D \to [0, \infty)$$

be continuous. If V is positive definite and

$$\triangle V(n, x(n)) \leq 0,$$

for $(n, x) \in [n_0, \infty) \times D$, and for $x \neq 0$, then V is called *Lyapunov function* for the system (6.3.1).

If the inequality is strict, that is $\triangle V(n, x(n)) < 0$ then V is said to be a *strict Lyapunov function*.

Definition 6.3.4. Let D be an open set in $\mathbb{R}^n$ containing 0. Let

$$V : [0, \infty) \times D \to [0, \infty).$$

We say $V(n, x)$ is:

(a) *decrescent* in D if there is a wedge W_1 such that

$$V(n, x) \leq W_1(|x|) \quad \text{for all } x \in D,$$

(b) *radially unbounded* if $V(n, x) \to \infty$ as $|x| \to \infty$. This means that given an $M > 0$, there is an $N > 0$ such that $V(n, x) > M$ for all $n \in [0, \infty)$ provided that $|x| > N$,

(c) both positive definite and decrescent if and only if

$$W_1(|x|) \leq V(n, x) \leq W_2(|x|).$$

We provide the following example.

Example 6.12. Consider the function

$$V(n.x) = \frac{1}{2}(x_1^2(n) + x_1^2(n)), \quad n \geq 0, \quad x = [x_1, x_2]^T \in \mathbb{R}^2.$$

Then

$$V(n, x) = \frac{1}{2}(x_1^2(n) + x_1^2(n)) = \frac{1}{2}\left(\sqrt{x_1^2(n) + x_1^2(n)}\right)^2 = \frac{1}{2}(r(x))^2 \overset{def}{=} W(r(x)),$$

where $W(r) = \frac{1}{2}r^2$ is a wedge. Recall that for $x = [x_1, x_2, \cdots, x_k]^T$ in $\mathbb{R}^k$, $|x| = \sum_{i=1}^{k} |x_i|$ is equivalent to $r(x) = \sqrt{x_1^2 + x_2^2 + \cdots + x_k^2}$. □

In the next theorem, we as ask that the function V is decrescent.

Theorem 6.3.1. *Let $D \subset \mathbb{R}^n$ be an open set containing the origin, and let $V(n, x) : [0, \infty) \times D \to [0, \infty)$ be a given continuously differentiable function in x satisfying*

$$V(n, 0) = 0, \tag{6.3.2}$$

$$W_1(|x|) \leq V(n, x), \tag{6.3.3}$$

and

$$\triangle V(n, x) \leq 0, \quad \text{for all } n \geq 0, \quad \text{and} \quad x \neq 0, \tag{6.3.4}$$

then the zero solution of (6.3.1) *is stable.*

Proof. From (6.3.4) we have $\triangle V(n, x) \leq 0$, V is continuous in x, $V(n, 0) = 0$, and $W_1(|x|) \leq V(n, x)$. Let $\epsilon > 0$ and $n_0 \geq 0$ be given. We must find δ such that $|x_0| < \delta$ and $n \geq n_0$ imply $|x(n, n_0, x_0)| < \epsilon$. (Throughout these proofs we assume ϵ is small enough so that $|x(n, n_0, x_0)| < \epsilon$ implies that $x \in D$.) As V is continuous in x and $V(n, 0) = 0$ there is a $\delta > 0$ such that $|x_0| < \delta$ implies $V(n_0, x_0) < W_1(\epsilon)$. Thus, if $n \geq n_0$ and $|x_0| < \delta$ and $x = x(n, n_0, x_0)$, we have from (6.3.3) that

$$W_1(|x(n)|) \leq V(n, x) \leq V(n_0, x_0) < W_1(\epsilon),$$

or $|x(n)| < \epsilon$ as required. □

The next example will show that $\triangle V(n, x) \leq 0$ is not enough to drive solutions to zero.

Example 6.13 (Raffoul). Let $g : [0, \infty) \to (0, \beta]$ with $g(0) = 1$, and $0 < \beta \leq 1$. Consider the non-autonomous difference equation

$$x(n + 1) = \left[g(n + 1)/g(n)\right]x(n). \tag{6.3.5}$$

It is clear that $x(n) = g(n)$ is a solution of (6.3.5). Our goal is to construct a function

$$V(n, x) = a(n)x^2(n)$$

such that $\Delta V(n, x) = -\alpha(n)x^2(n)$, where $a(n), \alpha(n) > 0$ for $n \in \mathbb{Z}^+$, and $\sum_{s=0}^{\infty} \alpha(s) < \infty$. That is $\Delta V(n, x) \leq 0$. Along the solutions of (6.3.5) we have

$$\Delta V(n, x) = \left[a(n+1)\frac{g^2(n+1)}{g^2(n)} - a(n)\right]x^2(n).$$

By setting

$$a(n+1)\frac{g^2(n+1)}{g^2(n)} - a(n) = -\alpha(n),$$

we get the difference equation

$$a(n+1) - \frac{g^2(n)}{g^2(n+1)}a(n) = -\frac{g^2(n)}{g^2(n+1)}\alpha(n). \qquad (6.3.6)$$

Using the variation of parameters formula given by (2.2.6), Eq. (6.3.6) has the solution, after some simplification,

$$a(n) = \left(\prod_{i=0}^{n-1}\frac{g^2(i)}{g^2(i+1)}\right)a(0) - \sum_{r=0}^{n-1}\prod_{i=r+1}^{n-1}\frac{g^2(i)}{g^2(i+1)}\frac{g^2(r)}{g^2(r+1)}\alpha(r)$$

$$= \left[a(0)g^2(0) - \sum_{r=0}^{n-1}g^2(r)\alpha(r)\right]/g^2(n)$$

$$\geq \left[a(0) - \beta^2\sum_{r=0}^{n-1}\alpha(r)\right]/g^2(n).$$

Since $0 < g \leq 1$, and $\sum_{s=0}^{\infty}\alpha(s) < \infty$, we may choose $a(0)$ so large to imply that $a(n) > 1$ for all $n \geq 1$. Thus we have shown that $V \geq 0$ and ΔV is negative definite are not enough to drive solutions to tend to zero. Notice that V is not decrescent. That is, there is no wedge W_2 with $V(n, x) \leq W_2(|x|)$. This shows condition (6.3.3) is necessary. $\qquad \square$

In the upcoming theorem we as ask that the function V be both positive definite and decrescent in order to obtain (US).

Theorem 6.3.2. *Let $D \subset \mathbb{R}^n$ be an open set containing the origin, and let $V(n, x) : [0, \infty) \times D \to [0, \infty)$ be a given continuously differentiable function in x satisfying*

$$V(n, 0) = 0,$$
$$W_1(|x|) \leq V(n, x) \leq W_2(|x|),$$

and

$$\Delta V(n, x) \leq 0, \quad \text{for all } n \geq 0, \quad \text{and} \quad x \neq 0,$$

then the zero solution of (6.3.1) *is uniformly stable.*

Proof. For a given ϵ, from the stability part, we select a $\delta > 0$ such that $W_2(\delta) < W_1(\epsilon)$ where $W_1(|x|) \leq V(n, x) \leq W_2(|x|)$. If $n_0 \geq 0$, we have

$$W_1(|x(n)|) \leq V(n, x) \leq V(n_0, x_0)$$
$$\leq W_2(|x_0|) < W_2(\delta) < W_1(\epsilon),$$

or $|x(n)| < \epsilon$ as required. This completes the proof. □

The next theorem shows that for (UAS) we must ask that V be positive definite, decrescent, and $\Delta V(n, x)$ negative definite.

Theorem 6.3.3. *Let $D \subset \mathbb{R}^n$ be an open set containing the origin, and let $V(n, x) : [0, \infty) \times D \to [0, \infty)$ be a given continuously differentiable function in x satisfying*

$$V(n, 0) = 0,$$
$$W_1(|x|) \leq V(n, x) \leq W_2(|x|),$$

and

$$\Delta V(n, x) \leq -W_3(|x|), \quad \text{for all } n \geq 0, \quad \text{and} \quad x \neq 0,$$

then the zero solution of (6.3.1) *is uniformly asymptotically stable.*

Proof. Let $\epsilon = 1$, and find δ of uniform stability and call it η. Let γ be given. We must find $T > 0$ such that

$$|x_0| < \eta, \quad n_0 \geq 0, \quad \text{and} \quad n \geq n_0 + T$$

imply $|x(n, n_0, x_0)| < \gamma$. Pick $\mu > 0$ with $W_2(\mu) < W_1(\gamma)$, so that there is $n_1 \geq n_0$ with $|x(n_1)| < \mu$, then, for $n \geq n_1$, we have

$$W_1(|x(n)|) \leq V(n, x) \leq V(n_1, x_1)$$
$$\leq W_2(|x_1|) < W_2(\delta) < W_1(\gamma),$$

or $|x(n_1)| < \gamma$. Since $\Delta V(n, x) \leq -W_3(|x|)$, so as long as $|x(n)| > \mu$, then $\Delta V(n, x) \leq -W_3(\mu)$; thus

$$V(n, x(n)) \leq V(n_0, x_0) - \sum_{s=n_0}^{n-1} W_3(|x(s)|)$$
$$\leq W_2(|x_0|) - W_3(\mu)(n - n_0)$$
$$\leq W_2(\eta) - W_3(\mu)(n - n_0),$$

which vanishes at

$$n = n_0 + \frac{W_2(\eta)}{W_3(\mu)} \geq n_0 + T,$$

where $T \geq \frac{W_2(\eta)}{W_3(\mu)}$. Hence, if $T > \frac{W_2(\eta)}{W_3(\mu)}$, then $|x(n)| > \mu$ fails, and we have $|x(n)| < \gamma$ for all $n \geq n_0 + T$. This proves (UAS). This completes the proof. $\square$

We display the following example.

Example 6.14. Consider the second-order difference equation

$$x(n+2) + e(n)x(n) = 0,$$

where e satisfies the condition that there is a constant $\alpha \in (0, 1)$ so that

$$\alpha \leq e(n) \leq 1 \quad \text{and} \quad \Delta e(n) \leq 0.$$

Let $x_1(n) = x(n)$ and $x_2(n) = x(n+1)$. Then we arrive at the system

$$x_1(n+1) = x_2(n), \quad x_2(n+1) = e(n)x_1(n).$$

Let

$$V(n, x) = e(n)|x_1(n)| + |x_2(n)|,$$

where $x = \begin{pmatrix} x_1 \\ x_2 \end{pmatrix}$. Then along the solutions we have

$$
\begin{aligned}
\Delta V(n, x) &= e(n+1))|x_1(n+1)| + |x_2(n+1)| - e(n)|x_1(n)| - |x_2(n)| \\
&= e(n+1))|x_2(n)| + e(n)|x_1(n)| - e(n)|x_1(n)| - |x_2(n)| \\
&= (e(n+1) - 1)\,|x_2(n)| \\
&= \Delta e(n)|x_2(n)| + (e(n) - 1)\,|x_2(n)| \\
&\leq 0.
\end{aligned}
$$

Due to the condition on $e(n)$ we have that

$$V(n, x) = e(n)|x_1(n)| + |x_2(n)| \geq \alpha\,(|x_1(n)| + |x_2(n)|) = W_1(|x|),$$

and

$$V(n, x) = e(n)|x_1(n)| + |x_2(n)| \leq (|x_1(n)| + |x_2(n)|) = W_2(|x|).$$

Thus, the zero solution is uniformly stable by Theorem 6.3.2. $\square$

Next we address instability of (6.3.1).

Definition 6.3.5. The zero solution of (6.3.1) is unstable if there is an $\varepsilon > 0$, and $n_0 \geq 0$, such that for any $\delta > 0$ there is an x_0 with $|x_0| < \delta$ and there is an $n_1 > n_0$ such that $|x(n_1, n_0, x_0)| \geq \varepsilon$.

Theorem 6.3.4 (Raffoul). *Suppose there exists a continuous Lyapunov function* $V : \mathbb{Z}^+ \times D \to [0, \infty)$ *which is locally Lipschitz in x such that*

$$W_1(|x|) \leq V(n, x) \leq W_2(|x|), \tag{6.3.7}$$

and along the solutions of (6.3.1) *we have*

$$\Delta V(n, x) \geq W_3(|x|). \tag{6.3.8}$$

Then the zero solution of (6.3.1) *is unstable.*

Proof. Suppose not, then for $\epsilon = \min\{1, d(0, \partial D)\}$ we can find a $\delta > 0$ such that $|x_0| < \delta$ and $n \geq 0$ imply that $|x(n, 0, x_0)| < \epsilon$. We may pick x_0 in such a way so that $|x_0| = \delta/2$ and find $\gamma > 0$ with $W_2(\gamma) = W_1\delta/2)$. Then for $x(n) = x(n, 0, x_0)$ we have $\Delta V(n, x) \geq 0$ so that

$$W_2(|x(n)|) \geq V(n, x(n)) \geq V(0, x_0) \geq W_1(\delta/2) = W_2(\gamma)$$

from which we conclude that $\gamma \leq |x(n)|$ for $n \geq 0$. Thus

$$\Delta V(n, x) \geq W_3(|x(n)|) \geq W_3(\gamma).$$

Thus,

$$W_2(|x(n)|) \geq V(n, x(n)) \geq V(0, x_0) + n W_3(\gamma),$$

from which we conclude that $|x(n)| \to \infty$, which is a contradiction. This completes the proof. $\square$

We provide the following example.

Example 6.15. Consider the non-autonomous system

$$x(n + 1) = a(n)x(n) + b(n)y(n)$$
$$y(n + 1) = -a(n)x(n) - b(n)y(n).$$

Clearly, $(0, 0)$ is an equilibrium solution. Assume there is a $\delta > 0$ so that

$$2|a(n)| - 1 \leq -\delta, \ 2|b(n)| - 1 \leq -\delta,$$

and

$$\lim_{n \to \infty} (2|a(n)| - 1) \neq 0, \ \lim_{n \to \infty} (2|b(n)| - 1) \neq 0.$$

Let

$$V(n, x, y) = |x(n)| + |y(n)|.$$

Then V is positive definite and decrescent on $\mathbb{N} \times \mathbb{R}^2$. Set $V(x, y) = x^2 + y^2$. It readily follows that

$$\Delta V(x) = |x(n+1)| + |y(n+1)| - |x(n)| - |y(n)|$$
$$\leq -\delta(|x(n)| + |y(n)|)$$

It is easy to see that

$$W_1(|(x, y)|) = W_2(|(x, y)|) = |x(n)| + |y(n)|, \quad \text{and}$$
$$W_3(|(x, y)|) = \delta(|x(n)| + |y(n)|).$$

It follows from Theorem 6.3.3 that the zero solution is (UAS). $\qquad \square$

6.3.1 Exercises

Exercise 6.20. Let $D \subset \mathbb{R}^n$ be an open set containing the origin, and let $V(n, x) : [0, \infty) \times D \to [0, \infty)$ be a given continuously differentiable function satisfying

$$V(n, 0) = 0,$$
$$W_1(|x|) \leq V(n, x),$$

and

$$\Delta V(n, x) \leq -W_2(|x|), \quad \text{for all } n \geq 0, \text{ and } x \neq 0,$$

then the zero solution of (6.3.1) is asymptotically stable.

Exercise 6.21. Use the function

$$V(n, x) = e(n)x_1^2(n) + x_2^2(n)$$

to show the zero solution of the system in Example 6.14 is uniformly stable under the condition that $\alpha \leq e(n) \leq 1$, for $\alpha \in (0, 1)$, and $\Delta e(n) \leq 0$.

Exercise 6.22. Use the function

$$V(n, x) = |x_1(n)| + |x_2(n)|$$

to show the zero solution of the system

$$x_1(n+1) = \frac{1}{2}x_1(n) + e(n)x_2(n), \quad x_2(n+1) = \frac{1}{2}x_1(n) - \frac{1}{2}x_2(n)$$

is uniformly stable under the condition that $0 \leq e(n) \leq \frac{1}{2}$.

Exercise 6.23. Let $D = \{x \in \mathbb{R} : |x| \leq 1, x \neq 0\}$. Define a function $V(n, x):$ $[0, \infty) \times D \to [0, \infty)$ to show the zero solution of

$$x(n + 1) = e(n)x^3(n)$$

is uniformly stable, where $e(n) \geq 0$, $e(n + 1) \leq 1$, for all $n \geq 0$.

Exercise 6.24. Use a Lyapunov function to show the zero solution of the system

$$x_1(n + 1) = \alpha(n)x_2(n), \quad x_2(n + 1) = \beta(n)x_1(n)$$

is uniformly asymptotically stable under the conditions that $|\alpha(n)| < 1$ and $|\beta(n)| < 1$, for either $|\alpha(n)| \leq |\beta(n)|$ or $|\beta(n)| \leq |\alpha(n)|$, for all $n \geq 0$.

Exercise 6.25. Use a Lyapunov function to show the zero solution of the scalar non-linear Volterra difference equation

$$x(n + 1) = a(n)x(n) + b(n)\frac{x(n)}{1 + \sum_{s=0}^{n-1} x^2(s)}, \quad n \geq 0,$$

is (US) for

$$|a(n)| + |b(n)| \leq 1,$$

and (UAS) for

$$|a(n)| + |b(n)| < 1.$$

Exercise 6.26. Show the function

$$V(n, x_1, x_2) = x_1^2(n) + (1 + n^2)x_2^2(n)$$

is positive definite in $\mathbb{Z}^+ \times \mathbb{R}^2$, but not decrescent.

Exercise 6.27. Show the function

$$V(n, x_1, x_2) = x_1^2(n) + \frac{1}{1 + n^2}x_2^2(n)$$

is not positive definite in $\mathbb{Z}^+ \times \mathbb{R}^2$, but decrescent.

Exercise 6.28. Consider the non-autonomous system

$$x(n + 1) = -a(n)x(n) + b(n)y(n)$$
$$y(n + 1) = a(n)x(n) + b(n)y(n).$$

Assume there is a $\delta > 0$ so that

$$2a^2(n) - 1 \leq -\delta, \quad 2b^2(n) - 1 \leq -\delta,$$

and

$$\lim_{n\to\infty} (2a^2(n) - 1) \neq 0, \quad \lim_{n\to\infty} (2b^2(n) - 1) \neq 0.$$

Use a quadratic Lyapunov function to show the zero solution is (UAS).

Exercise 6.29. Consider the non-autonomous system

$$x(n+1) = a(n)x(n) - b(n)y(n)$$
$$y(n+1) = a(n)x(n) + b(n)y(n).$$

Assume there is a $\delta > 0$ so that

$$2a^2(n) - 1 \geq \delta, \quad 2b^2(n) - 1 \geq \delta.$$

Use a quadratic Lyapunov function to show the zero solution is unstable.

6.4 Connection between Lyapunov functions and eigenvalues

Consider the homogeneous system

$$x(n+1) = Ax(n) \tag{6.4.1}$$

where A is an $k \times k$ constant matrix. We are mainly concerned with the existence of a Lyapunov function for system (6.4.1) in relation to the eigenvalues of the matrix A. We shall show later that if all the eigenvalues of A have a magnitude of less than one, then there is a strict Lyapunov function. The next definition plays a crucial role in our future results.

Definition 6.4.1. Let $x \in \mathbb{R}^k$ and B be an $k \times k$ constant matrix. We say B is a *positive definite* (*negative definite*), if

$$x^T Bx > 0 \, (< 0) \quad \text{for } x \neq 0.$$

Example 6.16. Let

$$B = I = \begin{pmatrix} 1 & 0 & \cdots & 0 \\ 0 & 1 & \ddots & \vdots \\ \vdots & \ddots & \ddots & \vdots \\ 0 & \cdots & 0 & 1 \end{pmatrix} \quad \text{and} \quad x = \begin{pmatrix} x_1 \\ x_2 \\ \vdots \\ x_k \end{pmatrix}.$$

Then

$$x^T B x = (x_1, x_2, \ldots, x_n) \begin{pmatrix} 1 & 0 & \cdots & 0 \\ 0 & 1 & \ddots & \vdots \\ \vdots & \ddots & \ddots & \vdots \\ 0 & \cdots & 0 & 1 \end{pmatrix} \begin{pmatrix} x_1 \\ x_2 \\ \vdots \\ x_k \end{pmatrix}$$

$$= x_1^2 + x_2^2 + \ldots + x_k^2 > 0 \quad \text{for } x \neq 0.$$

Thus the $k \times k$ identity matrix I is positive definite.

To introduce the concept that we are about to discuss, we assume the existence of a positive definite symmetric $k \times k$ constant matrix B and define the scalar function

$$V(x) = x^T(n) B x(n). \tag{6.4.2}$$

Then $V(x)$ is positive definite and moreover, along the solutions of (6.4.1) we have that

$$\Delta V(x) = x^T(n+1) B x(n+1) - x^T(n) B x(n)$$
$$= (Ax(n))^T B x - x^T(n) B x(n)$$
$$= x^T(n) A^T B A x(n) - x^T(n) B x(n)$$
$$= x^T(n) \left(A^T B A - B \right) x(n). \tag{6.4.3}$$

In order to get any form of stability using (6.4.3), we need to have $A^T B A - B = -C$ where C is a positive definite symmetric matrix. We notice that

$$\left(A^T B A - B \right)^T = (BA)^T A - B^T$$
$$= A^T B A - B,$$

which implies that $A^T B A - B$ is symmetric.

Remark 6.1. If (6.4.2) holds, then there are positive constants α_1 and α_2 such that

$$\alpha_1 x^T x \leq x^T B x \leq \alpha_2 x^T x. \tag{6.4.4}$$

Then $W_1(|x|) = \alpha_1 x^T x$ and $W_2(|x|) = \alpha_2 x^T x$ could serve as the two wedges such that

$$\alpha_1 x^T x \leq V(x) \leq \alpha_2 x^T x.$$

The next theorem is concerned with the existence of positive definite symmetric matrix B, provided all the eigenvalues of the matrix A lie inside the unit circle. Consider the autonomous linear system of difference equations

$$x(n+1) = Ax(n) \tag{6.4.5}$$

where A is an $k \times k$ constant matrix.

Theorem 6.4.1 (Raffoul). *For a given positive definite matrix C, the equation*

$$A^T BA - B = -C \tag{6.4.6}$$

can be solved for a positive definite symmetric matrix B if and only if all eigenvalues of A lie inside the unit circle.

Proof. Assume (6.4.6) and let $V = x^T Bx$. Then $\triangle V < 0$ for $x \neq 0$. Hence the zero solution of (6.4.5) is asymptotically stable by Theorem 6.1.2, and hence it follows from Theorem 5.4.3 and Corollary 5.1 that all the eigenvalues of the matrix A lie inside the unit circle.

Suppose all the eigenvalues of A lie inside the unit circle and define the matrix B by

$$B = \sum_{n=0}^{\infty} (A^T)^n C A^n.$$

Since all the eigenvalues of A reside inside the unit circle, we have $|A^n| \leq Ka^{-\delta n}$ for positive constants K, a and δ, such that $0 < a < 1$. Hence, the infinite sum converges (geometric series). It is clear from the definition of B, the matrix B is symmetric. Next we show it is positive definite.

$$x^T (A^T)^n C A^n x = (A^n x)^T C(A^n x) = y^T Cy,$$

and if $x \neq 0$, then $y \neq 0$; as C is positive definite $y^T Cy > 0$ for $x \neq 0$. Thus B is positive definite. Finally,

$$\begin{aligned}
A^T BA - B &= A^T \sum_{n=0}^{\infty} (A^T)^n C A^n A - \sum_{n=0}^{\infty} (A^T)^n C A^n \\
&= \sum_{n=0}^{\infty} (A^T)^{n+1} C A^{n+1} - \sum_{n=0}^{\infty} (A^T)^n C A^n \\
&= \sum_{n=0}^{\infty} \triangle_n \left((A^T)^n C A^n \right) \\
&= (A^T)^n C A^n \Big|_{n=0}^{\infty} = \text{zero matrix} - C \\
&= -C.
\end{aligned}$$

This completes the proof. $\qquad\square$

We have the immediate corollary.

Corollary 6.1. *Suppose there exists a positive definite symmetric matrix B such that*

$$A^T BA - B = -I. \tag{6.4.7}$$

Then the zero solution of (6.4.5) is (UAS).

Proof. Let V be given by (6.4.2). Then along the solutions of (6.4.5) we have from (6.4.3) that

$$\Delta V(x) = -x^T(n)x(n) := -W_3(|x|).$$

The two wedges $W_1(|x|)$ and $W_2(|x|)$ are readily made available by Remark 6.1, and hence the result follows from Theorem 6.3.3. This completes the proof. $\square$

When exploring stability and boundedness for non-linear non-autonomous systems, the effective approach is to study equations that are related in some way to the linear equations whose behavior is known to be covered by the known theory. Thus, we consider the perturbed non-linear system

$$x(n+1) = Ax(n) + g(n, x(n)), \quad n \geq 0 \tag{6.4.8}$$

where A is an $k \times k$ constant matrix and $g : \mathbb{Z}^+ \times \mathbb{R}^k \to \mathbb{R}^k$ is continuous in x with $g(n, 0) = 0$. The perturbation g is assumed small in some sense. The next theorem pertains to the stability of the perturbed non-linear difference equation given by (6.4.8).

Theorem 6.4.2 (Raffoul). *Suppose there exists a positive definite symmetric matrix B such that*

$$A^T BA - B = -I, \tag{6.4.9}$$

$$\lim_{x \to 0} \frac{|g(n, x)|}{|x|} = 0, \text{ uniformly in } n \tag{6.4.10}$$

then the zero solution of (6.4.8) is uniformly asymptotically stable.

Proof. We write x for $x(n)$ and we define the Lyapunov function V by $V = x^T Bx$. Using (6.4.9)–(6.4.10), we have along the solutions of (6.4.8) that

$$\Delta V = (Ax + g(n, x))^T B(Ax + g(n, x)) - x^T Bx$$
$$= x^T(A^T BA - B)x + 2x^T A^T Bg(n, x) + g^T(n, x)Bg(n, x)$$
$$= |x|^2 \left(-1 + 2\frac{x^T}{|x|} A^T B \frac{g(n, x)}{|x|} + \frac{g^T(n, x)}{|x|} B \frac{g(n, x)}{|x|} \right)$$
$$\leq |x|^2 \left(-1 + 2|A^T B| \frac{|g(n, x)|}{|x|} + \frac{|g^T(n, x)|}{|x|} B \frac{|g(n, x)|}{|x|} \right).$$

Due to condition (6.4.10), there are ξ_1, $\xi_2 > 0$ such that

$$2|A^T B| \frac{|g(n, x)|}{|x|} \leq \xi_1, \quad \text{and} \quad \frac{|g^T(n, x)|}{|x|} B \frac{|g(n, x)|}{|x|} \leq \xi_2,$$

where

$$-1 + \xi_1 + \xi_2 \leq -\xi, \quad \xi \in (0, 1),$$

for sufficiently small $|x|$ and $x \neq 0$. Consequently, we have

$$\triangle V \leq -\xi |x|^2.$$

Since B is symmetric and positive definite, from (6.4.4) we have that

$$\alpha^2 x^T x \leq V(x) \leq \beta^2 x^T x, \qquad (6.4.11)$$

from which we get

$$-|x|^2 \leq -\frac{1}{\beta^2} V(x) \leq -\frac{\alpha^2}{\beta^2} |x|^2.$$

As a consequence

$$\triangle V \leq -\frac{\xi}{\beta^2} V(x),$$

which has the solution

$$V(x(n)) \leq V(x_0) \left(1 + \frac{-\xi}{\beta^2}\right)^{(n-n_0)}.$$

Since β can be taken as large as needs to be and $\xi \in (0, 1)$ we have $\zeta = 1 - \frac{\xi}{\beta^2} \in (0, 1)$. From this we obtain

$$V(x(n)) \leq V(x_0) (\zeta)^{(n-n_0)}.$$

Using (6.4.11) we arrive at

$$\alpha^2 |x|^2 \leq V(x) \leq V(x_0) (\zeta)^{(n-n_0)}, \quad n \geq n_0.$$

It follows

$$\alpha^2 |x|^2 \leq x_0^T B x_0 (\zeta)^{(n-n_0)} \leq |x_0^T| |B| |x_0| (\zeta)^{(n-n_0)}.$$

For $\epsilon > 0$ and $|x_0| < \delta$ we have

$$|x| \leq \frac{1}{\alpha} (\delta |B| \delta)^{\frac{1}{2}} = \frac{\delta}{\alpha} |B|^{\frac{1}{2}} < \epsilon,$$

for $\delta = \frac{\epsilon\alpha}{|B|^{1/2}}$. Hence the zero solution is uniformly stable. Let $\delta = 1$ from the uniform stability so that for $|x(n_0)| = |x_0| < \delta$, need to show that for $n \geq n_0 + T(\epsilon)$, $|x(n)| < \epsilon$. Set $T(\epsilon) = \frac{2}{\ln(\zeta)} \ln\left(\frac{\alpha\epsilon}{|B|^{1/2}}\right)$. Then

$$|x| \leq \frac{|B|^{1/2}}{\alpha} (\zeta)^{\frac{n-n_0}{2}}$$

$$\leq \frac{|B|^{1/2}}{\alpha} (\zeta)^{\frac{T(\epsilon)}{2}}$$

$$< \epsilon.$$

This completes the proof. $\qquad\qquad\qquad\qquad\qquad\qquad\qquad\qquad\qquad$ $\square$

Remark 6.2. The hypothesis of Theorem 6.4.2 implies the zero solution of (6.4.8) is exponentially stable.

6.4.1 Exercises

Exercise 6.30. Let

$$A = \begin{pmatrix} 1 & -5 \\ 1/4 & -1 \end{pmatrix}.$$

(a) Show that the eigenvalues of A lie inside the unit circle, and find a symmetric positive definite matrix B so that $A^T B A - B = -I$.

(b) Write down a Lyapunov function that implies the zero solution of the system

$$x(n+1) = Ax(n), \quad x(n_0) = x_0, \quad n \geq n_0 \geq 0,$$

is (UAS).

Exercise 6.31. Repeat Exercise 6.30 for the matrix

$$A = \begin{pmatrix} 1/2 & 1/2 \\ 1/2 & 1/3 \end{pmatrix}.$$

Exercise 6.32. Construct a positive definite Lyapunov function and show the zero solution of the system

$$x(n+1) = \begin{pmatrix} 2 & 1 \\ 0 & -2 \end{pmatrix} x(n),$$

is unstable.

6.5 Exponential stability

In this section we employ non-negative definite Lyapunov functionals to obtain sufficient conditions that guarantee exponential stability of the zero solution of a non-linear discrete system

$$x(n+1) = f(n, x(n)), \quad n \geq 0 \qquad (6.5.1)$$
$$x(n_0) = x_0, \quad n_0 \geq 0$$

where $x(n) \in \mathbb{R}^k$, $f(n, x(n)) : \mathbb{Z}^+ \times \mathbb{R}^k \to \mathbb{R}^k$ is a given non-linear function satisfying $f(n, 0) = 0$ for all $n \in \mathbb{Z}^+$. We assume that $f(n, x)$ has the required conditions so that solutions exist for all $n \geq 0$. Throughout this section, $||x||$ is the Euclidean norm of the vector $x(n) \in \mathbb{R}^k$. From this point forward, when a function is written without its argument, the argument is assumed to be n. In addition, the function $V(n, x)$ is always assumed to be a positive definite, see Definition 6.3.2.

We restate the following definition regarding exponential stability.

Definition 6.5.1. The zero solution of system (6.5.1) is said to be exponentially stable if any solution $x(n, n_0, x_0)$ of (6.5.1) satisfies

$$||x(n, n_0, x_0)|| \leq C\Big(||x_0||, n_0\Big)a^{-\delta(n-n_0)}, \quad \text{for all } n \geq n_0,$$

where a is constant with $a > 1$, $C : \mathbb{R}^+ \times \mathbb{Z}^+ \to \mathbb{R}^+$, and δ is a positive constant. The zero solution of (6.5.1) is said to be uniformly exponentially stable if C is independent of n_0.

For motivational purpose, assume we have a positive definite function $V(n, x) : \mathbb{Z}^+ \times D \to \mathbb{R}^+$, where $D \subset \mathbb{R}^k$ is an open set containing the origin. Suppose along the solutions of (6.5.1) we have for a constant $\alpha \in (0, 1)$, that

$$\triangle V(n, x) \leq -\alpha V(n, x).$$

Then, by Section 2.1 we have the solution

$$V(n, x) = (1 - \alpha)^{n-n_0} V(n_0, x_0).$$

Now, in most cases there exists a positive integer r so that $||x||^r \leq V(n, x)$, then we end up with

$$||x|| \leq \big((1 - \alpha)^{n-n_0} V(n_0, x_0)\big)^{\frac{1}{r}},$$

which implies exponential stability of the zero solution. Our interest in this section is to introduce unconventional types of Lyapunov functions that satisfy the property

$$\triangle V(n, x) \leq -w(|x|) + e(n)$$

where w is a wedge and the function $e(n)$ decays exponentially, and yet still obtain exponential stability of the zero solution. We provide an example.

Example 6.17. Consider the non-linear system

$$x(n+1) = -\frac{x(n)}{2} + y^2(n)$$

$$y(n+1) = -\frac{y(n)}{2} - x(n)y(n).$$

In this example, we shall display a Lyapunov function and show that indeed the zero solution is (ES). In the computations to follow, we will suppress the independent variable n. Let

$$D = \{(x, y) \neq (0, 0) : |y| < \frac{1}{\sqrt{2}}\}$$

and consider the set D the Lyapunov function

$$V(x, y) = x^2(n) + y^2(n).$$

Then, $V(x^*, y^*) = V(0, 0) = 0$, and $V(x, y) > 0$, for $(x, y) \neq 0$. Moreover,

$$\Delta V(x, y) = x^2(n+1) + y^2(n+1) - x^2(n) - y^2(n)$$

$$= (-\frac{x}{2} + y^2)^2 + (-\frac{y}{2} - xy)^2 - x^2 - y^2$$

$$= -\frac{3}{4}x^2 - \frac{3}{4}y^2 + y^4 + x^2y^2$$

$$= x^2(-\frac{3}{4} + y^2) + y^2(-\frac{3}{4} + y^2)$$

$$= (-\frac{3}{4} + y^2)(x^2 + y^2)$$

$$\leq -\frac{1}{4}(x^2 + y^2)$$

$$= -\frac{1}{4}V(x, y).$$

This shows the zero solution $(0, 0)$ is (ES). Actually,

$$x^2(n) + y^2(n) \rightarrow (0, 0),$$

exponentially as $n \rightarrow \infty$ at the convergence rate

$$\sqrt{x^2(n) + y^2(n)} \leq \left((1 - \frac{1}{4})^{n-n_0} V(x_0, y_0) \right)^{\frac{1}{2}}$$

$$= \left((1 - \frac{1}{4})^{n-n_0} (x_0^2 + y_0^2) \right)^{\frac{1}{2}}, \quad n \geq n_0$$

provided that $(x_0, y_0) \in D$. $\hspace{6cm}$ $\square$

In whats to follow we preset a series of theorems regarding the role of Lyapunov functions with exponential stability. We begin with the following theorem.

Theorem 6.5.1. *Let a be a constant with a > 1. Let $D \subset \mathbb{R}^k$ be an open set containing the origin, and let $V(n, x) : \mathbb{Z}^+ \times D \to \mathbb{R}^+$ be a given function satisfying*

$$\lambda_1 ||x||^p \leq V(n, x) \leq \lambda_2 ||x||^q, \tag{6.5.2}$$

and

$$\Delta V(n, x) \leq -\lambda_3 ||x||^r + ka^{-\delta n}, \tag{6.5.3}$$

for some positive constants λ_1, λ_2, λ_3, p, q, r, k and δ. Moreover, if for some positive constants α and γ,

$$0 < \frac{\lambda_3}{\lambda_2^{r/q}} \leq \alpha < 1 \tag{6.5.4}$$

such that

$$V(n, x) - V^{r/q}(n, x) \leq \gamma a^{-\delta n} \tag{6.5.5}$$

with

$$\delta > -\frac{\ln(1 - \lambda_3/\lambda_2^{r/q})}{\ln(a)}, \tag{6.5.6}$$

then the zero solution of (6.5.1) is uniformly exponentially stable.

Proof. First note that in view of (6.5.4), the constant δ, which is given by (6.5.6) is positive. Let $M = -\dfrac{\ln(1 - \lambda_3/\lambda_2^{r/q})}{\ln(a)}$. Then taking the Δ of the function $V(n, x)a^{M(n-n_0)}$ we have

$$\Delta\left(V(n, x)a^{M(n-n_0)}\right) = \left[V(n + 1, x)a^M - V(n, x)\right]a^{M(n-n_0)}.$$

For $x \in D$, using (6.5.3) we get

$$\Delta\left(V(n, x)a^{M(n-n_0)}\right) \leq \left[-\lambda_3 ||x||^r a^M + V(n, x)a^M \right.$$
$$\left. + ka^{-\delta n}a^M - V(n, x)\right]a^{M(n-n_0)}. \tag{6.5.7}$$

From condition (6.5.2) we have $||x||^q \geq V(n, x)/\lambda_2$ and consequently, $-||x||^r \leq -[\frac{V(n,x)}{\lambda_2}]^{r/q}$. Thus, inequality (6.5.7) becomes

$$\Delta\left(V(n, x)a^{M(n-n_0)}\right) \leq \left[-a^M(\lambda_3/\lambda_2^{r/q})V^{r/q}(n, x) + V(n, x)a^M\right.$$
$$\left. + ka^{-\delta n}a^M - V(n, x)\right]a^{M(n-n_0)}$$
$$= \left[-a^M(\lambda_3/\lambda_2^{r/q})V^{r/q}(n, x) + (a^M - 1)V(n, x)\right.$$
$$\left. + ka^{-\delta n}a^M\right]a^{M(n-n_0)}.$$

Since

$$M = -\frac{\ln(1 - \lambda_3/\lambda_2^{r/q})}{\ln(a)}, \quad \text{we have } a^M - 1 = a^M(\lambda_3/\lambda_2^{r/q}),$$

and hence the above inequality reduces to

$$\Delta\left(V(n, x)a^{M(n-n_0)}\right) \leq \left[(a^M - 1)\left(V(n, x) - V^{r/q}(n, x)\right)\right.$$
$$\left. + ka^{-\delta n}a^M\right]a^{M(n-n_0)}. \tag{6.5.8}$$

By invoking condition (6.5.5), inequality (6.5.8) takes the form

$$\Delta\left(V(n, x)a^{M(n-n_0)}\right) \leq \left((a^M - 1)\gamma + ka^M\right)a^{-\delta n}a^{M(n-n_0)}$$
$$\leq \left((a^M - 1)\gamma + ka^M\right)a^{-\delta n+\delta n_0}a^{M(n-n_0)}$$
$$= La^{(M-\delta)(n-n_0)},$$

where $L = (a^M - 1)\gamma + ka^M$. Summing the above inequality from n_0 to $n - 1$ we obtain,

$$V(n, x)a^{M(n-n_0)} - V(n_0, x_0) \leq La^{-(M-\delta)n_0} \sum_{s=n_0}^{n-1} a^{(M-\delta)s}$$
$$= \frac{La^{-(M-\delta)n_0}}{a^{(M-\delta)} - 1}\left[a^{(M-\delta)n} - a^{(M-\delta)n_0}\right]$$
$$= \frac{L}{a^{(M-\delta)} - 1}\left[a^{(M-\delta)(n-n_0)} - 1\right].$$

Since $M < \delta$ and $V(n_0, x_0) \leq \lambda_2||x_0||^q$, the above inequality reduces to

$$V(n, x)a^{M(n-n_0)} \leq \lambda_2||x_0||^q + \frac{L}{1 - a^{(M-\delta)}}.$$

Set $B(||x_0||) = \lambda_2||x_0||^q + \frac{L}{1-a^{(M-\delta)}}$. Then

$$V(n, x) \le B(||x_0||)a^{-M(n-n_0)}. \tag{6.5.9}$$

From condition (6.5.2), we have $\lambda_1||x||^p \le V(n, x)$, which implies that

$$||x|| \le \{\frac{V(n, x)}{\lambda_1}\}^{1/p}. \tag{6.5.10}$$

Combining (6.5.9) and (6.5.10), we arrive at

$$||x|| \le \{\frac{B(||x_0||)}{\lambda_1}\}^{1/p}a^{-\frac{M}{p}(n-n_0)} = C(||x_0||)a^{-\frac{M}{p}(n-n_0)}.$$

Hence, the zero solution of (6.5.1) is uniformly exponentially stable. This completes the proof. $\qquad\square$

Example 6.18. Consider the non-linear difference equation

$$x(n + 1) = \sigma x(n) + Rx^{1/3}(n)a^{-ln}, \tag{6.5.11}$$

$a > 1$ and l are constants with $l > -\frac{1}{3}\frac{\ln(1-\lambda_3/\lambda_2^{r/q})}{\ln(a)}$, where $\lambda_1 = \lambda_2 = 1$, $\lambda_3 = 1 - (\sigma^2 + \frac{4}{3}|\sigma||R| + \frac{R^2}{3})$, $p = 2$, and $q = r = 2$. If

$$\sigma^2 + \frac{4}{3}|\sigma||R| + \frac{R^2}{3} < 1,$$

then the zero solution of (6.5.11) is uniformly exponentially stable.

To see this, let $V(n, x) = x^2(n)$. By calculating $\Delta V(n, x)$ along the solutions of (6.5.11), we obtain

$$\Delta V(n, x) = x^2(n + 1) - x^2(n)$$
$$= (\sigma x + Rx^{1/3}a^{-ln})^2 - x^2$$
$$\le \sigma^2 x^2 + 2|\sigma||R|x^{4/3}a^{-ln} + R^2x^{2/3}a^{-2ln} - x^2.$$

To further simplify $\Delta V(n, x)$, we make use of Young's inequality, which says that if $1/e + 1/f = 1$, then

$$wz < \frac{w^e}{e} + \frac{z^f}{f}.$$

Thus, for $e = 3/2$ and $f = 3$, we have

$$2|\sigma||R|x^{4/3}a^{-ln} \le 2|\sigma||R|\Big[\frac{(x^{4/3})^{3/2}}{3/2} + \frac{a^{-3ln}}{3}\Big]$$

$$= \frac{4}{3}|\sigma||R|x^2 + \frac{2}{3}|\sigma||R|a^{-3ln}.$$

Similarly, if we let $e = 3$ and $f = 3/2$, we have

$$R^2 x^{2/3} a^{-2ln} \leq R^2 \left[\frac{(x^{2/3})^3}{3} + \frac{a^{-3ln}}{3/2} \right]$$

$$= \frac{R^2 x^2}{3} + \frac{2}{3} R^2 a^{-3ln}.$$

Thus,

$$\Delta V(n, x) \leq \left[\sigma^2 + \frac{4}{3}|\sigma||R| + \frac{R^2}{3} - 1 \right] x^2 + \left[\frac{2}{3}|\sigma||R| + \frac{2}{3}R^2 \right] a^{-3ln}$$

$$\leq -\left[1 - \left(\sigma^2 + \frac{4}{3}|\sigma||R| + \frac{R^2}{3} \right) \right] x^2$$

$$+ \left[\frac{2}{3}|\sigma||R| + \frac{2}{3}R^2 \right] a^{-3ln}.$$

One can easily check that conditions (6.5.2)–(6.5.6) of Theorem 6.5.1 are satisfied with $3l = \delta$ and

$$k = \frac{2}{3}|\sigma||R| + \frac{2}{3}R^2.$$

Hence the zero solution of (6.5.11) is uniformly exponentially stable.

In the next theorem we show that the zero solution of (6.5.1) is exponentially stable. □

The next theorem generalizes Theorem 6.5.1 by assuming λ_i, $i = 1, 2, 3$ are functions of the independent variable n.

Theorem 6.5.2. *Let a be a constant with $a > 1$. Let $D \subset \mathbb{R}^k$ be an open set containing the origin, and let $V(n, x) : \mathbb{Z}^+ \times D \to \mathbb{R}^+$ be a given function satisfying*

$$\lambda_1(n)||x||^p \leq V(n, x) \leq \lambda_2(n)||x||^q, \tag{6.5.12}$$

and

$$\Delta V(n, x) \leq -\lambda_3(n)||x||^r + ka^{-\delta n}, \tag{6.5.13}$$

for some positive constants p, q, r, k, δ, and positive functions $\lambda_1(n)$, $\lambda_2(n)$ and $\lambda_3(n)$, where $\lambda_1(n)$ is a non-decreasing sequence. Moreover, suppose for some positive constants α and γ,

$$0 < \frac{\lambda_3(n)}{\lambda_2^{r/q}(n)} \leq \alpha < 1 \tag{6.5.14}$$

such that

$$V(n, x) - V^{r/q}(n, x) \leq \gamma a^{-\delta n} \qquad (6.5.15)$$

with

$$\delta > \inf_{n \in \mathbb{Z}^+} - \frac{\ln(1 - \lambda_3(n)/\lambda_2^{r/q}(n))}{\ln(a)}. \qquad (6.5.16)$$

Then the zero solution of (6.5.1) *is exponentially stable.*

Proof. First note that in view of (6.5.14), δ which is given by (6.5.16) is positive. Taking the difference of the function $V(n, x)a^{M(n-n_0)}$ with

$$M = \inf_{n \in \mathbb{Z}^+} - \frac{\ln(1 - \lambda_3(n)/\lambda_2^{r/q}(n))}{\ln(a)},$$

we have

$$\Delta\left(V(n, x)a^{M(n-n_0)}\right) = \left[V(n+1, x)a^M - V(n, x)\right]a^{M(n-n_0)}.$$

By a similar argument as in Theorem 6.5.1 we obtain,

$$V(n, x) \leq B(\|x_0\|, \lambda_2(n_0))a^{-M(n-n_0)}, \qquad (6.5.17)$$

where $B(\|x_0\|, \lambda_2(n_0)) = \lambda_2(n_0)\|x_0\|^q + \frac{L}{1-a^{(M-\delta)}}$. From condition (6.5.12) and the fact that $\lambda_1(n)$ is non-decreasing we have,

$$\|x\| \leq \{\frac{V(n, x)}{\lambda_1(n)}\}^{1/p} \leq \{\frac{V(n, x)}{\lambda_1(n_0)}\}^{1/p}. \qquad (6.5.18)$$

Combining (6.5.17) and (6.5.18) we obtain

$$\|x\| \leq \{\frac{B(\|x_0\|, \lambda_2(n_0))}{\lambda_1(n_0)}\}^{1/p}a^{-\frac{M}{p}(n-n_0)}$$

$$= C(\|x_0\|, n_0)a^{-\frac{M}{p}(n-n_0)}.$$

Hence, the zero solution of (6.5.1) is exponentially stable. This completes the proof. □

The next corollary is an immediate consequence of Theorem 6.5.2.

Corollary 6.2. *Suppose the hypotheses of Theorem 6.5.2 holds where the non-decreasing condition on $\lambda_1(n)$ is replaced by*

$$\lambda_1(n) \geq a^{-Nn} \quad \text{for all } n \geq n_0 \geq 0 \text{ with } 0 < N < M,$$

then the zero solution of (6.5.1) *is exponentially stable.*

The next theorem does not require an upper bound on the Lyapunov function.

Theorem 6.5.3. *Let a be a constant with $a > 1$. Let $D \subset \mathbb{R}^k$ be an open set containing the origin, and let $V(n, x) : \mathbb{Z}^+ \times D \to \mathbb{R}^+$ be a given function satisfying*

$$\lambda_1 \|x\|^p \leq V(n, x), \tag{6.5.19}$$

and

$$\Delta V(n, x) \leq -\lambda_2 V(n, x) + ka^{-\delta n}, \quad 0 < \lambda_2 < 1 \tag{6.5.20}$$

for some positive constants λ_1, λ_2, p, k and δ. Then, the zero solution of (6.5.1) is exponentially stable.

Proof. Let ϵ be a fixed number satisfying

$$0 < \varepsilon < \min\{\frac{-\ln(1 - \lambda_2)}{\ln(a)}, \delta\}.$$

Note that $\frac{-\ln(1-\lambda_2)}{\ln(a)} > 0$, since $0 < \lambda_2 < 1$ and $a > 1$. Taking $\Delta\left(a^{\epsilon n} V(n, x)\right)$ along the solutions of (6.5.1) and utilizing conditions (6.5.19) and (6.5.20), we obtain

$$\Delta\left(V(n, x) a^{\epsilon n}\right) = a^{\epsilon n}\left[a^{\epsilon}(1 - \lambda_2)V(n, x) - V(n, x) + ka^{-\delta n}\right]$$
$$\leq ka^{\epsilon n} a^{-\delta n}.$$

Thus,

$$V(n, x)a^{\epsilon n} \leq a^{\epsilon n_0} V(n_0, x_0) + k \sum_{s=n_0}^{n-1} a^{(\epsilon - \delta)s}$$
$$\leq a^{\epsilon n_0} V(n_0, x_0) + \frac{k}{a^{(\epsilon - \delta)} - 1}[a^{(\epsilon - \delta)n} - a^{\epsilon - \delta)n_0}]$$
$$\leq a^{\epsilon n_0} V(n_0, x_0) + \frac{ka^{(\epsilon - \delta)n_0}}{1 - a^{(\epsilon - \delta)}}$$
$$\leq a^{\epsilon n_0}\left(V(n_0, x_0) + \frac{k}{1 - a^{(\epsilon - \delta)}}\right).$$

Therefore,

$$|x|^p \leq \frac{1}{\lambda_1}\left(V(n_0, x_0) + \frac{k}{1 - a^{(\epsilon - \delta)}}\right)a^{-\epsilon(n - n_0)}, \quad \text{for all } n \geq n_0.$$

This completes the proof. □

Example 6.19. Let a be a constant such that $a > 1$ and consider the non-linear difference equation

$$x(n + 1) = \sigma x + Rx^{1/3} + a^{\gamma_1 n} \sin(x), \qquad (6.5.21)$$

where $\gamma_1 > 0$ and

$$a^{-\gamma_2}\left(|\sigma| + \frac{1}{3}\right) < 1.$$

Suppose

$$\gamma_2 < M = -\frac{\ln(a^{-\gamma_2}(|\sigma| + \frac{1}{3}))}{\ln(a)},$$

and if $\gamma_1 - \gamma_2 \leq -\eta$ for some positive constant η, with

$$\eta > -\frac{\ln(a^{-\gamma_2}(|\sigma| + \frac{1}{3}))}{\ln(a)}$$

and $\gamma_2 > 0$, then the zero solution of (6.5.21) is exponentially stable.
To see this, let

$$V(n, x) = a^{-\gamma_2 n}|x(n)|.$$

By calculating $\Delta V(n, x)$ along the solutions of (6.5.21) we obtain

$$\Delta V(n, x) = a^{-\gamma_2(n+1)}|x(n + 1)| - a^{-\gamma_2 n}|x(n)|$$
$$\leq \left(|\sigma||x| + |R||x|^{1/3} + a^{\gamma_1 n}\right)a^{-\gamma_2(n+1)} - a^{-\gamma_2 n}|x(n)|.$$

Using Young's inequality with $e = 3/2$ and $f = 3$, we obtain

$$|R||x|^{1/3} \leq \frac{2}{3}|R|^{3/2} + |x|/3.$$

Thus,

$$\Delta V(n, x) \leq a^{-\gamma_2 n}\left[a^{-\gamma_2}(|\sigma| + \frac{1}{3}) - 1\right]|x|$$
$$+ \frac{2}{3}a^{-\gamma_2(n+1)}|R|^{3/2} + a^{-\gamma_2(n+1)+\gamma_1 n}$$
$$\leq a^{-\gamma_2 n}\left[a^{-\gamma_2}(|\sigma| + \frac{1}{3}) - 1\right]|x|$$
$$+ \frac{2}{3}a^{-\gamma_2(n+1)}a^{\gamma_1 n}|R|^{3/2} + a^{-\gamma_2(n+1)+\gamma_1 n}$$
$$= a^{-\gamma_2 n}\left[a^{-\gamma_2}(|\sigma| + \frac{1}{3}) - 1\right]|x|$$
$$+ a^{-\gamma_2}(\frac{2}{3}|R|^{3/2} + 1)a^{(\gamma_1 - \gamma_2)n}$$

$$= -\left[1 - a^{-\gamma_2}(|\sigma| + \frac{1}{3})\right]a^{-\gamma_{2n}}|x|$$
$$+ a^{-\gamma_2}(\frac{2}{3}|R|^{3/2} + 1)a^{-\eta n}.$$

Thus, the conditions of Corollary 6.2 are satisfied with

$$N = \gamma_2, \ \lambda_1(n) = \lambda_2(n) = a^{-\gamma_{2n}},$$
$$\lambda_3(n) = \left[1 - a^{-\gamma_2}(|\sigma| + \frac{1}{3})\right]a^{-\gamma_{2n}},$$
$$p = q = r = 1, \delta = \eta, \ \text{and} \ k = a^{-\gamma_2}\left(\frac{2}{3}|R|^{3/2} + 1\right).$$

Hence the zero solution of (6.5.21) is exponentially stable. □

6.5.1 Exercises

Exercise 6.33. For $n \geq n_0 \geq 0$, consider the non-linear system

$$x(n+1) = -\frac{x(n)}{2} - x(n)y(n)$$
$$y(n+1) = -\frac{y(n)}{2} + x^2(n).$$

Refer to Example 6.17 and show the origin is (ES).

Exercise 6.34. For $n \geq n_0 \geq 0$, specify a set D so that solutions starting inside the set D, the scalar difference equation

$$x(n+1) = \frac{1}{4}x(n) + x^2(n)$$

is (ES).

Exercise 6.35. For $n \geq n_0 \geq 0$, consider the non-linear difference equation

$$x(n+1) = Rx^{1/3}(n)2^{-ln}, \tag{6.5.22}$$

where $l > -\frac{1}{3}\frac{\ln(1-\lambda_3)}{\ln(2)}$, with $\lambda_3 = 1 - \frac{R^2}{3}$. Show that if

$$\frac{R^2}{3} < 1,$$

then the zero solution of (6.5.22) is uniformly exponentially stable.

6.6 l_p-stability

The notion of l_p-Stability is better explained when we consider a functional difference equation. A functional difference equation is a difference equation in which the derivative $x(n + 1)$ of an unknown function x has a value at n that is related to x as a function of some other function at n. A general first-order functional difference equation is therefore given by

$$x(n + 1) = f(n, x(n), x(u(n))).$$

An example of a functional non-linear difference equation is the *Volterra difference equation*

$$x(n + 1) = a(n)x(n) + \sum_{s=0}^{n-1} b(n, s) f(s, x(s)),$$

where the function f is continuous in x, and a and b are known functions. When $f(n, x(n)) = x(n)$ then the equation is called *linear Volterra difference equation*. In this section we revisit the notion of l_p-stability and prove theorems under which it occurs. We begin by considering the general non-autonomous non-linear discrete system

$$x(n + 1) = G(n, x(s);\ 0 \le s \le n) \overset{def}{=} G(n, x(\cdot)) \qquad (6.6.1)$$

where $G : \mathbb{Z}^+ \times \mathbb{R}^k \to \mathbb{R}^k$ is continuous in x and $G(n, 0) = 0$. Let $C(n)$ denote the set of functions $\phi : [0, n] \to \mathbb{R}$ and $\|\phi\| = \sup\{|\phi(s)| : 0 \le s \le n\}$.

We say that $x(n) = x(n, n_0, \phi)$ is a solution of (6.6.1) with a bounded initial function $\phi : [0, n_0] \to \mathbb{R}^k$ if it satisfies (6.6.1) for $n > n_0$ and $x(j) = \phi(j)$ for $j \le n_0$.

Definition 6.6.1. The zero solution of (6.6.1) is stable (S) if for each $\varepsilon > 0$, there is a $\delta = \delta(n_0, \varepsilon) > 0$ such that $[n_0 \ge 0, \phi \in C(n_0), \|\phi\| < \delta]$ imply $|x(n, n_0, \phi)| < \varepsilon$ for all $n \ge n_0$. It is uniformly stable (US) if it is stable and δ is independent of n_0. It is asymptotically stable (AS) if it is (S) and $|x(n, n_0, \phi)| \to 0$, as $n \to \infty$.

Definition 6.6.2. The zero solution of system (6.6.1) is said to be l_p-stable if it is stable and if $\sum_{n=n_0}^{\infty} \|x(n, n_0, \phi)\|^p < \infty$ for positive p.

We have the following elementary theorem.

Theorem 6.6.1. *If the zero solution of (6.6.1) is exponentially stable (Definition 6.5.1), then it is also l_p-stable.*

Proof. Since the zero solution of (6.6.1) is exponentially stable, we have by the above definition that

$$\sum_{n=n_0}^{\infty} \|x(n, n_0, \phi)\| \le [C(\|\phi\|, n_0)]^p \sum_{n=n_0}^{\infty} a^{p\eta(n-n_0)}$$

$$= [C(||\phi||, n_0)]^P a^{-n_0 p \eta} \sum_{n=n_0}^{\infty} a^{p \eta n}$$

$$= [C(||\phi||, n_0)]^P / (1 - a^{p \eta}),$$

which is finite. This completes the proof. □

We caution that l_p-stability is not uniform with respect to p, as the next example shows. Also, it shows that (AS) does not imply l_p-stability for all p. In Chapter 1, we considered the difference equation

$$x(n+1) = \frac{n}{n+1} x(n), \quad x(n_0) = x_0 \neq 0, \quad n_0 \geq 1$$

and showed its solution is given by

$$x(n) := x(n, n_0, x_0) = \frac{x_0 n_0}{n}.$$

Clearly its zero solution is (US) and (AS). However, for $n_0 = n$, we have

$$x(2n, n, x_0) = \frac{x_0 n}{2n} \rightarrow \frac{x_0}{2} \neq 0$$

which implies that the zero solution is not (UAS). Moreover,

$$\sum_{n=n_0}^{\infty} ||x(n, n_0, x_0)||^P \leq \sum_{n=n_0}^{\infty} |(\frac{x_0 n_0}{n})|^P = |x_0|^P (n_0)^P \sum_{n=n_0}^{\infty} (\frac{1}{n})^P,$$

which diverges for $0 < p \leq 1$ and converges for $p > 1$.

The next example shows that asymptotic stability does not necessary imply l_p-stability for any $p > 0$. Let $g : [0, \infty) \rightarrow (0, \infty)$ with $\lim_{n \to \infty} g(n) = \infty$. Consider the non-autonomous difference equation

$$x(n+1) = [g(n)/g(n+1)]x(n), \quad x(n_0) = x_0, \tag{6.6.2}$$

which has the solution $x(n, n_0, x_0) = \frac{g(n_0)}{g(n)} x_0$. It is obvious that as $n \rightarrow \infty$ the solution tends to zero, for fixed initial n_0 and the zero solution is indeed asymptotically stable. On the other hand

$$\sum_{n=n_0}^{\infty} ||x(n, n_0, x_0)||^P = [g(n_0)x_0]^P \sum_{n=n_0}^{\infty} \left(\frac{1}{g(n)}\right)^P, \tag{6.6.3}$$

which may not converge for any $p > 0$. For example, if we take

$$g(n) = \log(n+2),$$

then from (6.6.3) we have

$$\sum_{n=n_0}^{\infty} ||x(n, n_0, x_0)||^p = [log(n_0 + 2)]^p ||x_0||^p \sum_{n=n_0}^{\infty} \left(\frac{1}{log(n + 2)}\right)^p,$$

which is known to diverge for all $p \geq 0$.

The next theorem relates l_p-stability to Lyapunov functionals.

Theorem 6.6.2. *Let $D \subset \mathbb{R}^k$ be an open set containing the origin, and let $V(n, x) : \mathbb{Z}^+ \times D \to \mathbb{R}^+$ (see Definition 6.1.1). If along the solutions of (6.6.1), V satisfies $\Delta V \leq -c||x||^p$, for some positive constants c and p, then the zero solution of (6.6.1) is l_p-stable.*

Proof. Set the solution $x(n) := x(n, n_0, \phi)$. The hypothesis of the theorem implies the zero solution is stable. Thus, for $n \geq n_0$ there is a positive constant M such that $||x(n, n_0, \phi)|| \leq M$. For $n \geq n_0$ we set

$$L(n) = V(n, x(n)) + c \sum_{s=n_0}^{n-1} ||x(s)||^p.$$

Then for all $n \geq n_0$ we have

$$\Delta L(n) = \Delta V(n, x) + c||x||^p$$
$$\leq -c||x||^p + c||x||^p = 0.$$

Therefore, $L(n)$ is decreasing and hence $0 \leq L(n) \leq L(n_0) = V(n_0, \phi)$, $n \geq n_0$. This implies that $0 \leq L(n) = V(n, x) + c \sum_{s=n_0}^{n-1} ||x(s)||^p \leq V(n_0, \phi)$, $n \geq n_0$ so that

$$0 \leq V(n, x) \leq -c \sum_{s=n_0}^{n-1} ||x(s)||^p + V(n_0, \phi).$$

As a consequence,

$$\sum_{s=n_0}^{n-1} ||x(s, n_0, \phi)||^p \leq V(n_0, \phi)/c, \; n \geq n_0.$$

Letting $n \to \infty$ on both sides of the above inequality gives

$$\sum_{n=n_0}^{\infty} ||x(n, n_0, \phi)||^p \leq V(n_0, \phi)/c < \infty.$$

This completes the proof. $\qquad \qquad \square$

In the next two examples we show that the l_p-stability depends on the type of Lyapunov functional that is being used. Moreover, there will be a price to pay if you want to obtain l_p-stability for higher values of p. We offer the next two corollaries as examples of l_p-stability.

Corollary 6.3. *Consider the scalar Volterra difference equation*

$$x(n+1) = a(n)x(n) + \sum_{s=0}^{n-1} b(n,s)f(s,x(s)) \qquad (6.6.4)$$

with f being continuous in x and there exists a constant λ_1 such that $|f(n,x)| \le \lambda_1|x|$. Assume there exists a positive α such that

$$|a(n)| + \lambda \sum_{s=n+1}^{\infty} |b(s,n)| - 1 \le -\alpha, \qquad (6.6.5)$$

and for some positive constant λ which is to be specified later, we have

$$\lambda_1 \le \lambda, \qquad (6.6.6)$$

then the zero solution of (6.6.4) is l_1-stable.

Proof. Define the Lyapunov functional V by

$$V(n,x) = |x(n)| + \lambda \sum_{j=0}^{n-1} \sum_{s=n}^{\infty} |b(s,j)||x(j)|.$$

We have along the solutions of (6.6.4) that

$$\Delta V(n) = |x(n+1)| + \lambda \sum_{j=0}^{n} \sum_{s=n+1}^{\infty} |b(s,j)||x(j)|$$

$$- |x(n)| - \lambda \sum_{j=0}^{n-1} \sum_{s=n}^{\infty} |b(s,j)||x(j)|$$

$$\le |a(n)||x(n)| + \lambda_1 \sum_{s=0}^{n-1} |b(n,s)||x(s)|$$

$$+ \lambda \sum_{j=0}^{n-1} \sum_{s=n+1}^{\infty} |b(s,j)||x(j)| + \lambda \sum_{s=n+1}^{\infty} |b(s,n)||x(n)|$$

$$- |x(n)| - \lambda \sum_{j=0}^{n-1} |b(n,j)|x(j)| - \lambda \sum_{j=0}^{n-1} \sum_{s=n+1}^{\infty} |b(s,j)||x(j)|$$

$$\leq \big(|a(n)| + \lambda \sum_{s=n+1}^{\infty} |b(s,n)| - 1\big)|x(n)| + (\lambda_1 - \lambda)\sum_{s=0}^{n-1} |b(n,s)||x(s)|$$

$$\leq \big(|a(n)| + \lambda \sum_{s=n+1}^{\infty} |b(s,n)| - 1\big)|x(n)|$$

$$\leq -\alpha|x(n)|.$$

This implies the zero solution is stable and l_1-stable by Theorem 6.6.2. This completes the proof. □

Corollary 6.4. *Consider (6.6.4) and assume f is continuous in x with $|f(n,x)| \leq \lambda_1 x^2$. Assume there exists a positive constant α such that*

$$a^2(n) + \lambda \sum_{s=n+1}^{\infty} |b(s,n)| + \lambda_1|a(n)| \sum_{s=0}^{n} |b(n,s)| - 1 \leq -\alpha, \qquad (6.6.7)$$

and for some positive constant λ which is to be specified later, we have

$$\lambda_1|a(n)| + \lambda_1^2 \sum_{s=0}^{n-1} |b(n,s)| - \lambda \leq 0. \qquad (6.6.8)$$

Then the zero solution of (6.6.4) is l_2-stable.

Proof. Define the Lyapunov functional V by

$$V(n,x) = x^2(n) + \lambda \sum_{j=0}^{n-1}\sum_{s=n}^{\infty} |b(s,j)|x^2(j).$$

We have along the solutions of (6.6.4) that

$$\Delta V(n) = \big(a(n)x(n) + \sum_{s=0}^{n-1} b(n,s)f(s,x(s))\big)^2 - x^2(n)$$

$$+ \lambda x^2(n)\sum_{s=n+1}^{\infty} |b(s,n)| - \lambda \sum_{s=0}^{n-1} |b(n,s)|x^2(s) - x^2(n)$$

$$\leq a^2(n)x^2(n) + 2\lambda_1|a(n)||x(n)|\sum_{s=0}^{n-1} |b(n,s)||x(s)|$$

$$+ \big(\sum_{s=0}^{n-1} b(n,s)f(s,x(s))\big)^2$$

$$+ \lambda x^2(n)\sum_{s=n+1}^{\infty} |b(s,n)| - \lambda \sum_{s=0}^{n-1} |b(n,s)|x^2(s) - x^2(n).$$

As a consequence of $2zw \leq z^2 + w^2$, for any real numbers z and w we have

$$2\lambda_1 |a(n)||x(n)| \sum_{s=0}^{n-1} |b(n,s)||x(s)| \leq \lambda_1 |a(n)| \sum_{s=0}^{n-1} |b(n,s)|(x^2(n) + x^2(s)).$$

Also, using Schwartz inequality we obtain

$$\left(\sum_{s=0}^{n-1} b(n,s)f(s,x(s))\right)^2 = \sum_{s=0}^{n-1} |b(n,s)|^{1/2}|b(n,s)|^{1/2}|f(s,x(s))|$$

$$\leq \sum_{s=0}^{n-1} |b(n,s)| \sum_{s=0}^{n-1} |b(n,s)|f^2(s,x(s))$$

$$\leq \lambda_1^2 \sum_{s=0}^{n-1} |b(n,s)| \sum_{s=0}^{n-1} |b(n,s)|x^2(s).$$

Putting all together, we get

$$\Delta V(n) \leq \left(a^2(n) + \lambda \sum_{s=n+1}^{\infty} |b(s,n)| + \lambda_1 |a(n)| \sum_{s=0}^{n} |b(n,s)| - 1\right)x^2(n)$$

$$+ \left(\lambda_1 |a(n)| + \lambda_1^2 \sum_{s=0}^{n-1} |b(n,s)| - \lambda\right) \sum_{s=0}^{n-1} |b(n,s)|x^2$$

$$\leq -\alpha x^2(n).$$

This implies the zero solution is stable and l_2-stable by Theorem 6.6.2. This completes the proof. $\qquad\square$

A quick comparison of (6.6.5) with (6.6.7) and (6.6.6) with (6.6.8) reveals that the conditions for the l_2-stability are more stringent than of the conditions for l_1-stability.

6.6.1 Exercises

Exercise 6.36. This exercise is related to Corollary 6.3. Consider the scalar Volterra difference equation

$$x(n+1) = a(n)x(n) + \sum_{s=0}^{n} b(n,s)f(s,x(s)) \qquad (6.6.9)$$

with f being continuous in x and there exists a constant λ_1 such that $|f(n,x)| \leq \lambda_1 |x|$. Let $V(n,x)$ be defined as in Corollary 6.3. Show that for $\lambda_1 \leq \lambda$, and if there exists a positive α such that

$$|a(n)| + \lambda \sum_{s=n+1}^{\infty} |b(s,n)| + \lambda_1|b(n,n)| - 1 \leq -\alpha,$$

then the zero solution of (6.6.9) is l_1-stable.

Exercise 6.37. Display a functional values of $a(n)$, $b(n,s)$, and $f(n,x(n))$ so that all the conditions of Corollary 6.3 are met.

Exercise 6.38.

$$x(n+1) = \sum_{s=0}^{n-1} c(n,s)f(x(s)) + g(n,x(n))$$

for all integers $n \geq 0$ and for integers, $0 \leq s \leq n$. The functions g and f are continuous in x and satisfy $|f(x)| \leq \delta|x|$, and $|g(n,x)| \leq \lambda(n)|x|$, where $\lambda : \mathbb{Z}^+ \to (0,1)$. Define the functional V by

$$V(n,x) = |x(n)| + \delta \sum_{s=0}^{n-1} \sum_{u=0}^{\infty} |c(u,s)||x(s)|.$$

Assume for positive constant α, we have

$$\lambda(n) + \delta \sum_{u=n+1}^{\infty} |c(u,n)| - 1 \leq -\alpha.$$

Show that

$$\Delta V(n,x) \leq \left[\lambda(n) + \delta \sum_{u=n+1}^{\infty} |c(u,n)| - 1\right]|x(n)|$$

$$\leq -\alpha|x(n)|,$$

and hence the zero solution is l_1-stable.

Chapter 7

New variation of parameters

7.1 Introduction

In the case of non-linear problems, whether in differential or difference equations, it is difficult and, in some cases, impossible to invert the problem and obtain a suitable mapping that can be effectively used in fixed point theory to qualitatively analyze its solutions. In this chapter, we consider the existence of a positive sequence and utilize it as a product factor to obtain a new variation of parameters formula. Then, we will use the obtained new variation of parameters formula and revert to the contraction principle to arrive at results concerning boundedness, periodicity, and stability. For background on normed spaces, Banach spaces, and the contraction mapping principle, we refer readers to Appendix A. Let us emphasize again that Appendix A must be read before attempting to read this chapter.

7.2 Scalar equations

Throughout this chapter, we use t for n as the independent variable, where t is an integer that will be specified according to the context. For motivational purpose, we consider the linear difference equation

$$x(t+1) = a(t)x(t), \ x(t_0) = x_0, \ t \geq t_0 \geq 0. \tag{7.2.1}$$

It is clear that the solution of (7.2.1) is given by

$$x(t) = x_0 \prod_{s=t_0}^{t-1} a(s), \tag{7.2.2}$$

provided that $a(t) \neq 0$ for all $t \in \mathbb{Z}^+$. Just as a reminder, we adopt the convention that for any sequence $x(k)$

$$\sum_{k=a}^{b} x(k) = 0 \quad \text{and} \quad \prod_{k=a}^{b} x(k) = 1 \quad \text{whenever } a > b.$$

Difference Equations and Applications. https://doi.org/10.1016/B978-0-44-331492-6.00013-3

Let $v(t)$ be a sequence such that $v : \mathbb{Z}^+ \to \mathbb{R}$ with $v(t) \neq 0$ for all $t \in \mathbb{Z}^+$. Multiply both sides of (7.2.1) by the product factor $\prod_{s=t_0}^{t} v^{-1}(s)$ to obtain

$$x(t+1) \prod_{s=t_0}^{t} v^{-1}(s) = a(t)x(t) \prod_{s=t_0}^{t} v^{-1}(s).$$

Thus, the above expression can be written in the compact form

$$\Delta\left[x(t) \prod_{s=t_0}^{t-1} v^{-1}(s) \right] = \left[(a(t) - v(t))x(t) \right] \prod_{s=t_0}^{t} a^{-1}(s). \qquad (7.2.3)$$

Summing Eq. (7.2.3) from t_0 to $t-1$ gives

$$x(t) = x_0 \prod_{s=t_0}^{t-1} v(s) + \sum_{r=t_0}^{t-1} (a(r) - v(r))x(r) \prod_{s=r+1}^{t-1} v(s). \qquad (7.2.4)$$

Note that (7.2.4) reduces to (7.2.2) if we set $v(t) = a(t)$ in (7.2.4). To obtain asymptotic stability of the zero solution of (7.2.1) using (7.2.2) one would have to assume that

$$\prod_{s=t_0}^{t} a(s) \to 0, \text{ as } t \to \infty.$$

On the other hand, if we use (7.2.4) instead, then such requirement is not necessary. But instead, we would have to ask that

$$\prod_{s=t_0}^{t} v(s) \to 0, \text{ as } t \to \infty.$$

Such technique of inversion is of more importance when the right hand of (7.2.1) is either totally non-linear or totally delayed. To see this, we consider the non-linear difference equation

$$x(t+1) = f(t, x(t)), \; x(t_0) = x_0, \qquad (7.2.5)$$

where the function $f : \mathbb{Z}^+ \times \mathbb{R} \to \mathbb{R}$ is continuous. We begin by restating some of the definitions regarding stability.

Definition 7.2.1. We say $x(t) := x(t, t_0, x_0)$ is a solution of (7.2.5) if $x(t_0) = x_0$ and satisfies (7.2.5) for $t \geq t_0 \geq 0$.

Definition 7.2.2. The zero solution of (7.2.5) is stable if for any $\epsilon > 0$ and any integer $t_0 \geq 0$ there exists a $\delta > 0$ such that $|x_0| \leq \delta$ implies $|x(t, t_0, x_0)| \leq \epsilon$ for $t \geq t_0$.

Definition 7.2.3. The zero solution of (7.2.5) is asymptotically stable if it is stable and $|x(t, t_0, x_0)| \to 0$ as $t \to \infty$.

For more on stability we refer to Chapter 5 and Chapter 6. We begin with the following Lemma. Its proof follows along the lines of the derivation of (7.2.4).

Lemma 7.1. *If $x(t)$ is a solution of* (7.2.5) *on an interval $\mathbb{Z}^+ \cap [0, T]$ and satisfies the initial condition $x(t_0) = x_0$, $t_0 \geq 0$, then $x(t)$ is a solution of the summation equation if and only if*

$$x(t) = x_0 \prod_{s=t_0}^{t-1} v(s) + \sum_{r=t_0}^{t-1} \left(f(r, x(r)) - v(r)x(r) \right) \prod_{s=r+1}^{t-1} v(s), \qquad (7.2.6)$$

where $v : \mathbb{Z}^+ \to \mathbb{R}$ with $v(t) \neq 0$ for all $t \in \mathbb{Z}^+$.

Next, we will use (7.2.6) to define a mapping on the proper space and show the zero solution is (AS). Let $\mathcal{C}$ be the set of all real-valued bounded sequences. Define the space

$$S = \{\phi : [0, \infty) \to \mathbb{R} / \phi \in \mathcal{C}, |\phi(t)| \leq L, \phi(t) \to 0, \text{ as } t \to \infty\}.$$

Then

$$(S, || \cdot ||)$$

is a complete metric space under the uniform metric

$$\rho(\phi_1, \phi_2) = ||\phi_1 - \phi_2||,$$

where

$$||\phi|| = \sup_{t \in \mathbb{Z}^+} \{|\phi(t)|\}.$$

Assume

$$f(t, 0) = 0, \qquad (7.2.7)$$

and the function f is *locally Lipschitz* on the set S. That is, for any ϕ_1 and $\phi_2 \in S$, we have

$$|f(t, \phi_1) - f(t, \phi_2)| \leq \lambda(t) ||\phi_1 - \phi_2||, \qquad (7.2.8)$$

for $\lambda : [0, \infty) \to (0, \infty)$. Assume for $\phi \in S$ and positive constant L, we have that

$$\left| x_0 \prod_{s=t_0}^{t-1} v(s) \right| + L \sum_{r=t_0}^{t-1} \left(|v(r)| + \lambda(r) \right) \left| \prod_{s=r+1}^{t-1} v(s) \right| \leq L. \qquad (7.2.9)$$

Note that (7.2.9) implies that

$$\sum_{r=t_0}^{t-1} \left(|v(r)| + \lambda(r)\right) \left| \prod_{s=r+1}^{t-1} v(s) \right| \leq \alpha < 1.$$

The next theorem offers results about stability and boundedness.

Theorem 7.2.1. *Assume* (7.2.7)–(7.2.9). *Suppose there exists a positive constant k such that*

$$\left| \prod_{s=t_0}^{t-1} v(s) \right| \leq k, \qquad (7.2.10)$$

then the unique solution of (7.2.5) *is bounded and its zero solution is stable.*
If, in addition,

$$\prod_{s=t_0}^{t-1} v(s) \to 0, \qquad (7.2.11)$$

then the zero solution of (7.2.5) *is asymptotically stable.*

Proof. For $\phi \in S$, define the mapping $\mathfrak{P} : S \to S$ by

$$(\mathfrak{P}\phi)(t) = x_0 \prod_{s=t_0}^{t-1} v(s) + \sum_{r=t_0}^{t-1} \left(f(r, \phi(r)) - \phi(r)v(r)\right) \prod_{s=r+1}^{t} v(s). \qquad (7.2.12)$$

It is clear that $(\mathfrak{P}\phi)(t_0) = x_0$. Now for $\phi \in S$, we have that

$$\left|(\mathfrak{P}\phi)(t)\right| \leq |x_0|k + \sum_{r=t_0}^{t-1} \left(\lambda(r)|\phi(r)| + |\phi(r)||v(r)|\right) \left| \prod_{s=r+1}^{t} v(s) \right|.$$

Consequently,

$$\|\mathfrak{P}\phi\| \leq |x_0|k + \sum_{r=t_0}^{t-1} \left(|v(r)| + \lambda(r)\right) \left| \prod_{s=r+1}^{t} v(s) \right| \|\phi\|.$$

Or,

$$\|\mathfrak{P}\phi\| \leq |x_0|k + \alpha\|\phi\| \leq L. \qquad (7.2.13)$$

Since $\mathfrak{P}$ is continuous we have that $\mathfrak{P} : S \to S$. Next we show that $\mathfrak{P}$ is a contraction.

For $\phi_1, \phi_2 \in S$, we have from (7.2.12) that

$$\left|(\mathfrak{P}\phi_1)(t) - (\mathfrak{P}\phi_2)(t)\right| \leq \sum_{r=t_0}^{t-1} \left(|v(r)| + \lambda(r)\right) \left| \prod_{s=r+1}^{t} v(s) \right| \|\phi_1 - \phi_2\|$$

$$\leq \alpha ||\phi_1 - \phi_2||.$$

This shows that $\mathfrak{P}$ is a contraction. By Banach's contraction mapping principle, $\mathfrak{P}$ has a unique fixed point $x \in S$ which is bounded. Moreover, the unique fixed point is a solution of (7.2.5) on $[0, \infty)$. Next we show the zero solution is stable. Let x be the unique solution. Then (7.2.13) implies that

$$||x|| \leq |x_0|k + \alpha ||\phi|| \leq L.$$

Thus, for $\varepsilon > 0$ we chose $\delta = \varepsilon \frac{1-\alpha}{k}$, so that for $|x_0| < \delta$, we have

$$(1 - \alpha)||x|| \leq |x_0|k < \delta k.$$

Or

$$||x|| \leq \varepsilon.$$

Left to prove that

$$(\mathfrak{P}\varphi)(t) \to 0, \text{ as } t \to \infty.$$

We have already proven that the zero solution of (7.2.5) is stable. Let δ be the one from stability such that $|x_0| < \delta$ and define

$$S^* = \left\{ \varphi : \mathbb{Z}^+ \to \mathbb{R} \mid \varphi(t_0) = x_0 , ||\varphi|| \leq \epsilon \text{ and } \varphi(t) \to 0 \text{ as } t \to \infty \right\}. \quad (7.2.14)$$

Let $\mathfrak{P}$ be given by (7.2.12) and define $\mathfrak{P} : S^* \to S^*$. The map $\mathfrak{P}$ is contraction and it maps from S^* into itself.

We next show that $(\mathfrak{P}\varphi)(t)$ goes to zero as t goes to infinity.

The first term on the right of (7.2.12) goes to zero due to condition (7.2.11). Left to show that

$$\left| \sum_{r=t_0}^{t-1} \left(f(r, x(r)) - v(r)\phi(r) \right) \prod_{s=r+1}^{t-1} v(s) \right| \to 0, \text{ as } t \to \infty.$$

Let $\varphi \in S^*$ then $|\varphi(t)| \leq \epsilon$. Also, since $\varphi(t) \to 0$ as $t \to \infty$, there exists a $t_1 > 0$ such that for $t > t_1$, $|\varphi(t)| < \epsilon_1$ for $\epsilon_1 > 0$. Due to condition (7.2.11) there exists a $t_2 > t_1$ such that for $t > t_2$ implies that

$$\left| \prod_{s=t_1}^{t} v(s) \right| < \frac{\epsilon_1}{\alpha\epsilon}.$$

Thus for $t > t_2$, we have

$$\left| \sum_{r=t_0}^{t-1} \left(f(r, x(r)) - v(r)\phi(r) \right) \prod_{s=r+1}^{t-1} v(s) \right|$$

$$\leq \sum_{r=t_0}^{t-1} \left(\lambda(r) + v(r)\right)|\phi(r)| \left|\prod_{s=r+1}^{t-1} v(s)\right|$$

$$\leq \sum_{r=t_0}^{t_1-1} \left(\lambda(r) + v(r)\right)|\phi(r)| \left|\prod_{s=r+1}^{t-1} v(s)\right|$$

$$+ \sum_{r=t_1}^{t-1} \left(\lambda(r) + v(r)\right)|\phi(r)| \left|\prod_{s=r+1}^{t-1} v(s)\right|$$

$$\leq \epsilon \sum_{r=t_0}^{t_1-1} \left(\lambda(r) + v(r)\right)|\phi(r)| \left|\prod_{s=r+1}^{t-1} v(s)\right| + \epsilon_1 \alpha$$

$$\leq \epsilon \sum_{r=t_0}^{t_1-1} \left(\lambda(r) + v(r)\right)| \prod_{s=r+1}^{t_1-1} v(s) \left|\prod_{s=t_1}^{t-1} v(s)\right| + \epsilon_1 \alpha$$

$$\leq \epsilon \left|\prod_{s=t_1}^{t-1} v(s)\right| \sum_{r=t_0}^{t_1-1} \left(\lambda(r) + v(r)\right) \left|\prod_{s=r+1}^{t_1-1} v(s)\right| + \epsilon_1 \alpha$$

$$\leq \epsilon \alpha |\prod_{s=t_1}^{t-1} v(s)| + \epsilon_1 \alpha$$

$$\leq \epsilon_1 + \epsilon_1 \alpha.$$

Since ϵ_1 is arbitrary small, this shows that $(\mathfrak{P}\varphi)(t) \to 0$ as $t \to \infty$. As $\mathfrak{P}$ has a unique fixed point, say x it implies the asymptotic stability of the zero solution of (7.2.11). This completes the proof. □

7.2.1 Contraction versus large contraction

Now we consider a particular non-linear equation and perform a rewrite of it so it can be inverted in the traditional way. This implication makes the contraction mapping principle ineffective in obtaining any meaningful results about the solutions. Let

$$f(t, x) = -a(t)x^3 + l(t, x),$$

where $l(t, x)$ is continuous in x and satisfies a smallness condition to be specified later. Thus, we consider

$$x(t + 1) = -a(t)x^3 + l(t, x). \tag{7.2.15}$$

To make the case for Large Contraction, we consider

$$x(t + 1) = -a(t)x + a(t)(x - x^3) + l(t, x). \tag{7.2.16}$$

Then by the variation of parameters formula we have

$$x(t) = x_0 \prod_{s=t_0}^{t-1} a(s) + \sum_{r=t_0}^{t-1} \left(a(r)(x(r) - x^3(r)) + l(t, x(r)) \right) \prod_{s=r+1}^{t-1} a(s).$$

$$(7.2.17)$$

It is naive to believe that every map can be defined so that it is a contraction, even with the strictest conditions. To see this, we let

$$g(x) = x - x^3.$$

Then, then for $x, y \in \mathbb{R}$ with $|x|, |y| \le \frac{\sqrt{3}}{3}$ we have

$$|g(x) - g(y)| = |x - x^3 - y + y^3| \le |x - y| \left(1 - \frac{x^2 + y^2}{2} \right).$$

It is clear from the above inequality that the contraction constant tends to one as $x^2 + y^2 \to 0$. Consequently, the regular contraction mapping principle failed to produce any results. To get around it, we let $v(t)$ be a sequence such that $v : \mathbb{Z}^+ \to \mathbb{R}$ with $v(t) \ne 0$ for all $t \in \mathbb{Z}^+$. By similar steps as in the development of (7.2.4) we arrive at the variation of parameters formula

$$x(t) = x_0 \prod_{s=t_0}^{t-1} v(s) + \sum_{r=t_0}^{t-1} \left(v(r)x(r) - a(t)x^3(r) + l(t, x(r)) \right) \prod_{s=r+1}^{t-1} v(s).$$

$$(7.2.18)$$

Thus, one can show that the function

$$f(x) = v(r)x(r) - a(t)x^3(r),$$

is a contraction on some bounded and small set provided a and v have small magnitudes. To better illustrate our intention we set $l(t, x) = 0$ in (7.2.15). Then from the above variation of parameters formula, we arrive at

$$x(t) = x_0 \prod_{s=t_0}^{t-1} v(s) + \sum_{r=t_0}^{t-1} \left(v(r)x(r) - a(t)x^3(r) \right) \prod_{s=r+1}^{t-1} v(s). \qquad (7.2.19)$$

Assume for $\phi \in S$ and positive constant L, we have that

$$\left| x_0 \prod_{s=t_0}^{t-1} v(s) \right| + \sum_{r=t_0}^{t-1} (L|v(r)| + L^3|a(r)|) \left| \prod_{s=r+1}^{t-1} v(s) \right| \le L, \qquad (7.2.20)$$

and

$$\sum_{r=t_0}^{t-1} (|v(r)| + 3L^2|a(r)|) \left| \prod_{s=r+1}^{t-1} v(s) \right| \le \alpha < 1. \qquad (7.2.21)$$

The next theorem offers results about stability and boundedness.

Theorem 7.2.2. *Assume* (7.2.7), (7.2.10), (7.2.20) *and* (7.2.21). *Then the unique solution of* (7.2.15) *with* $l(t, x) = 0$ *is bounded and its zero solution is stable.*

If, in addition, (7.2.11) *holds, then the zero solution of* (7.2.15) *is asymptotically stable.*

Proof. For $\phi \in S$, define the mapping $\mathfrak{P} : S \to S$, by

$$(\mathfrak{P}\phi)(t) = x_0 \prod_{s=t_0}^{t-1} v(s) + \sum_{r=t_0}^{t-1} \left(v(r)\phi(r) - \phi^3(r)a(r)\right) \prod_{s=r+1}^{t} v(s). \quad (7.2.22)$$

It is clear that $(\mathfrak{P}\phi)(t_0) = x_0$. Now for $\phi \in S$, we have that

$$\left|(\mathfrak{P}\phi)(t)\right| \le |x_0|k + \sum_{r=t_0}^{t-1} \left(|v(r)||\phi(r)| + |\phi^3(r)||a(r)|\right) \left| \prod_{s=r+1}^{t} v(s) \right|$$

$$\le \left| x_0 \prod_{s=t_0}^{t-1} v(s) \right| + \sum_{r=t_0}^{t-1} \left(L|v(r)| + L^3|a(r)|\right) \left| \prod_{s=r+1}^{t-1} v(s) \right|.$$

Thus,

$$\|\mathfrak{P}\phi\| \le L.$$

Since $\mathfrak{P}$ is continuous we have that $\mathfrak{P} : S \to S$. Next we show that $\mathfrak{P}$ is a contraction. For $\phi_1, \phi_2 \in S$, (7.2.12) implies

$$\left|(\mathfrak{P}\phi_1)(t) - (\mathfrak{P}\phi_2)(t)\right|$$

$$\le \sum_{r=t_0}^{t-1} \left(|v(r)||\phi_1(r) - \phi_2(r)|\right.$$

$$+ \sum_{r=t_0}^{t-1} |a(r)||\phi_1(r) - \phi_2(r)|(\phi_1^2(r) + |\phi_1(r)\phi_2(r)| + \phi_1^2(r))) \left| \prod_{s=r+1}^{t} v(s) \right|$$

$$\le \sum_{r=t_0}^{t-1} \left(|v(r)| + 3L^2|a(r)|\right) \left| \prod_{s=r+1}^{t-1} v(s) \right| \|\phi_1 - \phi_2\|$$

$$\le \alpha\|\phi_1 - \phi_2\|.$$

This shows that $\mathfrak{P}$ is a contraction. By Banach's contraction mapping principle, $\mathfrak{P}$ has a unique fixed point $x \in S$ which is bounded. The proof for stability and asymptotic stability follows along the lines of the proof of Theorem 7.2.1. $\quad \square$

For the rest of this section, we set $l(t, x) = 0$ in (7.2.17) and use large contraction that we define next, and prove parallel theorem to Theorem 7.2.2. We saw before that the function or map $g(x) = x - x^3$ does not define a contraction. To get around such an issue, we use the notion of large contraction. Next, we state the contraction mapping principle, in which the regular contraction is replaced with large contraction. Then, based on the notion of large contraction, we introduce a theorem to obtain boundedness results in which large contraction is substituted for regular contraction.

Definition 7.2.4. Let $(\mathcal{M}, d)$ be a metric space and $B : \mathcal{M} \to \mathcal{M}$. The map B is said to be large contraction if $\phi, \varphi \in \mathcal{M}$, with $\phi \neq \varphi$ then $d(B\phi, B\varphi) \leq d(\phi, \varphi)$ and if for all $\varepsilon > 0$, there exists a $\delta \in (0, 1)$ such that

$$[\phi, \varphi \in \mathcal{M}, d(\phi, \varphi) \geq \varepsilon] \Rightarrow d(B\phi, B\varphi) \leq \delta d(\phi, \varphi).$$

The next theorem is an alternative to the regular contraction mapping principle, in which we substitute large contraction for regular contraction.

Theorem 7.2.3. *Let $(\mathcal{M}, \rho)$ be a complete metric space and B be a large contraction. Suppose there are an $x \in \mathcal{M}$ and an $L > 0$ such that $\rho(x, B^n x) \leq L$ for all $n \geq 1$. Then B has a unique fixed point in $\mathcal{M}$.*

Next we state a remarkable theorem by Adivar and Raffoul that generalizes the concept of *large contraction*. Its proof can be found in [1]. The theorem provides easily checked sufficient conditions under which a mapping is a large contraction. Several authors have published it in their work without the proper citations.

Consider the mapping H defined by

$$H(x(u)) = x(u) - h(x(u)). \tag{7.2.23}$$

Let $\alpha \in (0, 1]$ be a fixed real number and define the set $\mathbb{M}_\alpha$ by

$$\mathbb{M}_\alpha = \{\phi : \phi \in C(\mathbb{R}, \mathbb{R}) \text{ and } \|\phi\| \leq \alpha\}. \tag{7.2.24}$$

H.1. $h : \mathbb{R} \to \mathbb{R}$ is continuous on $[-\alpha, \alpha]$ and differentiable on $(-\alpha, \alpha)$,
H.2. The function h is strictly increasing on $[-\alpha, \alpha]$,
H.3. $\sup_{t \in (-\alpha, \alpha)} h'(t) \leq 1$.

Theorem 7.2.4 (Adivar-Raffoul). *(Classifications of Large Contraction Theorem) Let $h : \mathbb{R} \to \mathbb{R}$ be a function satisfying (H.1–H.3). Then the mapping H in (7.2.23) is a large contraction on the set $\mathbb{M}_\alpha$.*

Example 7.1. Let $\alpha \in (0, 1)$ and $k \in \mathbb{N}$ be fixed elements and $u \in (-1, 1)$.

   (i) The condition (H.2) is not satisfied for the function $h_1(u) = \frac{1}{2k} u^{2k}$.
   (ii) The function $h_2(u) = \frac{1}{2k+1} u^{2k+1}$ satisfies (H.1–H.3).

Proof. Since $h_1'(u) = u^{2k-1} < 0$ for $-1 < u < 0$, the condition (H.2) is not satisfied for h_1. Evidently, (H.1–H.2) hold for h_2. (H.3) follows from the fact that $h_2'(u) \leq \alpha^{2k}$ and $\alpha \in (0, 1)$. $\qquad\square$

We have the following lemma. Define the mapping

$$H(x) = x - x^3. \tag{7.2.25}$$

Lemma 7.2. *Let* $\| \cdot \|$ *denote the supremum norm. If*

$$\mathbb{M} = \left\{ \phi : \mathbb{Z} \to \mathbb{R} \mid \phi(0) = \phi_0, \text{ and } \|\phi\| \le \frac{\sqrt{3}}{3} \right\},$$

then the mapping H *defined by* (7.2.25) *is a large contraction on the set* $\mathbb{M}$.

Proof. Let $\alpha = \frac{\sqrt{3}}{3}$ and $h(x) = x^3$. Then, clearly h satisfies (H.1–H.2). Moreover, $\sup_{x \in (-\alpha, \alpha)} h'(x) = 1$, which satisfies H.3. Hence by Theorem 7.2.4 defines a large contraction. $\square$

For $\psi \in \mathbb{M}$, we define the map $B : \mathbb{M} \to \mathbb{M}$ by

$$(B\psi)(t) = \psi_0 \prod_{s=0}^{t-1} a(s) + \sum_{s=0}^{t-1} \left(a(s) H(\psi(s)) \prod_{u=s+1}^{t-1} a(u) \right). \tag{7.2.26}$$

Lemma 7.3. *Assume for all* $t \in \mathbb{Z}$

$$|\psi_0| \left| \prod_{s=0}^{t-1} a(s) \right| + \frac{2\sqrt{3}}{9} \sum_{s=0}^{t-1} \left| \prod_{u=s}^{t-1} a(u) \right| \le \frac{\sqrt{3}}{3}. \tag{7.2.27}$$

If H *is a large contraction on* $\mathbb{M}$, *then so is the mapping* B.

Proof. It is easy to see that

$$|H(x(t))| = |x(t) - x(t)^3| \le \frac{2\sqrt{3}}{9} \text{ for all } x \in \mathbb{M}.$$

By Lemma 7.2, H is a large contraction on $\mathbb{M}$. Hence, for $x, y \in \mathbb{M}$ with $x \ne y$, we have $\|Hx - Hy\| \le \|x - y\|$. Hence,

$$|Bx(t) - By(t)| \le \sum_{s=0}^{t-1} |H(x(s)) - H(y(s))| \left| \prod_{u=s}^{t-1} a(u) \right|$$

$$\le \frac{2\sqrt{3}}{9} \sum_{s=0}^{t-1} \left| \prod_{u=s}^{t-1} a(u) \right| \|x - y\|$$

$$= \|x - y\|.$$

Taking supremum norm over the set $[0, \infty)$, we get that $\|Bx - By\| \le \|x - y\|$. For a given $\varepsilon \in (0, 1)$, suppose $x, y \in \mathbb{M}$ with $\|x - y\| \ge \varepsilon$. Then for $\delta =$

$\min\{1 - \varepsilon^2/16, 1/2\}$, which implies that $0 < \delta < 1$. Hence, for all such $\varepsilon > 0$ we know that

$$[x, y \in \mathbb{M}, \|x - y\| \geq \varepsilon] \Rightarrow \|Hx - Hy\| \leq \delta\|x - y\|.$$

Therefore, using (7.2.27), one easily verifies that

$$\|Bx - By\| \leq \delta\|x - y\|.$$

The proof is complete. ☐

We arrive at the following theorem in which we prove boundedness.

Theorem 7.2.5. *Assume* (7.2.27). *Then* (7.2.15) *has a unique solution in* $\mathbb{M}$ *which is bounded.*

Proof. $(\mathbb{M}, \|\cdot\|)$ is a complete metric space of bounded sequences. For $\psi \in \mathbb{M}$ we must show that $(B\psi)(t) \in \mathbb{M}$. From (7.2.26) and the fact that

$$|H(x(t))| = |x(t) - x(t)^3| \leq \frac{2\sqrt{3}}{9} \text{ for all } x \in \mathbb{M},$$

we have

$$|(B\psi)(t)| \leq |\psi_0|\left|\prod_{s=0}^{t-1} a(s)\right| + \frac{2\sqrt{3}}{9}\sum_{s=0}^{t-1}\left|\prod_{u=s}^{t-1} a(u)\right|$$

$$\leq \frac{\sqrt{3}}{3}.$$

This shows that $(B\psi)(t) \in \mathbb{M}$. Lemma 7.2 implies the map B is a large contraction and hence by Theorem 7.2.3, the map B has a unique fixed point in $\mathbb{M}$ which is a solution of (7.2.15). This completes the proof. ☐

7.2.2 Periodic solutions

In this section we apply our new method to linear or non-linear difference equations to show the existence of periodic solutions without the requirement of some classic conditions. We need the following theorem concerning Banach's spaces for the results of this section.

Theorem 7.2.6. *Let* X *be the space of* T-*periodic functions. Then* X *is a Banach space when it is endowed with the maximum norm*

$$\|x\|_\infty = \max_{t \in [0, T-1]} |x(t)|.$$

Proof. Let X be the space of T-periodic functions. To show X is a Banach space, we need to show the following:

1) The space X is a normed vector space. 2) The space X is complete, meaning that every Cauchy sequence in X converges to a limit point in X. First we verify X is a normed vector space. For every $f \in X$, we define $\|f\|_\infty = \max_{t \in [0, T-1]} |f(t)|$, which defines a norm on X called the maximum norm. Let X be the space of T-periodic functions with the maximum norm. For verifying the closure, we assume f and g are T-periodic functions in X, then their sum $f + g$ is also T-periodic, and we can define the maximum norm as $\|f + g\|_\infty = max(\|f\|_\infty, \|g\|_\infty)$. This demonstrates that X is closed under addition. For the scalar multiplication, we assume f is a T-periodic function in X, and a is a scalar, then af is also a T-periodic function, and we may define the maximum norm as $\|af\|_\infty = |a| \|f\|_\infty$. This shows that X is closed under scalar multiplication. Moreover, the zero function is T-periodic which exists in X. Additionally, for every f in X, there exists $-f$ in X such that $f + (-f) = 0$. Now we turn our attention toward showing completeness. That is to establish that X is a Banach space, we need to show that it is complete. In other words, we need to demonstrate that every Cauchy sequence in X converges to a limit within X. Assume the existence of a Cauchy sequence $\{f_n\}$ in X, such that $f_n(t) \to f(t)$, as $n \to \infty$, for any $t \in \mathbb{R}$. We are left to prove that f is T-periodic and $f_n \to f$ uniformly.

$$|f(t) - f_m(t)| = \lim_{n \to \infty} |f_n(t) - f_m(t)| \le \lim_{n \to \infty} \|f_n - f_m\|_\infty.$$

Using the fact that f_n is Cauchy for $\| \, \|_\infty$, we have that for any $\varepsilon > 0$, there is N_ε such that:

$$\lim_{n \to \infty} \|f_n - f_m\|_\infty < \varepsilon \text{ for } m > N_\varepsilon.$$

Hence

$$|f(t) - f_m(t)| < \varepsilon \text{ for } m > N_\varepsilon \text{ and any } t \in \mathbb{R}.$$

In other words

$$\|f(t) - f_m(t)\|_\infty \to 0 \text{ as } m \to \infty,$$

as required. Left to show the limit function f is T-periodic. Since $f_n(t) \to f(t)$ then $f_n(t + T) \to f(t)$, and because $f_n(t) = f_n(t + T)$, it follows that $f(t + T) = f(t)$, for all $t \in \mathbb{R}$. This completes the proof. $\qquad \square$

To better illustrate our approach, we consider the non-linear difference equation

$$x(t + 1) = a(t)x(t) + f(t, x(t)) \tag{7.2.28}$$

where f is continuous in x. Let T be an integer such that $T \geq 1$. We assume the periodicity condition

$$a(t + T) = a(t), \text{ and } f(t + T, \cdot) = f(t, \cdot). \tag{7.2.29}$$

Let BC be the space of bounded sequences $\phi : \mathbb{Z} \to \mathbb{R}$ with the maximum norm $||\cdot||$. Define $P_T = \{\phi \in BC, \phi(t+T) = \phi(t)\}$. Then P_T is a Banach space when it is endowed with the maximum norm

$$\|x\| = \max_{t \in [0, T-1]} |x(t)|.$$

Also, we assume that

$$\prod_{s=t-T}^{t-1} a(s) \neq 1. \tag{7.2.30}$$

Throughout this section we assume that $a(t) \neq 0$ for all $t \in [0, T-1]$. Let $x \in P_T$. Then Eq. (7.2.28) is equivalent to

$$\Delta\left[x(t) \prod_{s=t_0}^{t-1} a^{-1}(s)\right] = f(t, x(t)) \prod_{s=t_0}^{t} a^{-1}(s). \tag{7.2.31}$$

Summing Eq. (7.2.31) from $t - T$ to $t - 1$ and using the fact that $x(t - T) = x(t)$, gives

$$x(t) = \left(1 - \prod_{s=t-T}^{t-1} a(s)\right)^{-1} \sum_{r=t-T}^{t-1} f(r, x(r)) \prod_{s=r+1}^{t-1} a(s). \tag{7.2.32}$$

Define the mapping $\mathfrak{P}$ on P_T by

$$(\mathfrak{P}\phi)(t) = \left(1 - \prod_{s=t-T}^{t-1} a(s)\right)^{-1} \sum_{r=t-T}^{t-1} f(r, \phi(r)) \prod_{s=r+1}^{t-1} a(s). \tag{7.2.33}$$

One can easily verify that $(\mathfrak{P}\phi)(t + T) = (\mathfrak{P}\phi)(t)$, and hence $\mathfrak{P} : P_T \to P_T$.

Theorem 7.2.7. *Suppose $a(t) \neq 0$ for all $t \in [0, T-1]$ and assume (7.2.30). Suppose the function f is Lipschitz continuous with Lipschitz constant k. If*

$$k\left|\left(1 - \prod_{s=t-T}^{t-1} a(s)\right)^{-1}\right| \sum_{r=t-T}^{t-1} \left|\prod_{s=r+1}^{t-1} a(s)\right| \leq \alpha,$$

for $\alpha \in (0, 1)$, then Eq. (7.2.28) has a unique periodic solution.

Proof. The proof is easily obtained by direct application of contraction mapping principle on the set P_T. $\qquad\square$

Next, we use our new technique to avoid the requirement that $a(t) \neq 0$ for all $t \in [0, T-1]$ along with condition (7.2.30). Let $v(t)$ be a sequence such that $v : \mathbb{Z}^+ \to \mathbb{R}$ with $v(t) \neq 0$ for all $t \in [0, T-1]$. Assume (7.2.29) and for $v \in P_T$, multiply both sides of (7.2.28) by $\prod_{s=t_0}^{t} v^{-1}(s)$ to obtain

$$\Delta\left[x(t) \prod_{s=t_0}^{t-1} v^{-1}(s)\right] = \left[(a(t)x(t) - v(t)x(t) + f(t, x(t))\right] \prod_{s=t_0}^{t} v^{-1}(s).$$

(7.2.34)

Summing Eq. (7.2.34) from $t-T$ to t-1 and using the fact that $x(t-T) = x(t)$, gives

$$x(t) = \left(1 - \prod_{s=t-T}^{t-1} v(s)\right)^{-1} \sum_{r=t-T}^{t-1} [a(r)x(r) - v(r)x(r) + f(r, x(r))] \prod_{s=r+1}^{t-1} v(s).$$

(7.2.35)

Define the mapping $\mathfrak{P}$ on P_T by

$$(\mathfrak{P}\phi)(t) = \left(1 - \prod_{s=t-T}^{t-1} v(s)\right)^{-1} \sum_{r=t-T}^{t-1} [a(r)x(r) - v(r)x(r)$$

$$+ f(r, \phi(r)) \prod_{s=r+1}^{t-1} v(s).$$

(7.2.36)

One can easily verify that $(\mathfrak{P}\phi)(t+T) = (\mathfrak{P}\phi)(t)$, and hence $\mathfrak{P} : P_T \to P_T$.

Theorem 7.2.8. *Suppose* $v(t) \neq 0$ *for all* $t \in [0, T-1]$ *and assume*

$$\prod_{s=t-T}^{t-1} v(s) \neq 1.$$

(7.2.37)

Suppose the function f *is Lipschitz continuous with Lipschitz constant* k. *If*

$$\left|\left(1 - \prod_{s=t-T}^{t-1} v(s)\right)^{-1}\right| \sum_{r=t-T}^{t-1} [|a(r)| + |v(r)| + k] \left|\prod_{s=r+1}^{t-1} v(s)\right| \leq \alpha,$$

for $\alpha \in (0, 1)$, *then Eq. (7.2.28) has a unique periodic solution.*

Proof. The proof is easily obtained by direct application of the contraction mapping principle on the set P_T. □

Next we display an example.

Example 7.2. For positive constant k, we consider the difference equation

$$x(t+1) = (1 - (-1)^t)x(t) + \frac{kx}{1+x^2}.$$

(7.2.38)

It is clear that $a(t) = \left(1 - (-1)^t\right)$ is periodic of period $T = 2$ and $a(0) = 0$. Hence Theorem 7.2.7 can not be applied. On the other hand we may apply Theorem 7.2.8 by taking $v(t) = \frac{(-1)^t}{2}$, and sufficiently small k. $\qquad\square$

7.2.3 Neutral difference equations

A *neutral difference equation* is an equation that involves both past and present values of a sequence. These equations are used in various fields, including mathematics, engineering, and physics, to model systems with delayed or advanced effects. They are called neutral because they exhibit a neutral balance between past and present values.

In this section we extend the results of the previous sections to the neutral difference equation with functional delay

$$x(t+1) = a(t)x(t) + b(t)x(t - g(t)) + c(t)\Delta x(t - g(t)) \qquad (7.2.39)$$

where $a, b, c : \mathbb{Z} \to \mathbb{R}$, and $g : \mathbb{Z} \to \mathbb{Z}^+$. Moreover, we will discuss the concept of equi-boundedness.

If for some positive constant k, $|g| \le k$ then for any integer $t_0 \ge 0$, we define $\mathbb{Z}_0$ to be the set of integers in $[t_0 - k, t_0]$. If g is unbounded then $\mathbb{Z}_0$ will be the set of integers in $(-\infty, t_0]$. We assume the existence of a given bounded initial sequence $\psi(t) : \mathbb{Z}_0 \to \mathbb{R}$. We will use the summation by parts formula

$$\sum \left(Ex(t)\Delta z(t) \right) = x(t)z(t) - \sum z(t)\Delta x(t)$$

where E is defined as $Ex(t) = x(t+1)$.

Definition 7.2.5. We say $x(t) := x(t, t_0, \psi)$ is a solution of (7.2.39) if $x(t) = \psi(t)$ on $\mathbb{Z}_0$ and satisfies (7.2.39) for $t \ge t_0$.

Definition 7.2.6. The zero solution of (7.2.39) is stable if for any $\epsilon > 0$ and any integer $t_0 \ge 0$ there exists a $\delta > 0$ such that $|\psi(t)| \le \delta$ on $\mathbb{Z}_0$ implies $|x(t, t_0, \psi)| \le \epsilon$ for $t \ge t_0$.

Definition 7.2.7. The zero solution of (7.2.39) is asymptotically stable if it is stable and if for any integer $t_0 \ge 0$ there exists $r(t_0) > 0$ such that $|\psi(t)| \le r(t_0)$ on $\mathbb{Z}_0$ implies $|x(t, t_0, \psi)| \to 0$ as $t \to \infty$.

Definition 7.2.8. A solution $x(t, t_0, \psi)$ of (7.2.39) is said to be bounded if there exists a $B(t_0, \psi) > 0$ such that $|x(t, t_0, \psi)| \le B(t_0, \psi)$ for $t \ge t_0$.

Definition 7.2.9. The solutions of (7.2.39) are said to be equi-bounded if for any t_0 and any $B_1 > 0$, there exists a $B_2 = B_2(t_0, B_1) > 0$ such that $|\psi(t)| \le B_1$ on $\mathbb{Z}_0$ implies $|x(t, t_0, \psi)| \le B_2$ for $t \ge t_0$.

For the remaining of the section we assume that there is a positive constant k, $|g| \le k$.

Lemma 7.4. *If $x(t)$ is a solution of (7.2.39) and satisfies the initial condition $x(t) = \psi(t)$ for $t \in \mathbb{Z}_0$, then $x(t)$ is a solution of the summation equation if and only if*

$$x(t) = \left[x(t_0) - c(t_0 - 1)x(t_0 - g(t_0))\right] \prod_{s=t_0}^{t-1} v(s) + c(t - 1)x(t - g(t))$$

$$+ \sum_{r=t_0}^{t-1} \left[(a(r) - v(r))x(r) \prod_{s=r+1}^{t-1} v(s)\right]$$

$$+ \sum_{r=t_0}^{t-1} \left([b(r) - \phi(r)]x(r - g(r)) \prod_{s=r+1}^{t-1} v(s)\right), \; t \geq t_0 \qquad (7.2.40)$$

where

$$\phi(r) = c(r) - c(r - 1)v(r).$$

$[0, T]$ *where* $v : \mathbb{Z} \cap [-k, \infty) \to \mathbb{R}$ *with* $v(t) \neq 0$.

Multiply both sides of (7.2.39) by $\prod_{s=t_0}^{t} v^{-1}(s)$ and then notice the resulting expression is equivalent to

$$\Delta\left[x(t) \prod_{s=t_0}^{t-1} v^{-1}(s)\right] = \left[(a(t) - v(t))x(t) + b(t)x(t - g(t))\right.$$

$$\left. + c(t)\Delta x(t - g(t))\right] \prod_{s=t_0}^{t} v^{-1}(s).$$

Summing the above expression from t_0 to t-1 gives

$$x(t) \prod_{s=t_0}^{t-1} v^{-1}(s) - x(t_0) = \sum_{r=t_0}^{t-1} \left[(a(r) - v(r))x(r)\right.$$

$$\left. + b(r)x(r - g(r)) + c(r)\Delta x(r - g(r))\right] \prod_{s=t_0}^{r} v^{-1}(s).$$

Dividing both sides by $\prod_{s=t_0}^{t-1} v^{-1}(s)$, gives

$$x(t) = x(t_0) \prod_{s=t_0}^{t-1} v(s) + \sum_{r=t_0}^{t-1} \left[(a(r) - v(r))x(r) \prod_{s=r+1}^{t-1} v(s)\right]$$

$$+ \sum_{r=t_0}^{t-1} \left[b(r)x(r - g(r))\right.$$

$$+ c(r)\Delta x(r - g(r))\Big] \prod_{s=t_0}^{r} v^{-1}(s) \prod_{s=t_0}^{t-1} v(s)$$

$$= x(t_0) \prod_{s=t_0}^{t-1} v(s) + \sum_{r=t_0}^{t-1} \Big[(a(r) - v(r))x(r) \prod_{s=r+1}^{t-1} v(s) \Big]$$

$$+ \sum_{r=t_0}^{t-1} \Big[b(r)x(r - g(r)) \Big] \prod_{s=r+1}^{t-1} v(s)$$

$$+ \sum_{r=t_0}^{t-1} \Big[c(r)\Delta x(r - g(r)) \Big] \prod_{s=r+1}^{t-1} v(s).$$

Using summation by parts and after some calculations and simplification we arrive at (7.2.40).

Theorem 7.2.9. *Suppose* $v(t) \neq 0$ *for* $t \geq t_0$ *and* $v(t)$ *satisfies*

$$\Big| \prod_{s=t_0}^{t-1} v(s) \Big| \leq M$$

for $M > 0$. *Also, suppose that there is an* $\alpha \in (0, 1)$ *such that*

$$|c(t - 1)| + \sum_{r=t_0}^{t-1} |a(r) - v(r| \Big| \prod_{s=r+1}^{t-1} v(s) \Big|$$

$$+ \sum_{r=t_0}^{t-1} \Big[|b(r) - \phi(r)| \Big] \Big| \prod_{s=r+1}^{t-1} a(s) \Big| \leq \alpha, \ t \geq t_0. \tag{7.2.41}$$

Then solutions of (7.2.39) *are equi-bounded.*

Proof. Let B_1 and B_2 be two positive constants to be defined later in the proof and let $\psi(t)$ be a bounded initial function satisfying $|\psi(t)| \leq B_1$ on $\mathbb{Z}_0$. Define

$$S = \Big\{ \varphi : \mathbb{Z} \to \mathbb{R} | \ \varphi(t) = \psi(t) \text{ on } \mathbb{Z}_0 \text{ and } ||\varphi|| \leq B_2 \Big\},$$

where

$$||\varphi|| = \sup_{t \in \mathbb{Z}} |\varphi(t)|.$$

Then $(S, || \cdot ||)$ is a complete metric space.
Define mapping $P : S \to S$ by

$$(P\varphi)(t) = \psi(t) \text{ on } \mathbb{Z}_0$$

and

$$\left(P\varphi\right)(t) = \left[\psi(t_0) - c(t_0 - 1)\psi(t_0 - g(t_0))\right]\prod_{s=t_0}^{t-1} v(s) + c(t-1)\varphi(t - g(t))$$

$$+ \sum_{r=t_0}^{t-1}\left[(a(r) - v(r))\varphi(r)\prod_{s=r+1}^{t-1} v(s)\right]$$

$$+ \sum_{r=t_0}^{t-1}\left[(b(r) - \phi(r))\varphi(r - g(r))\prod_{s=r+1}^{t-1} v(s)\right], \quad t \geq t_0. \quad (7.2.42)$$

Let $B_1 > 0$ be given. Choose B_2 such that

$$|1 - c(t_0 - 1)|MB_1 + \alpha B_2 \leq B_2. \quad (7.2.43)$$

We first show that P maps from S to S. By (7.2.43)

$$|(P\varphi)(t)| \leq |1 - c(t_0 - 1)|MB_1 + \alpha B_2$$
$$\leq B_2 \quad \text{for} \, t \geq t_0.$$

Thus P maps from S into itself. We next show that P is a contraction under the supremum norm. Let $\zeta, \eta \in S$. Then

$$|(P\zeta)(t) - (P\eta)(t)| \leq \left(|c(t-1)| + \sum_{r=t_0}^{t-1}\left[|b(r) - \phi(r)|\right]\left|\prod_{s=r+1}^{t-1} v(s)\right|\right)||\zeta - \eta||$$

$$+ \sum_{r=t_0}^{t-1}|a(r) - v(r)|\left|\prod_{s=r+1}^{t-1} v(s)||\zeta - \eta||\right.$$

$$\leq \alpha||\zeta - \eta||.$$

This shows that P is a contraction. Thus, by the contraction mapping principle, P has a unique fixed point in S which solves (7.2.39). Hence solutions of (7.2.39) are equi-bounded. $\qquad\square$

Theorem 7.2.10. *Assume that the hypotheses of Theorem 7.2.9 hold. Then the zero solution of (7.2.39) is stable.*

Proof. Let $\epsilon > 0$ be given. Choose $\delta > 0$ such that

$$|1 - c(t_0 - 1)|M\delta + \alpha\epsilon \leq \epsilon. \quad (7.2.44)$$

Let $\psi(t)$ be a bounded initial function satisfying $|\psi(t)| \leq \delta$. Define the complete metric space S by

$$S = \left\{\varphi : \mathbb{Z} \to \mathbb{R} \, | \, \varphi(t) = \psi(t) \text{ on } \mathbb{Z}_0 \text{ and } ||\varphi|| \leq \epsilon\right\}.$$

Let $P : S \to S$ be defined by (7.2.42). Then, from the proof of Theorem 7.2.10 we have that P is a contraction map and for any $\varphi \in S$, $\|P\varphi\| \leq \epsilon$. $\square$

Hence the zero solution of (7.2.39) is stable.

Theorem 7.2.11. *Assume that the hypotheses of Theorem 7.2.9 hold. Also assume that*

$$\prod_{s=t_0}^{t-1} v(s) \to 0 \text{ as } t \to \infty. \tag{7.2.45}$$

Then the zero solution of (7.2.39) *is asymptotically stable.*

Proof. We have already shown that the zero solution of (7.2.39) is stable. Let $r(t_0)$ be the δ of stability of the zero solution.

Let $\psi(t)$ be any initial discrete function satisfying $|\psi(t)| \leq r(t_0)$. Define

$$S^* = \left\{ \varphi : \mathbb{Z} \to \mathbb{R} |\, \varphi(t) = \psi(t) \text{ on } \mathbb{Z}_0, \ \|\varphi\| \leq \epsilon \text{ and } \varphi(t) \to 0 \text{ as } t \to \infty \right\}.$$

Define $P : S^* \to S^*$ by (7.2.42). Then from Theorem 7.2.9, the map P is a contraction and it maps from S^* into itself.

Left to show that $(P\varphi)(t) \to 0$ as $t \to \infty$.

Let $\varphi \in S^*$. Then the first term on the right of (7.2.42) goes to zero. The second term on the right side of (7.2.42) goes to zero due condition (7.2.45) and the fact that $\varphi \in S^*$.

Now we show that the second term on the right side of (7.2.45) goes to zero as $t \to \infty$. Let $\varphi \in S^*$ then $|\varphi(t)| \leq \epsilon$. Also, since $\varphi(t) \to 0$ as $t \to \infty$, there exists a $t_1 > 0$ such that for $t > t_1$, $|\varphi(t)| < \epsilon_1$ for $\epsilon_1 > 0$. Due to condition (7.2.45) there exists a $t_2 > t_1$ such that for $t > t_2$ implies that

$$\left| \prod_{s=t_1}^{t} v(s) \right| < \frac{\epsilon_1}{\alpha \epsilon}.$$

Thus for $t > t_2$, we have

$$\left| \sum_{r=t_0}^{t-1} \left[a(r) - v(r) \right] \varphi(r) \prod_{s=r+1}^{t-1} v(s) \right|$$

$$\leq \sum_{r=t_0}^{t-1} \left| (a(r) - v(r)) \varphi(r) \prod_{s=r+1}^{t-1} v(s) \right|$$

$$\leq \sum_{r=t_0}^{t_1-1} \left| (a(r) - v(r)) \varphi(r) \prod_{s=r+1}^{t-1} v(s) \right| + \sum_{r=t_1}^{t-1} \left| (a(r) - v(r)) \varphi(r) \prod_{s=r+1}^{t-1} v(s) \right|$$

$$\leq \epsilon \sum_{r=t_0}^{t_1-1} \left| (a(r) - v(r)) \prod_{s=r+1}^{t-1} v(s) \right| + \epsilon_1 \alpha$$

$$\leq \epsilon \sum_{r=t_0}^{t_1-1} \left| [a(r) - v(r)] \prod_{s=r+1}^{t_1-1} v(s) \prod_{s=t_1}^{t-1} v(s) \right| + \epsilon_1 \alpha$$

$$\leq \epsilon \left| \prod_{s=t_1}^{t-1} v(s) \right| \sum_{r=t_0}^{t_1-1} \left| [a(r) - v(r)] \prod_{s=r+1}^{t_1-1} v(s) \right| + \epsilon_1 \alpha$$

$$\leq \epsilon \alpha \left| \prod_{s=t_1}^{t-1} v(s) \right| + \epsilon_1 \alpha$$

$$\leq \epsilon_1 + \epsilon_1 \alpha.$$

This shows that the second term of (7.2.42) goes to zero as t goes to infinity. Showing that the last term on the right side of (7.2.45) goes to zero as $t \to \infty$ is similar, and hence we omit. This implies that $(P\varphi)(t) \to 0$ as $t \to \infty$.

By the contraction mapping principle, P has a unique fixed point that solves (7.2.39) and goes to zero as t goes to infinity. This concludes that the zero solution of (7.2.39) is asymptotically stable. $\qquad\square$

Remark 7.1. If the delay function $g(t)$ is unbounded, then we may prove a similar theorem to Theorem 7.2.11 by making the additional requirement that $t - g(t) \to 0$, as $t \to \infty$.

Example 7.3. Solutions of the linear neutral difference equation

$$x((t+1) = \frac{2^{t+1}}{8(1+t)!} x(t-2) + \frac{2^{t+1}}{8(1+t)!} \Delta x(t-2), \ t \geq 0 \qquad (7.2.46)$$

are equi-bounded and the zero solution is asymptotically stable. To see this, we let $v(t) = \frac{1}{3(1+t)}$. Comparing terms, we see that $a(t) = 0$, $b(t) = c(t) = \frac{2^{t+1}}{8(1+t)!}$. Set $t_0 = 0$. Then (7.2.41) is equivalent to

$$|c(t-1)| + \sum_{r=0}^{t-1} |v(r)| \left| \prod_{s=r+1}^{t-1} v(s) \right|$$

$$+ \sum_{r=0}^{t-1} \left[|b(r) - \phi(r)| \right] \left| \prod_{s=r+1}^{t-1} v(s) \right|$$

$$\leq \frac{2^t}{8(t)!} + \sum_{r=0}^{t-1} \prod_{s=r}^{t-1} \frac{1}{3(1+s)} + \sum_{r=0}^{t-1} \frac{2^r}{8(1+r)!} \prod_{s=r+1}^{t-1} \frac{1}{3(1+s)}.$$

Now,

$$\sum_{r=0}^{t-1} \frac{2^r}{8(1+r)!} \prod_{s=r+1}^{t-1} \frac{1}{3(1+s)} \leq 1/3 \sum_{r=0}^{t-1} \frac{2^r}{8(1+r)!} \frac{1}{(r+2)(r+3)...(t)}$$

$$\leq 1/3 \sum_{r=0}^{t-1} \frac{2^r}{8t!}$$

$$\leq \frac{1}{24t!}(2^t - 1) \leq \frac{2^t}{24t!}.$$

Similarly, by estimating $\frac{1}{1+s} \leq 1$, for $s \geq 0$, we have that

$$\sum_{r=0}^{t-1} \prod_{s=r}^{t-1} \frac{1}{3(1+s)} \leq \sum_{r=0}^{t-1} (\frac{1}{3})^{t-r}$$

$$\leq (\frac{1}{3})^t \sum_{r=0}^{t-1} 3^r = (\frac{1}{3})^t [\frac{3^r}{2}]\big|_0^{t-1}$$

$$\leq \frac{1}{6}[1 - 2^{1-t}] \leq 1/6.$$

Combining the two inequalities we end up with

$$|c(t-1)| + \sum_{r=0}^{t-1} |v(r)| \left| \prod_{s=r+1}^{t-1} v(s) \right|$$

$$+ \sum_{r=0}^{t-1} \left[|b(r) - \phi(r)| \right] \left| \prod_{s=r+1}^{t-1} v(s) \right|$$

$$\leq \frac{2^t}{8(t)!} + \frac{1}{3} + \frac{2^t}{24t!}$$

$$\leq \frac{1}{4} + \frac{1}{6} + \frac{1}{12} = \frac{1}{2} < 1.$$

Hence (7.2.41) is satisfied. It is clear that condition (7.2.45) is satisfied for the specified value of v. This implies the zero solution is asymptotically stable, by Theorem 7.2.11. Left to show solutions are equi-bounded.

Since $t_0 = 0$, we have that $\mathbb{Z}_0 = [-2, 0]$.

Let $B_1 > 0$ be given and $\psi(t) : \mathbb{Z}_0 \to \mathbb{R}$ be a given initial function with $|\psi(t)| \leq B_1$. We need to choose B_2 so that (7.2.43) is satisfied. It is clear that $c(t_0 - 1) = c(-1) = \frac{1}{8}$, and hence $|1 - c(t_0 - 1)| = 1 - \frac{1}{8} = \frac{7}{8}$. In addition

$$\left| \prod_{s=0}^{t-1} v(s) \right| \leq M$$

is satisfied for $M = \frac{1}{3}$. From the above calculation for asymptotic stability, we see that $\alpha = \frac{1}{2}$. Now we choose B_2 such that

$$\frac{7}{24} B_1 \leq \frac{B_2}{2}.$$

Then, in our case, inequality (7.2.43) corresponds to

$$|1 - c(t_0 - 1)|M B_1 + \alpha B_2 \leq B_2.$$

Or equivalently,

$$\frac{7}{24} B_1 + \frac{B_2}{2} \leq B_2,$$

is satisfied.   □

It is worth mentioning that the results of Section 7.2.3 can be easily extended to the non-linear neutral difference equation

$$x(t+1) = a(t)x(t) + c(t)\Delta x(t - g(t)) + q(t, x(t), x(t - g(t))) \quad (7.2.47)$$

where $a(t), c(t)$ and $g(t)$ are defined as before. We assume that, $q(t, 0, 0) = 0$ for the stability and q is locally Lipschitz in x and y. That is, there is a $K > 0$ so that if $|x|, |y|, |z|$ and $|w| \leq K$ then

$$|q(x, y) - q(z, w)| \leq L|x - z| + E|y - w|$$

for some positive constants L and E.

Note that

$$|q(x, y)| = |q(x, y) - q(0, 0) + q(0, 0)|$$
$$\leq |q(x, y) - q(0, 0)| + |q(0, 0)|$$
$$\leq L|x| + E|y|.$$

7.2.4   Exercises

Exercise 7.1. Give all details on how (7.2.18) was obtained.

Exercise 7.2. Let the mapping $\mathfrak{P}$ be defined by (7.2.36). Show that

$$(\mathfrak{P}\phi)(t + T) = (\mathfrak{P}\phi)(t).$$

Exercise 7.3. Prove Theorem 7.2.7.

Exercise 7.4. Prove Theorem 7.2.8.

Exercise 7.5. Redevelop the content to Section 7.2.3 for the equation

$$x(t+1) = a(t)x(t) + b(t)x(t-K)$$

where $a, b : \mathbb{Z} \to \mathbb{R}$, and $K \in \mathbb{Z}^+$.

Project
Extend the contents of Chapter 7 to the system

$$x(t+1) = A(t)x(t) + g(t, x(t)),$$

where $A(t)$ is an $k \times k$ matrix and the function g is an $k \times 1$ vector function that is continuous in x, and g can be varied according to the content of the section being generalized.

Appendix A

Banach spaces

This appendix is devoted to introductory materials related to Cauchy sequences, metric spaces, contraction, compactness, contraction mapping principle, and Banach spaces. Materials in this section will be of use in several places in the book especially in Chapter 7. Throughout, the notation $C(I, \mathbb{R}^n)$ denotes the space of all continuous functions $f : I \to \mathbb{R}^n$, for an interval I that could be infinite.

Definition A.0.1. A pair (E, ρ) is a metric space if E is a set and $\rho : E \times E \to [0, \infty)$ such that when y, z, and u are in E then

(a) $\rho(y, z) \geq 0$, $\rho(y, y) = 0$, and $\rho(y, z) = 0$ implies $y = z$,
(b) $\rho(y, z) = \rho(z, y)$, and
(c) $\rho(y, z) \leq \rho(y, u) + \rho(u, z)$.

The next definitions are concerned with Cauchy sequences.

Definition A.0.2 (Cauchy sequence). A sequence $\{x_n\} \subseteq E$ is a Cauchy sequence if for each $\varepsilon > 0$ there exists an $N \in \mathbb{N}$ such that $n, m > N \implies \rho(x_n, x_m) < \varepsilon$.

Complete metric spaces play a major role when showing a fixed point belongs to the metric space of interest.

Definition A.0.3 (Completeness of metric space). A metric space (E, ρ) is said to be complete if every Cauchy sequence in E converges to a point in E.

Definition A.0.4. A set L in a metric space (E, ρ) is compact if each sequence in L has a subsequence with a limit in L.

Definition A.0.5. Let $\{f_n\}$ be a sequence of real functions with $f_n : [a, b] \to \mathbb{R}$.

(i) $\{f_n\}$ is uniformly bounded on $[a, b]$ if there exists $M > 0$ such that $|f_n(t)| \leq M$ for all $n \in \mathbb{N}$ and for all $t \in [a, b]$.
(ii) $\{f_n\}$ is equicontinuous at t_0 if for each $\varepsilon > 0$ $\delta > 0$ such that for all $n \in \mathbb{N}$, if $t \in [a, b]$ and $|t_0 - t| < \delta$ then $|f_n(t_0) - f_n(t)| < \varepsilon$. Also, $\{f_n\}$ is equicontinuous if $\{f_n\}$ is equicontinuous at each $t_0 \in [a, b]$.
(iii) $\{f_n\}$ is uniformly equicontinuous if for each $\varepsilon > 0$ there exists $d > 0$ such that for all $n \in \mathbb{N}$, if $t_1, t_2 \in [a, b]$ and $|t_1 - t_2| < \delta$ then $|f_n(t_1) - f_n(t_2)| < \varepsilon$.

Easy to see that $\{f_n\} = \{x^n\}$ is not an equicontinuous sequence of functions on $[0, 1]$ but each f_n is uniformly continuous.

Proposition 1 (Cauchy criterion for uniform convergence). *If $\{F_n\}$ is a sequence of bounded functions that is Cauchy in the uniform norm, then $\{F_n\}$ converges uniformly.*

Definition A.0.6. A real-valued function f defined on $E \subseteq \mathbb{R}$ is said to be Lipschitz continuous with Lipschitz constant K if $|f(x) - f(y)| \le K|x - y|$ for all $x, y \in E$.

It is easy to see that the function $f(x) = x^2$ is not Lipschitz on $\mathbb{R}$. This is due to the fact that for any x and y in $\mathbb{R}$ we have that $f(x) - f(y) = |x^2 - y^2| = |x + y||x - y|$ and so there is no constant K such that $|x^2 - y^2| \le K|x - y|$. Definition A.0.6 implies f is globally Lipschitz since the constant K is uniform for all x and y in $\mathbb{R}$.

Remark 1. It is an easy exercise that a Lipschitz continuous function is uniformly continuous. Also, if each f_n in a sequence of functions $\{f_n\}$ has the same Lipschitz constant, then the sequence is uniformly equicontinuous.

Lemma 1. *If $\{f_n\}$ is an equicontinuous sequence of functions on a closed bounded interval, then $\{f_n\}$ is uniformly equicontinuous.*

Proof. Suppose $\{f_n\}$ is equicontinuous defined on $[a, b]$ (which is contraction). Let $\varepsilon > 0$. For each $x \in K$, let $\delta_x > 0$ be such that $|y - x| < \delta_x \implies |f_n(x) - f_n(y)| < \varepsilon/2$ for all $n \in \mathbb{N}$. The collection $\{B(x, \delta_x/2) : x \in [a, b]\}$ is an open cover of $[a, b]$ so has a finite subcover $\{B(x_i, \delta_{x_i}/2) : i = 1, \dots, k\}$. Let $\delta = \min\{\delta_{x_i}/2 : i = 1, \dots, k\}$. Then, if $x, y \in [a, b]$ with $|x - y| < \delta$, there is some i with $x \in B(x_i, \delta_{x_i}/2)$. Since $|x - y| < \delta \le \delta_{x_i}/2$, we have $|x_i - y| \le |x_i - x| + |x - y| < \delta_{x_i}/2 + \delta_{x_i}/2 = \delta_{x_i}$. Hence $|x_i - y| < \delta_{x_i}$ and $|x_i - x| < \delta_{x_i}$. So, for any $n \in \mathbb{N}$ we have $|f_n(x) - f_n(y)| \le |f_n(x) - f_n(x_i)| + |f_n(x_i) - f_n(y)| < \varepsilon/2 + \varepsilon/2 = \varepsilon$. So, $\{f_n\}$ is uniformly equicontinuous. $\qquad\square$

The next theorem gives us the main method of proving compactness in the spaces in which we are interested.

Theorem A.0.1 (Ascoli-Arzelà). *If $\{f_n(t)\}$ is a uniformly bounded and equicontinuous sequence of real valued functions on an interval $[a, b]$, then there is a subsequence which converges uniformly on $[a, b]$ to a continuous function.*

Proof. Since $\{f_n(t)\}$ is equicontinuous on $[a, b]$, by Lemma 1, $\{f_n(t)\}$ is uniformly equicontinuous. Let $t_1, t_2, \dots$ be a listing of the rational numbers in $[a, b]$ (note, the set of rational numbers is countable, so this enumeration is possible). The sequence $\{f_n(t_1)\}_{n=1}^{\infty}$ is a bounded sequence of real numbers (since $\{f_n\}$ is uniformly bounded) so, it has a subsequence $\{f_{n_k}(t_1)\}$ converging to a number which we call $\phi(t_1)$. It will be more convenient to represent this subsequence without sub-subscripts, so we write f_k^1 for f_{n_k} and switch the index from k to n. So, the subsequence is written as $\{f_n^1(t_1)\}_{n=1}^{\infty}$. Now, the sequence $\{f_n^1(t_2)\}$ is bounded, so it has a convergent subsequence, say $\{f_n^2(t_2)\}$, with limit $\phi(t_2)$.

We continue in this way obtaining a sequence of sequences $\{f_n^m(t)\}_{n=1}^{\infty}$ (one sequence for each m) each of which is a subsequence of the previous. Furthermore, we have $f_n^m(t_m) \to \phi(t_m)$ as $n \to \infty$ for each $m \in \mathbb{N}$. Now, consider the "diagonal" functions defined $F_k(t) = f_k^k(t)$. Since $f_n^m(t_m) \to \phi(t_m)$, it follows that $F_r(t_m) \to \phi(t_m)$ as $r \to \infty$ for each $m \in \mathbb{N}$ (in other words, the sequence $\{F_r(t)\}$ converges pointwise at each t_m). We now show that $\{F_k(t)\}$ converges uniformly on $[a, b]$, by showing it is Cauchy in the uniform norm. Let $\varepsilon > 0$. Let $\delta > 0$ be as in the definition of uniformly equicontinuous for $\{f_n(t)\}$ applied with $\varepsilon/3$. Divide $[a, b]$ into p intervals where $p > \frac{b-a}{\delta}$. Let ξ_j be a rational number in the jth interval, for $j = 1, \ldots, p$. Remember, $\{F_r(t)\}$ converges at each of the points ξ_j, since they are rational numbers. So, for each j, there is $M_j \in \mathbb{N}$ such that $|F_r(\xi_j) - F_s(\xi_j)| < \varepsilon/3$ whenever $r, s > M_j$. Let $M = \max\{M_j : j = 1, \ldots, p\}$. If $t \in [a, b]$, then it is in one of the p intervals, say the jth. So, $|t - \xi_j| < \delta$ and so $|f_r^r(t) - f_r^r(\xi_j)| = |F_r(t) - F_r(\xi_j)| < \varepsilon/3$ for every r. Also, if $r, s > M$, then $|F_r(\xi_j) - F_s(\xi_j)| < \varepsilon/3$ (since M is the max of the M_i's). So, we have for $r, s > M$,

$$|F_r(t) - F_s(t)| = |F_r(t) - F_r(\xi_j) + F_r(\xi_j) - F_s(\xi_j) + F_s(\xi_j) - F_s(t)|$$
$$\leq |F_r(t) - F_r(\xi_j)| + |F_r(\xi_j) - F_s(\xi_j)| + |F_s(\xi_j) - F_s(t)|$$
$$\leq \frac{\varepsilon}{3} + \frac{\varepsilon}{3} + \frac{\varepsilon}{3} = \varepsilon.$$

By the Cauchy Criterion for convergence, the sequence $\{F_r(t)\}$ converges uniformly on $[a, b]$. Since each $F_r(t)$ is continuous, the limit function $\phi(t)$ is also continuous. $\qquad\square$

Remark 2. The Ascoli-Arzelà Theorem can be generalized to a sequence of functions from $[a, b]$ to $\mathbb{R}^n$. You apply the Ascoli-Arzelà to the first coordinate function to get a uniformly convergent subsequence. Then, apply the theorem again, this time to the corresponding subsequence of functions restricted to the second coordinate, getting a sub-subsequence, and so on.

The next criteria, known as Weierstrass M-test plays an important role in showing the existence of solutions.

Lemma 2 (Weierstrass M-test). *Let $\{f_n\}$ be a sequence of functions defined on a set E. Suppose for all $n, n = 1, \ldots$, there is a constant M_n such that $|f_n(t)| \leq M_n$, for all $t \in E$. If*

$$\sum_{n=1}^{\infty} M_n < \infty, \text{ then } \sum_{n=1}^{\infty} f_n(t)$$

converges absolutely and uniformly on the E.

We remark that the Weierstrass M-test can be easily generalized if the domain of the sequence of functions is a subset of Banach space endowed with an appropriate norm.

Here is an example of the Weierstrass M-test.

Example A.1. For $n = 1, 2, \ldots$ define the sequence of functions $\{f_n\}$ on $\mathbb{R}$ by $f_n(t) = \frac{1}{t^2+n^2}$. Then $|f_n(t)| = |\frac{1}{t^2+n^2}| \leq \frac{1}{n^2} := M_n$, for all $t \in \mathbb{R}$, and $n \geq 1$. Since the series $\sum_{n=1}^{\infty} \frac{1}{n^2}$ converges, then by the Weierstrass M-test the series $\sum_{n=1}^{\infty} \frac{1}{t^2+n^2}$ converges uniformly on $\mathbb{R}$. Moreover we can say more. As each term of the series is continuous and the convergence is uniform the sum function is also continuous. (As the uniform limit of continuous functions is continuous.) $\square$

Here is another example with a simple twist to it.

Example A.2. We prove the following series

$$\sum_{n=1}^{\infty} \frac{n^2 + x^4}{n^4 + x^2}$$

converges to a continuous function $f : \mathbb{R} \to \mathbb{R}$.

Let c be a positive constant. Then for all $x \in [-c, c]$ we see that

$$\left|\frac{n^2 + x^4}{n^4 + x^2}\right| \leq \frac{n^2 + x^4}{n^4} \leq \frac{1}{n^2} + \frac{c^4}{n^4} := M_n.$$

On the other hand the series

$$\sum_{n=1}^{\infty} M_n = \sum_{n=1}^{\infty} \frac{1}{n^2} + c^4 \sum_{n=1}^{\infty} \frac{1}{n^4}$$

converges, so Weierstrass M-test implies the series converges uniformly to a function f on the bounded interval $[-c, c]$. Each term in the series is continuous and since the uniform limit of continuous functions is continuous, we have the limit function f is continuous on $[-c, c]$ for every $c > 0$. Now since every $x \in \mathbb{R}$ lies in such an interval for sufficiently large c, it follows that f is continuous on $\mathbb{R}$. Note the series does not converge uniformly on $\mathbb{R}$, so we can not use the argument that the sum is continuous on $\mathbb{R}$ because the series converges uniformly on $\mathbb{R}$. $\square$

Banach space form an important class of metric spaces. We now define Banach space in several steps.

Definition A.0.7. A triple $(V, +, \cdot)$ is said to be a linear (or vector) space over a field F if V is a set and the following are true.

(i) Properties of $+$
    a. $+$ is a function from $V \times V$ to V. Outputs are denoted $x + y$.
    b. for all $x, y \in V$, $x + y = y + x$. ($+$ is commutative)

   c. for all $x, y, w \in V$, $x + (y + w) = (x + y) + w$. (+ is associative)
   d. there is a unique element of V which we denote 0 such that for all
      $x \in V$, $0 + x = x + 0 = x$. (additive identity)
   e. for each $x \in V$ there is a unique element of V which we denote $-x$
      such that $x + (-x) = -x + x = 0$. (additive inverse).
(ii) Scalar multiplication
   a. $\cdot$ is a function from $F \times V$ to V. Outputs are denoted $\alpha \cdot x$, or αx.
   b. for all $\alpha, \beta \in F$ and $x \in V$, $\alpha(\beta x) = (\alpha\beta)x$.
   c. for all $x \in V$, $1 \cdot x = x$.
   d. for all $\alpha, \beta \in F$ and $x \in V$, $(\alpha + \beta)x = \alpha x + \beta x$.
   e. for all $\alpha \in F$ and $x, y \in V$, $\alpha(x + y) = \alpha x + \alpha y$.

Commonly, the real numbers or complex numbers are the field in the above
definition. For our purposes, we only consider the field of real numbers $F = \mathbb{R}$.

Definition A.0.8 (Normed spaces). A vector space $(V, +, \cdot)$ is a **normed space**
if for each $x \in V$ there is a nonnegative real number $\|x\|$, called the **norm** of x,
such that for each $x, y \in V$ and $\alpha \in \mathbb{R}$

   (i) $\|x\| = 0$ if and only if $x = 0$
   (ii) $\|\alpha x\| = |\alpha|\|x\|$
   (iii) $\|x + y\| \leq \|x\| + \|y\|$.

Remark 3. A norm on a vector space always defines a metric $\rho(x, y) = \|x - y\|$
on the vector space. Given a metric ρ defined on a vector space, it is tempting
to define $\|v\| = \rho(v, 0)$. But this is not always a norm.

Definition A.0.9. A Banach space is a complete normed vector space. That is,
a vector space $(X, +, \cdot)$ with norm $\| \cdot \|$ for which the metric $\rho(x, y) = \|x - y\|$
is complete.

Example A.3. The space $(\mathbb{R}^n, +, \cdot)$ over the field $\mathbb{R}$ is a vector space (with the
usual vector addition, $+$ and scalar multiplication, $\cdot$) and there are many suitable
norms for it. For example, if $x = (x_1, x_2, \ldots, x_n)$ then

   (i) $\|x\| = \max_{1 \leq i \leq n} |x_i|$,
   (ii) $\|x\| = \sqrt{\sum_{i=1}^{n} x_i^2}$, or
   (iii) $\|x\| = \sum_{i=1}^{n} |x_i|$,
   (iv) $\|x\|_p = \left(\sum_{i=1}^{n} |x_i|^p \right)^{1/p}$, $p \geq 1$

are all suitable norms. Norm 2. is the Euclidean norm: the norm of a vector is its
Euclidean distance to the zero vector and the metric defined from this norm is
the usual Euclidean metric. Norm 3. generates the "taxi-cab" metric on $\mathbb{R}^2$ and
Norm 4. is the l^p norm.

   Throughout the book, it should cause no confusion to use $|\cdot|$, instead of $\|\cdot\|$
to denote a particular norm. $\qquad\qquad\qquad\qquad\qquad\qquad\qquad\qquad\qquad$ $\square$

Remark 4. Consider the vector space $(\mathbb{R}^n, +, \cdot)$ as a metric space with its metric defined $\rho(x, y) = \|x - y\|$ where $\|\cdot\|$ is any of the norms as in Example A.3. The completeness of this metric space comes directly from the completeness of $\mathbb{R}$, hence $(\mathbb{R}^n, \|\cdot\|)$ is a Banach space.

Remark 5. In the Euclidean space $\mathbb{R}^n$, compactness is equivalent to closed and bounded (Heine-Borel Theorem). In fact, the metrics generated from any of the norms in Example A.3 are equivalent in the sense that they generate the same topologies. Moreover, compactness is equivalent to closed and bounded in each of those metrics.

Example A.4. Let $C([a, b], \mathbb{R}^n)$ denote the space of all continuous functions $f : [a, b] \to \mathbb{R}^n$.

   (i) $C([a, b], \mathbb{R}^n)$ is a vector space over $\mathbb{R}$.
  (ii) If $\|f\| = \max_{a \leq t \leq b} |f(t)|$ where $|\cdot|$ is a norm on $\mathbb{R}^n$, then $(C([a, b], \mathbb{R}^n), \|\cdot\|)$ is a Banach space.
 (iii) Let M and K be two positive constants and define

$$L = \{f \in C([a, b], \mathbb{R}^n) : \|f\| \leq M; |f(u) - f(v)| \leq K|u - v|\}$$

then L is compact. $\qquad\square$

Proof. (of part 3.) Let $\{f_n\}$ be any sequence in L. The functions are uniformly bounded by M and have the same Lipschitz constant, K. So, the sequence is uniformly equicontinuous. By the Ascoli-Arzelà Theorem, there is a subsequence, $\{f_{n_k}\}$, that converges uniformly to a continuous function $f : [a, b] \to \mathbb{R}^n$. We now show that $f \in L$. Well, $|f_n(t)| \leq M$ for each $t \in [a, b]$, so $|f(t)| \leq M$ for each $t \in [a, b]$ and hence $\|f\| \leq M$. Now, fix $u, v \in [a, b]$ and fix $\varepsilon > 0$. Since $\{f_{n_k}\}$ converges uniformly to f, there is $N \in \mathbb{N}$ such that $|f_{n_k}(t) - f(t)| < \varepsilon/2$ for all $t \in [a, b]$ and all $k \geq N$. So, fix any $k \geq N$ and we have

$$\begin{aligned}
|f(u) - f(v)| &= |f(u) - f_{n_k}(u) + f_{n_k}(u) - f_{n_k}(v) + f_{n_k}(v) - f(v)| \\
&\leq |f(u) - f_{n_k}(u)| + |f_{n_k}(u) - f_{n_k}(v)| + |f_{n_k}(v) - f(v)| \\
&< \varepsilon/2 + K|u - v| + \varepsilon/2 = K|u - v| + \varepsilon.
\end{aligned}$$

Since $\varepsilon > 0$ was arbitrary, $|f(u) - f(v)| \leq K|u - v|$. Hence $f \in L$. We have demonstrated that $\{f_n\}$ has a subsequence converging to an element of L. Hence, L is compact. $\qquad\square$

Example A.5. Consider $\mathbb{R}$ as a vector space over $\mathbb{R}$ and define the metric $d(x, y) = \frac{|x-y|}{1+|x-y|}$. For each $x \in \mathbb{R}$, we can define $\|x\| = d(x, 0)$. Explain why $\|\cdot\|$ is not a norm on $\mathbb{R}$. $\qquad\square$

Example A.6. Let $\phi : [a, b] \to \mathbb{R}^n$ be continuous and let S be the set of continuous functions $f : [a, c] \to \mathbb{R}^n$ with $c > b$ and with $f(t) = \phi(t)$ for $a \leq t \leq b$. Define $\rho(f, g) = \|f - g\| = \sup_{a \leq t \leq c} |f(t) - g(t)|$ for $f, g \in S$. Then (S, ρ) is a complete metric space but not a Banach space since $f + g$ is not in S. $\qquad\square$

Example A.7. (S, ρ) be the space of continuous bounded functions $f :$ $(-\infty, 0] \to \mathbb{R}$ with $\rho(f, g) = \|f - g\| = \sup_{-\infty < t \leq 0} |f(t) - g(t)|$.

(i) Show that (S, ρ) is a Banach space.

(ii) The set $L = \{f \in S : \|f\| \leq 1, |f(u)f(v)| \leq |u - v|\}$ is not compact in (S, ρ). □

Proof. (of 2.) Consider the sequence of functions defined

$$f_n(t) = \begin{cases} 0 & \text{if } t \leq -n \\ \frac{t}{n} + 1 & \text{if } -n < t \leq 0. \end{cases}$$

Then, the sequence converges pointwise to $f = 1$, but $\rho(f_n, f) = 1$ for all $n \in \mathbb{N}$. So, there is no subsequence of $\{f_n\}$ converging in the norm $\| \cdot \|$ (i.e. converging uniformly) to f. □

Example A.8. Let (S, ρ) be the space of continuous functions $f : (-\infty, 0] \to \mathbb{R}^n$ with

$$\rho(f, g) = \sum_{n=1}^{\infty} 2^{-n} \rho_n(f, g) / \{1 + \rho_n(f, g)\}$$

where

$$\rho_n(f, g) = \max_{-n \leq s \leq 0} |f(s) - g(s)|$$

and $| \cdot |$ is the Euclidean norm on $\mathbb{R}^n$

(i) Then (S, ρ) is a complete metric space. The distance between all functions is bounded by 1.

(ii) $(S, +, \cdot)$ is a vector space over $\mathbb{R}$.

(iii) (S, ρ) is not a Banach space because ρ does not define a norm, since $\rho(x, 0) = \|x\|$ does not satisfy $\|\alpha x\| = |\alpha| \|x\|$.

(iv) Let M and K be given positive constants. Then the set

$$L = \{f \in S : \|f\| \leq M \text{ on } (-\infty, 0], |f(u) - f(v)| \leq K|u - v|\}$$

is compact in (S, ρ). □

Proof. (of 4.) Let $\{f_n\}$ be a sequence in L. It is clear that if $f_n \to f$ uniformly on compact subsets of $(-\infty, 0]$ then we have $\rho(f_n, f) \to 0$ as $n \to \infty$. Let's begin by considering $\{f_n\}$ on $[-1, 0]$. Then the sequence is uniformly bounded and equicontinuous and so there is a subsequence, say $\{f_n^1\}$ converging uniformly to some continuous f on $[-1, 0]$. Moreover the argument of Example A.4 shows that $|f(t)| \leq M$, and $|f(u) - f(v)| \leq K|u - v|$. Next we consider $\{f_n^1\}$ on $[-2, 0]$. Then the sequence is uniformly bounded and equicontinuous and so there is a subsequence, say $\{f_n^2\}$ converging uniformly, say, to some continuous

f on $[-2, 0]$. Continuing this way we arrive at $F_n = f_n^n$ which has a subsequence of $\{f_n\}$ and it converges uniformly on compact subsets of $(-\infty, 0]$ to a function $f \in L$. This proves L is compact. $\qquad\square$

The next result is stated in the form of a theorem that we leave its proof to the reader.

Theorem A.0.2. *Let $g : (-\infty, 0] \to [1, \infty)$ be a continuous strictly decreasing function with $g(0) = 1$ and $g(r) \to \infty$ as $r \to -\infty$. Let $(S, |\cdot|_g)$ be the space of continuous functions $f : (-\infty, 0] \to \mathbb{R}^n$ for which*

$$|f|_g := \sup_{-\infty < t \leq 0} \frac{|f(t)|}{|g(t)|}$$

exists. Then

  (i) *$(S, |\cdot|_g)$ is a Banach space.*
  (ii) *Let M and K be given positive constants. Then the set*

$$L = \{f \in S : \|f\| \leq M \text{ on } (-\infty, 0], |f(u) - f(v)| \leq K|u - v|\}$$

is compact in (S, ρ).

Definition A.0.10. Let (E, ρ) be a metric space and $D : E \to E$. The operator or mapping D is a contraction if there exists an $\alpha \in (0, 1)$ such that

$$\rho\Big(D(x), D(y)\Big) \leq \alpha\rho(x, y).$$

The next theorem is known by the name of Caccioppoli Theorem, or Banach's Contraction Mapping Principle. A proof can be found in many advanced analysis books.

Theorem A.0.3 (Contraction mapping principle). *Let (E, ρ) be a complete metric space and $D : E \to E$ a contraction operator. Then there exists a unique $\phi \in E$ with $D(\phi) = \phi$. Moreover, if $\psi \in E$ and if $\{\psi_n\}$ is defined inductively by $\psi_1 = D(\psi)$ and $\psi_{n+1} = D(\psi_n)$, then $\psi_n \to \phi$, the unique fixed point.*

Proof. Let $y_0 \in E$ and define a sequence $\{y_n\}$ in E by $y_1 = Dy_0$, $y_2 = Dy_1 = D(Dy_0) = D^2 y_0, \ldots, y_n = Dy_{n-1} = D^n y_0$. Next we show that $\{y_n\}$ is a Cauchy sequence. To see this, if $m > n$, then

$$\rho(y_n, y_m) = \rho(D^n y_0, D^m y_0)$$
$$\leq \alpha\rho(D^{n-1} y_0, D^{m-1} y_0)$$
$$\vdots$$
$$\leq \alpha^n \rho(y_0, y_{m-1})$$
$$\leq \alpha^n \big\{\rho(y_0, y_1) + \rho(y_1, y_2) + \ldots + \rho(y_{m-n-1}, y_{m-n})\big\}$$

$$\leq \alpha^n \{ \rho(y_0, y_1) + \alpha \rho(y_0, y_1) + \ldots + \alpha^{m-n-1} \rho(y_0, y_1) \}$$
$$\leq \alpha^n \rho(y_0, y_1) \{ 1 + \alpha + \ldots + \alpha^{m-n-1} \}$$
$$\leq \alpha^n \rho(y_0, y_1) \frac{1}{1 - \alpha}.$$

Thus, since $\alpha \in (0, 1)$, we have that

$$\rho(y_n, y_m) \to 0, \text{ as, } n \to \infty.$$

This shows the sequence $\{y_n\}$ is Cauchy. Since (E, ρ) is a complete metric space, $\{y_n\}$ has a limit, say y in E. Since the mapping D is continuous we have that

$$D(x) = D(\lim_{n \to \infty} y_n) = \lim_{n \to \infty} D(y_n) = \lim_{n \to \infty} y_{n+1} = y,$$

and y is a fixed point. Left to show y is unique. Let $x, y \in E$ such that $D(x) = x$ and $D(y) = y$. Then

$$0 \leq \rho(x, y) = \rho(D(x), D(y)) \leq \alpha \rho(x, y),$$

which implies that

$$0 \leq (1 - \alpha) \rho(x, y) \leq 0.$$

Since $1 - \alpha \neq 0$, we must have $\rho(x, y) = 0$ and hence $x = y$. This completes the proof. $\square$

Another form of the contraction mapping principle.

Theorem A.0.4 (Contraction mapping principle, Banach fixed point theorem). *Let (E, ρ) be a complete metric space and $P : E \to E$ such that P^m is a contraction for some fixed positive integer m. Then there is a unique $x \in E$ with $P(x) = x$.*

Bibliography

[1] M. Adivar, M. Islam, Y. Raffoul, Separate contraction and existence of periodic solutions in totally non-linear delay differential equations, Hacettepe Journal of Mathematics and Statistics 41 (1) (2012) 1–13.

[2] R. Ashegi, Bifurcations and dynamics of discrete predator-prey system, Journal of Biological Dynamics 8 (1) (2014) 161–186.

[3] R. Bartle, The Elements of Real Analysis, second edition, Wiley, New York, 1976.

[4] M. Braun, Difference Equations and Their Applications, Springer, 1992.

[5] H. Broer, F. Takens, Non-linear Discrete Dynamical Systems, Cambridge University Press, 2011.

[6] T.A. Burton, Integral equations, implicit functions, and fixed points, Proceedings of the American Mathematical Society 124 (1996) 2383–2390.

[7] T. Burton, Volterra Integral and Differential Equations, Academic Press, New York, 1983.

[8] T. Burton, Stability and Periodic Solutions of Ordinary and Functional Differential Equations, Academic Press, New York, 1985.

[9] T. Burton, T. Furumochi, Fixed points and problems in stability theory, Dynamic Systems and Applications 10 (2001) 89–116.

[10] S.K. Chakraborty, A Course in Discrete Mathematical Structures, Springer, 2011.

[11] M.R. Crisci, V.B. Kolmanovskii, A. Vecchio, Boundedness of discrete Volterra equations, Journal of Mathematical Analysis and Applications 211 (1997) 106–130.

[12] S. Elaydi, Periodicity and stability of linear Volterra difference systems, Journal of Mathematical Analysis and Applications 181 (1994) 483–492.

[13] S. Elaydi, An Introduction to Difference Equations, Springer, New York, 1999.

[14] S. Elaydi, S. Murakami, Uniform asymptotic stability in linear Volterra difference equations, Journal of Difference Equations 3 (1998) 203–218.

[15] P. Eloe, M. Islam, Y.N. Raffoul, Uniform asymptotic stability in non-linear Volterra discrete systems, Computers & Mathematics With Applications: Special Issue on Advances in Difference Equations IV 45 (2003) 1033–1039.

[16] A. Gelfond, Calculus of Finite Differences, Hindustan, Delhi, India, 1971.

[17] P. Glendinning, Difference Equations from Rabbits to Chaos, Oxford University Press, 1998.

[18] Y. Hino, S. Murakami, Total stability and uniform asymptotic stability for linear Volterra equations, Journal of the London Mathematical Society 43 (1991) 305–312.

[19] M. Islam, Y.N. Raffoul, Exponential stability in non-linear difference equations, Journal of Difference Equations and Applications 9 (2003) 819–825.

[20] M. Islam, E. Yankson, Boundedness and stability in non-linear delay difference equations employing fixed point theory, Electronic Journal on the Qualitative Theory of Differential Equations 26 (2005) 1–18.

[21] A. Jerri, Linear Difference Equations with Discrete Transform Methods, CRC Press, 2016.

[22] G.W. Kelley, A. Peterson, Difference Equations: An Introduction with Applications, Academic Press, 2000.

[23] G. Korn, T. Korn, Dynamical Systems and Difference Equations with Mathematica, CRC Press, 2002.

[24] L. Lakshmikantham, D. Trigiante, Theory of Difference Equations: Numerical Methods and Applications, Academic Press, New York, 1991.

[25] N. Linh, V. Phat, Exponential stability of non-linear time-varying differential equations and applications, Electronic Journal of Differential Equations 34 (2001) 1–13.

[26] M. Maroun, Y.N. Raffoul, Periodic solutions in non-linear neutral difference equations with functional delay, Journal of the Korean Mathematical Society 42 (2) (2005) 255–268.

[27] R. Medina, Asymptotic behavior of Volterra difference equations, Computers & Mathematics With Applications 41 (2001) 679–687.

[28] R. Medina, The asymptotic behavior of the solutions of a Volterra difference equation, Computers & Mathematics With Applications 181 (1994) 19–26.

[29] Y.N. Raffoul, Applied Mathematics for Scientists and Engineers, Taylor & Francis/Chapman and Hall Press, 2023.

[30] Y.N. Raffoul, Advanced Differential Equations, Elsevier/Academic Press, April 19, 2022.

[31] Y.N. Raffoul, Qualitative Theory of Volterra Difference Equations, Springer, Switzerland, 2018.

[32] W. Reid, A criterion of oscillation for generalized differential equations, The Rocky Mountain Journal of Mathematics 7 (1977) 799–806.

[33] P. Ribenboim, The Book of Prime Number Records, Springer-Verlag, New York, 1988.

[34] C. Richardson, An Introduction to the Calculus of Finite Differences, Van Nostrand, Princeton, 1954.

[35] A. Sarkovskii, Coexistence of cycles of a continuous map of a line into itself, Ukrainian Mathematical Journal 16 (1964) 153–158.

[36] D. Sherbert, Difference equations with applications, UMAP, Unit 332 (1979) 1–34.

[37] J. Smith, Mathematical Ideas in Biology, Cambridge Press, Cambridge, 1968.

[38] R. Smith, Sufficient conditions for stability of difference equations, Duke Mathematical Journal 33 (1966) 725–734.

[39] M. Trott, The Mathematica GuideBook for Programming, Springer-Verlag, New York, 2004, pp. 24–25, http://www.mathematicaguidebooks.org/.

[40] E. Weisstein, Logistic Map, From MathWorld–A Wolfram Web Resource, https://mathworld.wolfram.com/LogisticMap.

[41] S. Wolfram, A New Kind of Science, Wolfram Media, Champaign, IL, 2002, pp. 918–921 and 1098.

[42] A. Wouk, Difference equations and J-matrices, Duke Mathematical Journal (1953) 141–159.

[43] M. Yamaguti, H. Matano, Euler's finite difference scheme and chaos, Proceedings of the Japan Academy 55A (1979) 78–80.

[44] M. Yamaguti, S. Ushiki, Discretization and chaos, Comptes Rendus de L'Académie Des Sciences. Paris 290 (1980) 637–640.

[45] Z-transform, Chapter 4, https://theengineeringmaths.com/wp-content/uploads/2017/10/ztransforms-web.pdf.

Index

Symbols

l_2-stable, 282, 283
l_p-stability, 278
l_p-stable, 175
z-inverse, 111
Z-transform, 83

A

Abel's summation formula, 25
Ambient, 33
Amortization of a Loan, 33
Antidifference, 18, 21
Antidifference operator, 18
Applications, 32
Ascoli-Arzelà, 310
Asymptotic stability, 227
Asymptotically stable, 167, 168, 175, 180, 189,
        210, 215, 218, 220, 238–240, 243,
        244, 246, 250, 278, 287
Asymptotically stable focus, 205
Asymptotically stable node, 202
Autonomous, 235
Autonomous systems, 237
Auxiliary equation, 54

B

Banach space, 292, 309, 312–316
Bank account, 32
Binomial coefficient function, 12
Bounded, 187, 189, 194
Bounded solutions, 295
Boundedness, 240

C

Calculus, 1
Carrying capacity, 36
Casoratian, 50, 68, 70
Cauchy sequence, 309
Causal, 108
Cayley-Hamilton Theorem, 147
Cell division, 32

Characteristic equation, 54
Cobweb diagram, 170
Compact, 309, 310, 314–316
Complete metric, 295, 314–317
Complex integration, 118
Complex roots, 57, 58
Constant coefficients, 53
Constant solution, 165, 174, 178, 253
Continuous, 1, 235, 278
Contour, 118
Contraction, 301, 316, 317
Contraction mapping principle, 309
Convolution, 103, 111
Convolution method, 121
Critical, 229

D

Decrescent, 254
Degenerate case, 204
Degenerate node, 201
Delay block, 109
Delayed unit step sequence, 94
Delta difference operator, 2
Diagonal, 140
Diagonalizable, 152
Difference equations, 1, 27
Difference operator, 11
Differential equations, 1
Direct inversion, 111
Discrete, 1
Discrete Fourier transform, 85
Distinct eigenvalues, 143
Distinct roots, 54
Domain of attraction, 244
Double factorial, 8

E

Eigenpair, 141
Eigenvalues, 141
Eigenvectors, 141
Equi-bounded, 301
Equicontinuous, 309–311, 314, 315

Equilibrium points, 167
Equilibrium solution, 35, 165, 174, 253
Euler's formula, 57
Euler's method, 31
Exponential order, 83
Exponentially stable, 211, 218, 236, 267, 269, 277

F

Factorial function, 8, 95
Fibonacci sequence, 32, 33, 55
Final value theorem, 97
Finite difference, 1
First-order, 27
Fish populations, 173
Fixed point, 59, 165, 316, 317
Floquet multipliers, 223, 226, 232
Floquet theorem, 225
Floquet theory, 220
Forcing function, 61
Forward difference, 2
Functional difference equations, 278
Fundamental matrix, 158, 180
Fundamental set of solutions, 158
Fundamental solution, 49, 52, 56
Fundamental theorem of difference calculus, 21

G

Gamma function, 8
Generalized eigenvector, 145
Global asymptotic stability, 242
Globally asymptotically stable, 176, 250, 251
Golden ratio, 55
Gronwall's inequality, 191, 194

H

Higher-order difference equations, 48
Hill's equation, 233
Homogeneous, 5, 48, 59, 69
Homogeneous system, 161
Host-parasitoid model, 219

I

Indefinite sum, 18
Initial value theorem, 96
Integration by parts, 23
Intrinsic, 223
Invariant set, 248, 249
Inverse z-transform, 110

J

Jacobian, 209

Jacobian matrix, 207
Jordan matrix, 199

L

Laplace table, 125
Large contraction, 293, 294
LaSalle invariance principle, 248
Laurent series, 83
Left shifting, 92
Limit set, 248
Linear, 4, 6
Linear difference equations, 27
Linearity, 86
Linearization, 206, 218, 243, 246
Linearly dependent, 48
Linearly independent, 48, 152
Linearly independent eigenvectors, 142
Lipschitz, 287, 310
Locally Lipschitz, 287
Logistic discrete model, 35
Logistic growth model, 36, 169
Logistice equation, 38
Lower triangular, 140
Lyapunov function, 218, 243, 246, 254–257, 260, 269, 277
Lyapunov functional, 240, 281, 282

M

Matrix norm, 180
Mean Value Theorem, 168

N

Negative definite matrix, 262
Neutral, 299
New variation of parameters, 285
Newton's cooling law, 33
Non-autonomous, 174, 197, 253
Non-autonomous systems, 253
Non-homogeneous, 48, 59, 61, 69
Non-homogeneous periodic systems, 227
Non-homogeneous Riccati equation, 74
Non-homogeneous systems, 161
Non-linear, 74
Non-oscillatory, 229
Noncritical, 229
Norm, 175, 313

O

Open ball, 238
Orbit, 167, 199
Order, 3
Oscillate, 28
Oscillatory, 229

P

Partial fractions, 111
Periodic, 229
Periodic sequence, 103
Periodic solutions, 295
Periodic systems, 220
Perturbation term, 194
Perturbed linear systems, 191
Perturbed systems, 194
Poles, 118
Positive definite, 236, 254
Positive definite matrix, 262
Power series, 111
Power series method, 120
Predator-prey model, 213
Principal matrix, 158
Product factor, 44
Properties, 86
Putzer algorithm, 147

R

Rabbit breeding, 32
Radially bounded, 242
Radially unbounded, 241, 254
Ramp function, 132
Rationally homogeneous, 75
Reformed discrete logistic model, 40
Repeated roots, 55, 56
Residues, 111, 118
Revised host-parasitoid model, 220
Riccati equation, 73
Ricker equation, 173
Right shifting, 92, 126
Rotation matrix, 205

S

Saddle with reflection, 201
Sampling, 89, 131
Second order difference, 3
Shift operator, 2, 11
Shifting, 92, 126
Simple pole, 118
Sink, 200
Source, 201
Source with reflection, 201
Spectrum, 185
Stability, 174
Stability indicator, 167
Stable, 167, 175, 178, 180, 197, 236, 238, 240, 255–257, 260, 286
Stable center, 205
Stable node, 200, 204

Stable star, 202
Staircase diagram, 170, 237
Steady state, 136
Steady state solution, 35
Strict Lyapunov function, 236, 254
Summation by parts, 23
Summation theorem, 21
Symmetric, 151
System, 117, 124, 138, 174

T

Taylor series, 83, 206
Time-varying systems, 157
Tower of Hanoi, 34
Trajectory, 199
Transfer function, 127
Transformation, 138
Transients, 136
Two-sided z-transform, 84

U

Undetermined coefficients, 61
Uniform asymptotic stability, 182
Uniform metric, 287
Uniformly asymptotically stable, 175, 176, 184, 196, 257, 260, 261, 265
Uniformly bounded, 240, 309, 310, 314, 315
Uniformly exponentially stable, 175, 277
Uniformly l_p-stable, 176
Uniformly stable, 175, 182, 184, 187, 194, 258, 260, 267
Unit impulse sequence, 88
Unit step sequence, 88, 94
Unstable, 167, 168, 175, 178, 200, 201, 204, 218, 236, 259
Unstable degenerate node, 201
Unstable focus, 205
Unstable node, 201, 202, 204
Unstable star, 202
Upper triangular, 140

V

Variation of parameters, 27, 66, 161, 286
Vector space, 296
Verhulst process, 36, 169
Volterra difference equation, 278

W

Wedge, 253

Weierstrass M-test, 311

Y

Young's inequality, 272

Z

Zero-input response, 127
Zero-state response, 127